Benchmark Papers in Geology

Series Editor: Rhodes W. Fairbridge
Columbia University

PUBLISHED VOLUMES

RIVER MORPHOLOGY / *Stanley A. Schumm*
SLOPE MORPHOLOGY / *Stanley A. Schumm and M. Paul Mosley*
SPITS AND BARS / *Maurice L. Schwartz*
BARRIER ISLANDS / *Maurice L. Schwartz*
ENVIRONMENTAL GEOMORPHOLOGY AND LANDSCAPE CONSERVATION, VOLUME I: Prior to 1900 / VOLUME II: Urban Areas / VOLUME III: Non-Urban Regions / *Donald R. Coates*
TEKTITES / *Virgil E. Barnes and Mildred A. Barnes*
GEOCHRONOLOGY: Radiometric Dating of Rocks and Minerals / *C. T. Harper*
MARINE EVAPORITES: Origin, Diagenesis, and Geochemistry / *Douglas W. Kirkland and Robert Evans*
GLACIAL ISOSTASY / *John T. Andrews*
GLACIAL DEPOSITS / *Richard P. Goldthwait*
PHILOSOPHY OF GEOHISTORY: 1785–1970 / *Claude C. Albritton, Jr.*
GEOCHEMISTRY OF GERMANIUM / *Jon N. Weber*
GEOCHEMISTRY AND THE ORIGIN OF LIFE / *Keith A. Kvenvolden*
GEOCHEMISTRY OF WATER / *Yasushi Kitano*
GEOCHEMISTRY OF IRON / *Henry Lepp*
GEOCHEMISTRY OF BORON / *C. T. Walker*
SEDIMENTARY ROCKS: Concepts and History / *Albert V. Carozzi*
METAMORPHISM AND PLATE TECTONIC REGIMES / *W. G. Ernst*
SUBDUCTION ZONE METAMORPHISM / *W. G. Ernst*
PLAYAS AND DRIED LAKES: Occurrence and Development / *James T. Neal*
PLANATION SURFACES: Peneplains, Pediplains, and Etchplains / *George Adams*
SUBMARINE CANYONS AND DEEP-SEA FANS: Modern and Ancient / *J. H. McD. Whitaker*

BENCHMARK PAPERS IN GEOLOGY (continued)

ENVIRONMENTAL GEOLOGY / *Frederick Betz, Jr.*
LOESS: Lithology and Genesis / *Ian J. Smalley*
PERIGLACIAL DEPOSITS / *Cuchlaine A. M. King*
LANDFORMS AND GEOMORPHOLOGY: Concepts and History / *Cuchlaine A. M. King*
METALLOGENY AND GLOBAL TECTONICS / *Wilfred Walker*
HOLOCENE TIDAL SEDIMENTATION / *George deVries Klein*
PALEOBIOGEOGRAPHY / *Charles A. Ross*
MECHANICS OF THRUST FAULTS AND DÉCOLLEMENT / *Barry Voight*
WEST INDIES ISLAND ARCS / *P. H. Mattson*
CRYSTAL FORM AND STRUCTURE / *Cecil J. Schneer*

Additional volumes in preparation

Benchmark Papers
in Geology / 34

A BENCHMARK® Books Series

CRYSTAL FORM AND STRUCTURE

Edited by

CECIL J. SCHNEER

University of New Hampshire

Dowden, Hutchinson
& Ross, Inc.

STROUDSBURG, PENNSYLVANIA

For Mary

Benchmark Papers in Geology, Volume 34
Library of Congress Catalog Card Number: 76-15565
ISBN: 0-87933-181-X

79 78 77 1 2 3 4 5
Manufactured in the United States of America.

LIBRARY OF CONGRESS CATALOGING IN PUBLICATION DATA
Main entry under title:
Crystal form and structure
(Benchmark papers in geology/34)
Includes indexes.
1. Crystallography—Addresses, essays, lectures. I. Schneer, Cecil J., 1923–
QD921.C76 548 76-15565
ISBN 0-87933-181-X

Exclusive Distributor: **Halsted Press**
A Division of John Wiley & Sons, Inc.
ISBN: 0-470-98937-8

SERIES EDITOR'S PREFACE

The philosophy behind the "Benchmark Papers in Geology" is one of collection, sifting, and rediffusion. Scientific literature today is so vast, so dispersed, and, in the case of old papers, so inaccessible for readers not in the immediate neighborhood of major libraries that much valuable information has been ignored by default. It has become just so difficult, or so time consuming, to search out the key papers in any basic area of research that one can hardly blame a busy man for skimping on some of his "homework."

This series of volumes has been devised, therefore, to make a practical contribution to this critical problem. The geologist, perhaps even more than any other scientist, often suffers from twin difficulties—isolation from central library resources and immensely diffused sources of material. New colleges and industrial libraries simply cannot afford to purchase complete runs of all the world's earth science literature. Specialists simply cannot locate reprints or copies of all their principal reference materials. So it is that we are now making a concerted effort to gather into single volumes the critical material needed to reconstruct the background of any and every major topic of our discipline.

We are interpreting "geology" in its broadest sense: the fundamental science of the planet Earth, its materials, its history, and its dynamics. Because of training and experience in "earthy" materials, we also take in astrogeology, the corresponding aspect of the planetary sciences. Besides the classical core disciplines such as mineralogy, petrology, structure, geomorphology, paleontology, and stratigraphy, we embrace the newer fields of geophysics and geochemistry, applied also to oceanography, geochronology, and paleoecology. We recognize the work of the mining geologists, the petroleum geologists, the hydrologists, the engineering and environmental geologists. Each specialist needs his working library. We are endeavoring to make his task a little easier.

Each volume in the series contains an Introduction prepared by a specialist (the volume editor)—a "state of the art" opening or a summary of the object and content of the volume. The articles, usually some thirty to fifty reproduced either in their entirety or in significant extracts, are selected in an attempt to cover the field, from the key papers of the last century to fairly recent work. Where the original works are in foreign languages, we have endeavored to locate or commission translations. Geologists, because of their global subject, are often acutely aware of the oneness of our world. The selections cannot, therefore, be restricted to any one country, and whenever possible an attempt is made to scan the world literature.

To each article, or group of kindred articles, some sort of "highlight commentary" is usually supplied by the volume editor. This commentary should serve to bring that article into historical perspective and to emphasize its particular role in the growth of the field. References, or citations, wherever possible, will be reproduced in their entirety—for by this means the observant reader can assess the background material available to that particular author, or, if he wishes, he, too, can double check the earlier sources.

A "benchmark," in surveyor's terminology, is an established point on the ground, recorded on our maps. It is usually anything that is a vantage point, from a modest hill to a mountain peak. From the historical viewpoint, these benchmarks are the bricks of our scientific edifice.

RHODES W. FAIRBRIDGE

PREFACE

It is the purpose of this volume to place in context certain key papers in the development of our ideas on the relationship between the macroscopic external form of crystals and the microscopic internal structure, that ultimate mechanism by which, Robert Hooke wrote in 1660, nature accomplishes all her effects. It has proved impossible to be exhaustive or even approximately complete. Bravais's *Études cristallographiques* alone, for example, would exceed the space of this volume, and contain, moreover, little or nothing that is not central to the topic, succinct, or readily obtainable elsewhere. In collecting these papers, I have relied principally on these criteria: First, was the paper of influence? Second, was it representative of the state of the science when published? Third, are the ideas presented original and of intrinsic interest? Fourth, is the material readily available elsewhere? Also, in the words of A. Guinier, Chairman of the Teaching Commission of the International Union of Crystallography, is it "useful for the formation and the culture" of our students? Finally, recent work is included to illustrate modern ideas on the Pythagorean relationship between form and structure.

In every paper, reference is made to the subject of crystal growth, inseparable from the form that is its end result. However, the topic of crystal growth is very well covered by other review volumes; it is treated in these papers only insofar as it is related to the morphologic distributions that are the primary object of our concern.

I must acknowledge with gratitude the services of the translators, exonerating them at the same time from responsibility for errors in language or interpretation; also, the generosity of the publishers who agreed that articles such as von Laue's derivation of Wulff's theorem should be available in English. Dr. Robert Silverman of the University of New Hampshire Department of Mathematics very kindly corrected some of my errors. Authors of these articles were most generous with copies and with permission to publish, as well as helpful with advice and suggestions and uniformly patient. Finally, I am grateful to Nancy McCool and Linda Hoadley for their patience and the skill with which they assisted me in assembling this volume.

CECIL J. SCHNEER

CONTENTS

PART III: THE FIRST STRUCTURES

PART IV: BRAVAIS'S RULE AS NATURAL LAW

PART V: THE EXTENSION OF BRAVAIS'S RULE

PART VI: APPLICABILITY OF THERMODYNAMICS

PART VII: MORPHOLOGY IN PHASE SPACE

PART VIII: SPIRAL GROWTH

PART IX: PERIODIC BOND CHAINS

PART X: SURFACE ENERGY AS A FUNCTION OF STRUCTURE

CONTENTS BY AUTHOR

INTRODUCTION

> We rise from the conception of form to an understanding of the forces which gave rise to it . . . in the representation of form we see a diagram of forces in equilibrium, and in the comparison of kindred forms we discern the magnitude and the direction of the forces which have sufficed to convert the one form into the other.
>
> *d'Arcy Thompson*[1]

FORM (GEOMETRY) AND PERIODICITY

The earliest Pythagoreans explained the creation of the world by deriving form (substance) out of number (idea), a relationship between arithmetic and geometry that is preserved in modern languages, for example, when we speak of 4 or 8 as the *square* or *cube* of two. The Pythagoreans drew their numbers with dots: · for one, ··, ···, ····, etc. One, three, six, ten, etc., were triangular, ·, ∴, .∴.. This early association of number with form was not purely symbolic. It was essentially physics, because it was an account of phenomena and experience. Diogenes Laertius quotes this account of a pre-Platonic cosmology:

> The first principle of all things is the One. From the One came an indefinite Two. ... From the One and the Indefinite Two came numbers; and from numbers points; from points, lines; from lines, plane figures; from plane figures, solid figures; from solid figures, sensible bodies. The elements of these four are: fire, water, earth, air; ....[2]

These elements with the fifth, space, were identified with the five regular solids of Plato, the tetrahedron, octahedron, cube, icosahedron, and pentagonal dodecahedron. It was not simply mathematics as pure idea, but mathematical physics. It was an account of the origin of the world. ". . . out of them (the elements) comes to be a cosmos, animate, intelligent, spherical. . . ."

Serialization or periodization, the repetition of properties such as odd or even, square or cubic, was of great interest at the dawn of arithmetic, even though the Pythagoreans thought of this subject more as an auxiliary of their chief concern, harmony, than a subject in itself. The realization that the series of the squares 1, 4, 9, 16, . . . was generated from the sum of the series of odd numbers, 1, 3, 5, 7, . . . again confirmed the basis of Pythagoreanism—the identification of the world with number. This identification is a kind of mathematical physics. The discovery attributed to Pythagoras himself, that the musical harmonies of the octave, the fourth, the fifth, etc., were associated with simple integer ratios seemed a confirmation, by identifying sound with number, of the literal conviction that number was the world and the world was number.

These ideas were transmitted and shaped over centuries. They are the basis of Euclid's arithmetic and the solid geometry of his treatment of the Platonic solids in Book XIII. Through the *De Arithmetica et Musica* of the Roman Boethius (480–524), they passed to the Middle Ages. With the revival of Greek mathematics in the Renaissance, the association of geometric form and number illustrated in the first printed copies of Euclid (fifteenth century) by cubic lattices, for example, was a given. But the association of mathematics and physics, particularly for the early naturalists, came less and less to be regarded as God-given. Mathematics was more and more understood in Einstein's happy phrase as a "free creation of the human mind"; Alfred North Whitehead wrote, "the most original invention of mankind except for music."[3] The Pythagorean view went out with eighteenth-century rationalism. In the modern view, number and geometric form are symbols. They originate in mind and not in nature, which is to say, not in physics.

Karl Friedrich Gauss (1777–1855), whose concept of virtual work paved the way for J. Willard Gibbs (1839–1903) and Pierre Curie (1859–1906), and the idea of the specific surface energy of a crystal face, considered geometry as something more than pure abstraction.

> It is my deepest conviction that space-theory occupies an entirely different position with respect to *a priori* knowledge than pure number-theory. . . . even if number is the pure product of

DENIQVE & vnitas interdum ſolidus appellabitur numerus, etſi non proprie, quia eius latera ſunt tres vnitates, ipſam vnitatem ex mutua earum multiplicatione procreãtes.

DEFINIT autem & hic Euclides tantum numerum ſolidum rectangulum, cuius baſes oppoſitę ſunt parallelę, & qui continetur ſub tribus numeris, omiſſis infinitis alijs, de quibus Iordanus, ob cauſam in pręcedenti definitione datam; quia ſcilicet hi prorſus ęquales ſunt, & ſimiles cubis, & parallelepipedis Geometricis.

XVIII.

QVADRATVS numerus eſt, qui æqualiter æqualis. Vel qui ſub duobus æqualibus numeris continetur.

PLANVM illum numerum, qui ęqualiter ęqualis eſt, hoc eſt, qui ſecundum vnitates ſuas in longum & latum diſpoſitas, refert parallelogrammum rectangulum, cuius longitudo latitudini eſt ęqualis, ita vt omnia latera ſint ęqualia, vel qui ex multiplicatione mutua duorum numerorum ęqualium procreatur, atque adeo ſub illis continetur, vocat quadratum. Huiuſmodi eſt numerus 25. contentus ſub numeris ęqualibus 5. & 5. hoc eſt, ex eorum mutua multiplicatione genitus. Nam ſi eius vnitates in formam planam redigantur, referet perfectum quadratum Geometricum, in quolibet latere habens quinque vnitates, ideoq; ęqualiter ęqualis erit.

ALTERVTER autem numerorum ęqualium, ſub quibus quadratus numerus continetur, vel ex quorum multiplicatione producitur, latus quadrati à Geometris, radix vero ab Arithmeticis pleriſque appellatur.

CVBVS

XIX.

CVBVS vero, qui æqualiter æqualis æqualiter. Vel, qui ſub tribus æqualibus numeris continetur.

SOLIDVM quoque illum numerũ vocat cubum, qui eſt ęqualiter ęqualis ęqualiter, id eſt, cuius vnitates in longum, latum, atque profundum diſpoſita cubum Geometricum referunt, ita vt omnes eius dimenſiones, nimirum longitudo, latitudo, & altitudo, ſiue profunditas, ęquales ſint; Vel qui ex multiplicatione mutua trium ęqualium numerorum inter ſe producitur. Qualis eſt numerus 27. contentus ſub numeris ęqualibus 3. 3. 3. ſiue ex eorum multiplicatione mutua procreatus, cum ex 3. in 3. fiat numerus 9. & ex 9. in 3. gignatur numerus cubus 27. Omnes enim illius vnitates in formam ſolidam redactę referunt perfectum cubum Geometricum, exiſtuntque tam in longitudine, & latitudine, quam in profunditate, tres vnitates. Quare ipſe numerus 27. eſt ęqualiter ęqualis ęqualiter.

QVILIBET vero trium numerorum ęqualium, ſub quibus cubus continetur, vel ex quorum mutua multiplicatione procreatur, Geometris latus cubi, pleriſque autem Arithmeticis radix dicitur.

XX.

NVMERI proportionales ſunt, cũ primus ſecundi, & tertius quarti, æque multiplex eſt; vel eadem pars, vel eędem

B 2 partes:

Reprinted from Christopher Clavius, *Euclid*, edition of Grassi, Rome, 1589. This particular print was provided by the History of Science Collections, University of Oklahoma Libraries.

our mind, space has also a reality outside of our mind, the laws of which we cannot fully comprehend, *a priori*.[4]

Einstein and Whitehead would probably have agreed, although they might have insisted on separating geometry from physical space. In any event, metaphysical considerations are not determinant in science, although a historical background may be useful in laying bare the hidden assumptions that underlie modern ideas.

The Pythagorean arithmetic of Boethius has its modern counterpart in the concept of the Brillouin and Jones zones, developed for very different purposes. Yet the mathematical problem has not been developed in full generality. What is the geometry in three dimensions associated with a periodic distribution of points? What periodic distribution of points (e.g., the distribution of the electron density of a crystal) is associated with a particular polyhedral shape? The zone or band theory of solids has achieved general acceptance. Mineralogists and crystallographers, and even solid-state physicists studying periodic distributions considerably more complex than those of the metals of the early zone theories, might very well reconsider the metaphysics of number and space.

We are concerned in this volume with the relationship between the external morphology of the "free" crystal and the periodic distribution of matter within, i.e., the specific structure. Johannes Kepler (1571–1630), mathematician and neo-Platonic mystic, did not fail to see in the regular forms of natural crystals (see Paper 1 of this volume) the lattice solution for their genesis—*material necessity*, he called it. We now call it close-packing. Others, René Descartes (1596–1650), Robert Hooke (1635–1703), Domenico Guglielmini (1655–1710), Christian Huyghens (1629–1695), pursued the topic. It is interesting to speculate on the motives of Nicholaus Steno (1638–1687), a contemporary of Hooke's and a confirmed Cartesian, who rejected the concept of an internally periodic structure and, in doing so, succeeded in *measuring* interfacial angles instead of accepting them as angular permutations for cubic and hexagonal close-packings.

René Just Haüy (1743–1822) is credited with the discovery, anticipated in significant measure by others, such as Torbern Bergman (1735–1784), of the internal periodicity of crystals, the law of rational intercepts, and the observation of simple integer intercept ratios. "He made of crystallography," Cuvier wrote in his *Éloges,*[5] "a science as exact as astronomy." It was an exaggeration. It was Bravais, nearly half a century later, who replaced Haüy's unwieldly combination of polyhedral nuclei and *molécules intégrantes* with the 14 lattices. By a uniquely circular process of reasoning, Bravais hypothesized the association of lattice and morphology, and then proceeded to use morphology as the basis for hypothesizing symmetrical polyhedra associated with each lattice point. Bravais's theory made possible the later derivation of space groups by Artur Schoenflies (1853–1928), E. von Fedorov (1853–1919), and William Barlow (1845–1934); but we are here interested in the hypothetical associations that he drew between morphology and structure, which were in effect the source of the theory, and which it must be confessed, have not yet been tested definitively.

In the latter half of the nineteenth century, Gibbs and Curie established the principles of the thermodynamics of solids and of

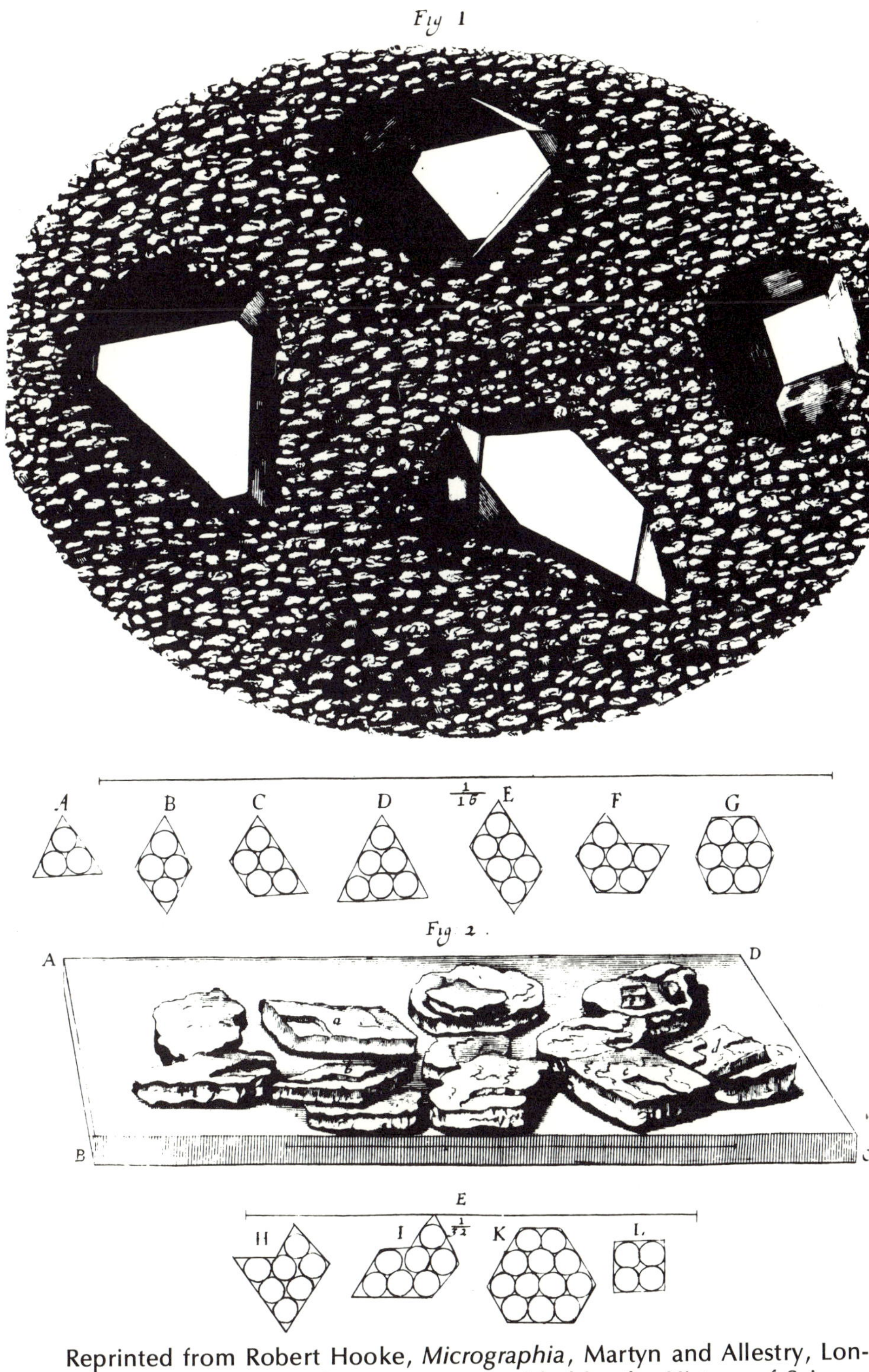

Reprinted from Robert Hooke, *Micrographia*, Martyn and Allestry, London, 1665. This particular print was provided by the History of Science Collections, University of Oklahoma Libraries.

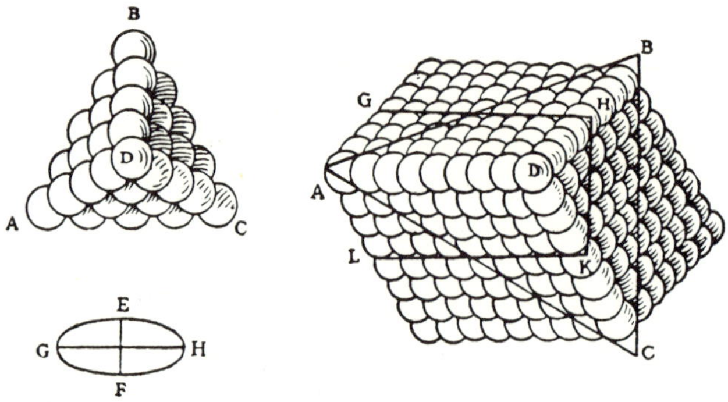

Internal molecular structure inferred from rhombic morphology of Iceland spar (calcite), and the explanation of birefringence. Reprinted from Christian Huyghens, *Traité de la lumière,* Van der Aa, Leiden, 1690. This particular print was provided by the History of Science Collections, University of Oklahoma Libraries.

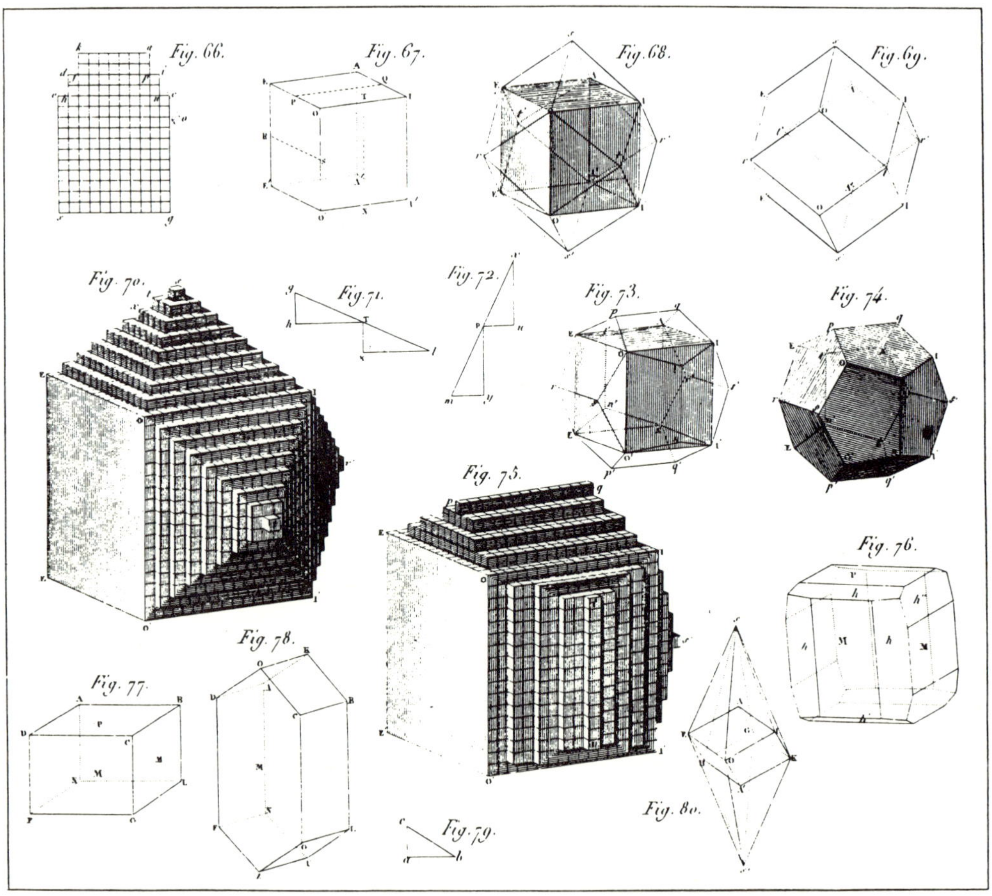

Reprinted from René Just Haüy, *Traité de cristallographie,* Bachelier et Huzard, Paris, 1822, vols. 1–2 and atlas. This particular print was provided by the History of Science Collections, University of Oklahoma Libraries.

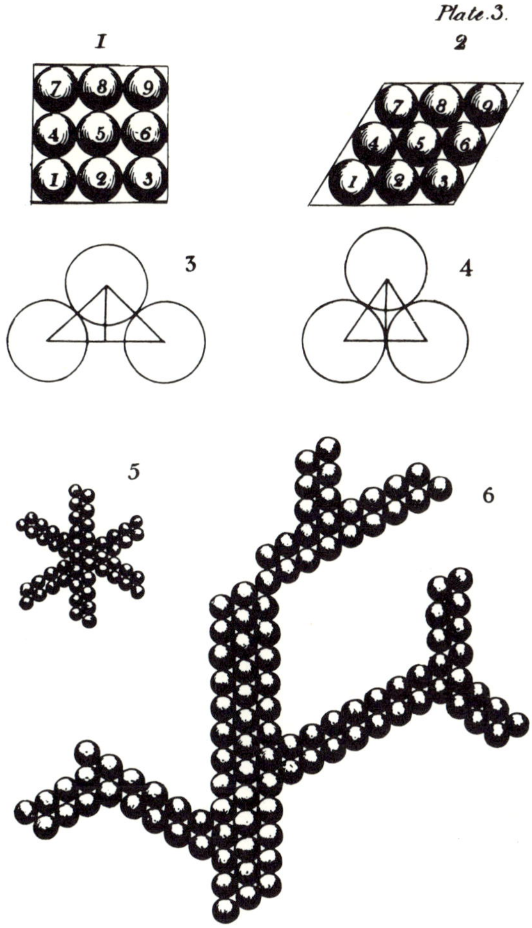

Representation of the "particles of water" in ice. Reprinted from John Dalton, *A New System of Chemical Philosophy*, Manchester, 1808.

crystal growth. These principles, it is clear, were independent of what Georges Friedel (1865–1933) was to refer to as "the reticular hypothesis" and, in fact, independent of kinetic-molecular and even atomic hypotheses of any sort. Gibbs's masterpiece, "On the Equilibrium of Heterogeneous Substances," published in the *Transactions of the Connecticut Academy of Sciences* for 1878, established the constraints within which any hypotheses, however based, must fit. The equations of Gibbs are the laws of physics for crystals. They provided for the first time a theoretical basis for experimentation on crystal growth. G. Wulff undertook a series of experiments on capillary constants, to relate Curie's specific surface energies to particular crystallographic planes. Out of this work emerged the immensely fruitful ideas of the Wulff point, the Wulff vector, and the Wulff diagram—the major modern conceptual tools for the treatment of problems in crystal growth and morphology.

The baton of French crystallography handed down by Bravais passed to Friedel, who saw more clearly than any of his generation the empirical basis of Haüy's periodicity, and attempted to place the laws of crystallography on separate but equally rigorous legs of *physics* (which is to say observations of the natural world) and on the rapidly developing mathematical science of symmetry groups. It seemed to Friedel that the often crude hypotheses of structure, such as those of William Barlow, resting as they did upon geometric parallels and apparently accidental coincidences, were mere speculation. Attacked also from the right by the German group theorists, such as Sohncke, Barlow nevertheless, with the chemist W. J. Pope (1870–1939), prepared the model of the internal structure of crystals with which today's X-ray diffraction is concerned. It was not a trivial observation in 1883 that the several atoms in a solid compound "... occupy distinct portions of space and do not lose their individuality."

With the advent of X-ray diffraction, morphological studies, with their capacity to *imply* lattice type and simple structure, were necessarily displaced from the position of principal method of investigation of structure. Nevertheless, it is important to keep in mind that the concept of a periodic distribution of atoms was already well advanced from the elementary speculations of Kepler and Hooke. More significantly, Barlow understood that the hypothetical interior distribution must have the symmetry of the external form—holohedral, hemihedral, tetartohedral, the structure might have less, but never more symmetry than the morphology. The X-ray structures revealed by more than half a century of the most intense investigation are the lineal descendants of the tetrahedra of zinc and sulfur atoms pointing alternatively up and down, of Barlow and Pope.

While physics owes to mineralogy the concepts of atomism and symmetry, mineralogy may owe to physics and particularly to Niels Bohr the opening of a new avenue of approach to the Pythagorean problem. The complementarity of wave and particle in the quantum theory is analogous to the complementarities of external form and internal structure and of number and geometry. After Bloch had shown in 1928 that waves in a crystal are modulated by the periodic structure,[6] Léon Brillouin demonstrated the geometric consequences, i.e., the polyhedral zones in $\bar{k}$-space exactly similar to the mineral polyhedra of the morphologist. Bravais had arrived at the lattice from an examination of the polyhedral boundary of periodically distributed material particles. Brillouin (and Jones) started from the lattice to arrive at polyhedral envelopes specifying field properties, magnitude, and direction, for the crystal energies.

The association drawn by Bravais between the stability of particular forms and their net reticular density is still today the fundamental theoretical principle of physical morphology. Fedorov considered it as "Now [1911] demonstrated by direct experiment." Yet the richness of detail of Bravais's original concept was only gradually appreciated. It is an avenue that has not yet been explored to the end. Paul Niggli (1888–1953), in attempting to take into account a more complex density, a depth of growing surface, rather than an ideal plane, very early entered upon this pathway. He began statistical studies to quantify the voluminous and exhaustive descriptive morphology of Goldschmidt's *Atlas der Kristallformen*, 18 quarto volumes published over a decade, including World War I, and illustrating every crystal mineral specimen known to science. In 1937, J. D. H. Donnay and D. Harker modified and extended the Bravais rule to the form in which it enters the textbook literature today.

The theory and practice of the growth of single crystals is a subject of major modern interest and not a few review volumes. It is also an industry with significant international economic effects. The mineralogist Niggli's concept of a stepped natural growth surface was to transform the entire theory of crystal growth when it attracted the attention of the theorists of the solid state. Frank's theory of spiral growth was a major development stemming from the defect theory of Frenkel, Buerger, Burgers, and others. It was the basis for Hartman and Perdok's conception of the dependence of morphology on periodic bond chains (PBC vectors). Yet the very success of Burton, Cabrera, and Frank's kinetics was a temporary setback for the development of a theory linking morphology and structure. The spiral growth theory cast doubt on the whole concept of an equilibrium morphology, since the external form of crystals appeared to be largely dependent on the random occurrence of defects.

In a summary review by A. F. Wells in 1946, the parallel was again drawn between the prominence of crystal planes and the intensity of X-ray diffraction. High diffraction intensities are obtained from morphologically significant planes, either growth planes or cleavages. A quantitative attempt to account for this parallelism was made by McLachlan in 1952. McLachlan has also introduced a fascinating wave-kinetic explanation to account for Kepler's snowflakes. Schneer (1966) added the concept of the equilibrium morphological assemblage and has applied Fourier transformation to pass from morphology to internal structure. The statistics of the assemblage are one means of treating the very real perturbations during growth, of environmental factors on morphology. It seems clear that the ultimate explanation of the formal parallellism between the polyhedral zones in $\bar{k}$-space, the intensities of wave diffraction, and the crystal morphology, will rest upon the single element that is common to all three, the internal structure. In his "Wave Propagation in Periodic Structures," Brillouin emphasized the generality of his results ". . . for any kind of waves (elastic, electromagnetic, or wave mechanical de Broglie waves) propagating in a medium with periodic structure . . . ."

At least two recent directions of research into the relation between form and structure are not represented in this volume. Physicists working with point emission spectroscopy are concerned with the equilibrium morphologies of very small crystals. If Gibbs was truly prescient, it may well be that these are the equilibrium forms for natural mineral crystals. The bounding planes of the seed are those of the final crystals. The advantage of this approach is that it is directly tied to experiment.

Crystallographers are also attempting computer simulation of crystal growth, i.e., programming alternative hypotheses with the aim of arriving at morphologies comparable to natural crystals. Yet as of this writing, nothing has been published.[7] The emergence of the high-speed digital computer has placed enormously sophisticated mathematical power into many hands. The computer itself, appropriately programmed, appears now to be capable of achieving a solution of the old Pythagorean problem: what is the geometric form corresponding to a given periodic distribution of points? It is hardly elegant, but it works up to a point. If the computer program is considered as a mathematical operator A, the electron density map F, and the morphology M, then

$$M = A(F)$$

The rule of Bravais was the function for the transformation from the lattice to the corresponding polyhedron, e.g., from the face-centered cubic lattice to the cubo-octahedron. The geometric structure factor was employed by Mott and Jones.[8] The Donnay-Harker rule suggests the function for the more complex assemblage of space-group equipoints. Modulated structure factors are employed to relate the electron densities themselves to morphology (McLachlan, Schneer). It may very well be that a natural science such as mineralogy will never be subject to a simple, direct mathematics, yet there is no reason to doubt that for crystallography a finite science, an elegant, direct thematics relating periodicity and form, is possible. We can only wait upon the mathematicians. In the meantime, the old tools of description, statistics, experiment, and the digital computer continue to generate fascinating if difficult questions.

REFERENCES

1. D.W. Thompson, *On Growth and Form*, Cambridge University Press, Cambridge, 1966, p. 270.
2. F.M. Cornford, *Plato and Parmenides*, Liberal Arts Press, New York, 1957, p. 3.
3. A.N. Whitehead, *Science and the Modern World*, New American Library, New York, 1945, p. 20.
4. Gauss to Bessel, 9 April, 1830, in Paul Stäckel and Friedrich Engel, *Die Theorie der Parallellinien von Euklid bis auf Gauss*, B. G. Teubner, Leipzig 1895, p. 227.
5. G. Cuvier, *Recueil des éloges historique*, V. III, Levrault, Paris.
6. F. Bloch, *Z. Phys.* **52**, 555 (1928).
7. C.J. Schneer, Computer Simulation of Crystal Growth, *Abstract with Programs*, 1973 Annual Meetings, Geol. Soc. Am., Boulder, Colo. 1973, p. 796.
8. N.F. Mott, and H. Jones, *The Theory of the Properties of Metals and Alloys*, Dover Publications, Inc., New York, 1958.

Part I

THE IDEA OF INTERNAL PERIODICITY

Editor's Comments on Papers 1 Through 4

1 **KEPLER**
Excerpts from *A New Year's Gift, or On a Hexagonal Snowflake*

2 **HAÜY**
Excerpts from *Elementary Treatise on Physics*

3 **BRAVAIS**
Excerpts from *The Crystal Considered as a Simple Assemblage of Points*

4 **BRAVAIS**
Excerpts from *The Crystal Considered as an Assemblage of Polyatomic Molecules*

In 1607, Johann Kepler's mathematical power applied to his Pythagorean preconceptions resulted in the first two of his three laws of planetary motion. It must have been in a relaxed mood, then, that he turned his glance from the heavens to the earth, determined to demonstrate that the Platonic solids ruled the sublunary world as well as the celestial. A snowflake landing on his sleeve was the point of departure for the charming essay reproduced here as Paper 1, which posed for the first time the central problem of crystallographic form and structure.[1]

Paper 2 consists of excerpts selected from the second edition of Haüy's *Traité Elémentaire de Physique*, 1805; they represent his own summary of a mature version of his ideas rather than their first or last form. In approximately this habit, Haüy's thesis dominated mineralogy, crystallography, and physical chemistry for better than half a century.

The molecular and atomistic hypotheses with which Haüy was primarily concerned were eliminated from mineralogy in the latter part of the nineteenth century by the application of new standards of rigor and the requirement of demonstration. Haüy's concept was recast as the law of rational intercepts, still today the fundamental law of morphological crystallography. Nevertheless, the banished *molécules intégrantes* persisted in the scientific underground, and were to emerge again to bright sunlight, transformed now into unit cells, which with atoms and ions, were destined to dominate solid-state theory. Haüy's laws of decrement, his nucleus and his subtractive molecules, as well as the distinction between primitive and secondary forms have not fared as well and have disappeared from the literature. His *rangées*, translated here as rows, and his diagrams of *laminae* of superposition (a much earlier idea of Romé de l'Isle and Bergman) are potent influences in the development of modern ideas. One has only to consider the *steps* and *kinks* of the series of papers, beginning in 1949, that were to establish the theory of spiral growth.

Cuvier wrote of Haüy that "he made of mineralogy a science as exact and as methodical as astronomy,"[2] but it was a statement more applicable to Bravais. It was Bravais (Papers 3 and 4) who, beginning in 1848, applied that mathematical rigor to the periodicity of crystals which Hessel had already applied to their external symmetry in 1830.[3] Haüy's classification of nuclear forms and integrant molecules was essentially arbitrary, however faithfully it might reproduce interfacial angles of real crystals. Bravais's derivation of the 14 lattices that bear his name was an achievement in pure mathematics.[4] The physical and chemical insight that led him to associate the points of his lattices with the centers of gravity of molecular motifs had two results: (1) by the consideration of the appropriate Hessel symmetries of the motifs, he assured the later derivation of the 230 space groups and the development of modern structure theory; (2) by the association of cleavage and morphology with net reticular density, he laid the foundation of molecular mechanics and prepared the way for the modern science of the solid state.

This first statement of the Bravais rule is based on the principle that the force between planes decreases with the interplanar spacing. Bravais makes a clear distinction between cleavage, which can depend only on intermolecular force, and growth, which must also be sensitive to the environment.

NOTES AND REFERENCES

1. See C. Schneer, "Kepler's New Year's Gift of a Snowflake," *Isis*, **51**, Pt. 4., No. 166, 531–545 (1960); also J. Kepler, *The Six-Cornered Snowflake*, Colin Hardie, trans. and ed., Oxford, Oxford, 1966.
2. *Recueil des éloges historique*, V. III, Levrault, Paris, p. 163.
3. J. F. C. Hessel, *Kristallometrie*, Leipzig, 1831; Hessel's work was apparently unknown to Bravais.
4. Some of this derivation had been previously performed by M. L. Frankenheim, *Acta Naturae Curiosorum*, V. XIX, 2nd pt., 1842, p. 483.

1

A NEW YEAR'S GIFT, OR ON A HEXAGONAL SNOWFLAKE

Johannes Kepler

This translation was prepared expressly for this Benchmark volume by Jean Schneer Silverman, from pp. 8–9 of Strena seu de Nive Sexangula, *Tampach, Frankfort, 1611*

[*Editor's Note:* In the original, material precedes this excerpt.]

For, in general, similar balls collected in a vessel arrange themselves in two ways, corresponding to two kinds of arrangement in a plane.

If one forces equal-sized balls moving on a horizontal plane, into a narrow space, so that they touch each other, they will collect into either a triangular or rectangular form; in the former case, six will surround each one, in the latter, four; in both cases the ratio of contact is the same, except for those on the edge. In the pentagonal form, equality cannot be retained, and the sexagonal is resolved into the triangular, so that there are, as mentioned, only two arrangements.

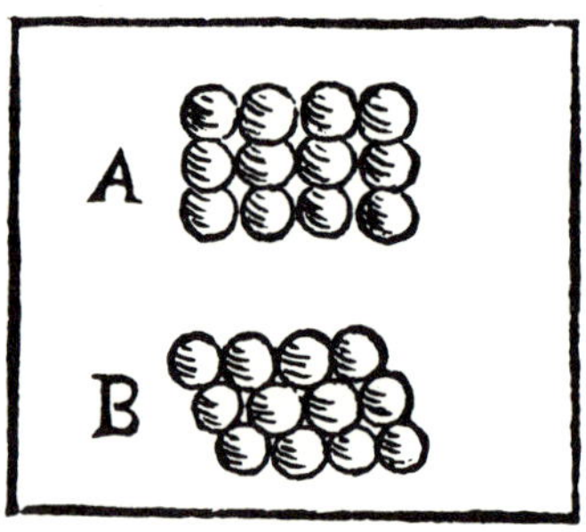

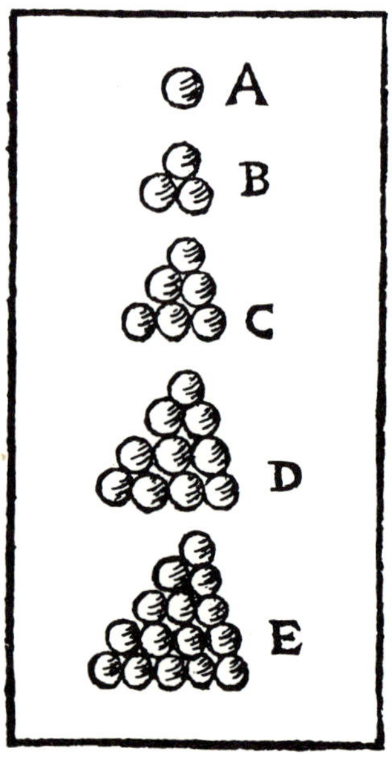

Reprinted from Johannes Kepler, "Two Planar Arrangements of Equal Spheres Found in Close-packing," *Strena seu de Nive Sexangula*, Tampach, Frankfort, 1611.

Now, if one proceeds to pack solids as closely as possible and sets rows upon rows, which were earlier arranged in a plane, they will be either square or triangular (A and B); if squares, the individual balls of the upper row will rest on those of the lower row; if triangular, the individual balls of the upper row will rest between four of the lower row. In the first case, each ball is touched by four surrounding it in the same plane and by one above it and one below it, and thus by six others altogether; this will be a cubic order. By pressure, cubes will be made, but it will not be the tightest compression. In the second case, not only is each ball touched by four surrounding it in the same plane, but also by four above it and four below it, and thus it will be touched by twelve altogether; by pressure, rhombs will be made from the balls. This order is like the octahedron and the pyramid [tetrahedron]. It has the tightest compression, so that by no other arrangement could more balls be put in the same vessel. Again, if triangular rows were constructed in a plane, then in a solid compression either ball of the upper row rests on each ball of the lower, again a loose compression, or each ball of the upper row rests between three of the lower. In the first case, each ball is touched by six surrounding ones in the same plane, by one above, and by one below, and thus by eight altogether. This order is like a prism, and, by pressure, columns are made with six rectangular sides and two hexagonal bases. In the second case, the same thing occurs as in the second case of the quadrilateral. For let B be the juncture of three balls. Above this is set one ball, A, for a vertex; then let there be another juncture of six balls, C, then another of ten, D, and another of fifteen, E. Always set a narrower row on a broader to make the figure of a pyramid. Although in this superposition, each upper ball rests between three of the lower ones; yet if the figure is inverted, so that not the vertex but the whole side [edge] of the pyramid [tetrahedron] is on top, as often as one removes one ball from the top, four will rest below in a square. And again one ball will be touched, as before, by twelve others, namely by six surrounding it in the same plane, three above and three below. Thus in the closest arrangement of solids the trigonal order cannot occur without squares, and vice versa.

[*Editor's Note:* This excerpt is from pages 8 and 9 of the original 24-page Latin material.]

2

ELEMENTARY TREATISE ON PHYSICS

René Just Haüy

This abridged translation was prepared expressly for this Benchmark volume by Cecil J. Schneer, from pp. 59–61, 65, 77, 78 of Traité élémentaire de physique, *2nd ed., Paris, 1860*

[*Editor's Note:* In the original, material precedes this excerpt.]

If one divides various original crystals of a single substance with strokes that correspond on all parts similarly situated, one is able to arrive at in this way a regular solid which is constant for all these crystals, even for those whose forms contrast most sharply . . . . We call *primitive forms* those solids which are each inscribed in all the crystals that belong to a single species, and *secondary forms*, those which differ from the primitive form.

110. The observed primitive forms are six in number: the tetrahedron, which in this case is always regular; the parallelepiped which is sometimes rhomboidal, sometimes cubic, etc.; the octahedron with a surface composed of triangles, which are, according to the species, equilateral, isosceles, or scalene; the regular hexahedral prism; the dodecahedron with equal and similar faces; and the dodecahedron composed of two hexagonal pyramids joined at their bases . . . .

[*Editor's Note:* Material omitted at this point.]

OF THE FORMS OF THE INTEGRANT MOLECULES

[Knowing the constant nuclear form for all the variety of crystals of the same species, it now remains to determine the various manners of arrangement of the molecules that envelope the nucleus.]

112. But before passing to the laws of such arrangement, we should know the molecules in question; and it is by the subdivision of the nucleus parallel to its different faces, and sometimes in still another direction, that we arrive at this knowledge.

To start, assume that the nucleus is a parallelepiped with no joints other than those parallel to these faces, and we choose, for example, the rhombohedron of carbonate of chalk [calcite]. Subdivision of this rhomboid by planes closer and closer together will give similar rhomboids of successively diminishing volume; and, in theory, if one continues beyond the point where the small solids are visible, one will be led to rhomboids of such a degree of tenuousness that one cannot further divide them without analysis, i.e., without breaking the bond of chemical principles which compose them. These rhomboids, obtained by whatever means at the limit of mechanical division, are what we call *integrant molecules* of carbonate of chalk, to distinguish them from the elementary molecules of the same substance which are in part those of chaux [lime], in part those of carbonic acid. . . .[a]

[*Editor's Note:* Material omitted at this point.]

OF THE LAWS TO WHICH THE STRUCTURE OF CRYSTALS IS SUBJECT

119. If one considers carefully the shapes of the laminae that cover successively the nucleus of a crystal, which we call *laminae of superposition*, we perceive that leaving the nucleus, they progressively decrease, sometimes on all sides at once, sometimes only on certain parts. But the quantity by which each layer is exceeded by the layer that preceded it can only arise from the omission of a certain sum of integrant molecules which the first lacks in order to arrive at the same level as the second; and because the boundaries of the decreasing layers are always straight lines parallel to one another on different layers, the result is that the differences of which we are speaking are measured by the subtraction of one or several rows of integrant molecules. Here then is the problem that arises for solution:

Given a secondary crystal, and the shape of its nucleus and of its integrant molecules also being known, and supposing further that each of the layers added to the nucleus is exceeded by the preceding in certain parts, by a quantity equal to one, two, three, etc., rows of molecules, determine from among the different laws of decrement those which would result in a form similar to that proposed, in the number, shape, and disposition of its faces, and in the measure of its plane and solid angles.

[*Editor's Note:* Material omitted at this point.]

GENERALITY OF THE THEORY

132. Here we enter into some details of the structure of secondary forms, the molecule of which is a tetrahedron or triangular prism; but we do not believe that we can conclude this article in any better way than by the exposition of a result which serves to tie this structure with that of the original forms of the parallelepiped. The connection in question consists of the fact that the tetrahedral or triangular prisms are always so distributed within the primitive form and within that of secondary crystals that, in considering little groups of two, four, or six, they make up parallelepipeds; as a consequence, the ranks subtracted by the effect of decrement are nothing other than the sum of these parallelepipeds.

[a] Haüy's ultimate particle in this respect is simply Aristotle's natural minimum applied to Lavoisier's "simple substances"; but to this relic of scholasticism he has added number and goniometry (*Editor's Note*).

[*Editor's Note:* Material omitted at this point.]

135. We have given the name *subtractive molecules* to these parallelepipeds composed of tetrahedra or triangular prisms, the rows of which measure the quantity of decrement that the layers of superposition experience. These parallelepipeds alone are required to conclude the calculations; and the kind of anatomy which these little solids undergo beyond this when one attempts to arrive at the true form of the integrant molecule is an affair of pure observation, which is foreign to the theory. The parallelepiped here represents unity; whatever is beyond this unity, the compound fractions of its subdivisions, is of little importance. At the center of this conformity between the results given by the various forms of integrant molecules, the theory has the advantage of being able to generalize its object, in leading back to a single element that multitude of forms which by their diversity seem hardly susceptible of coming together in a common point.

[*Editor's Note:* Material omitted at this point.]

Plate II

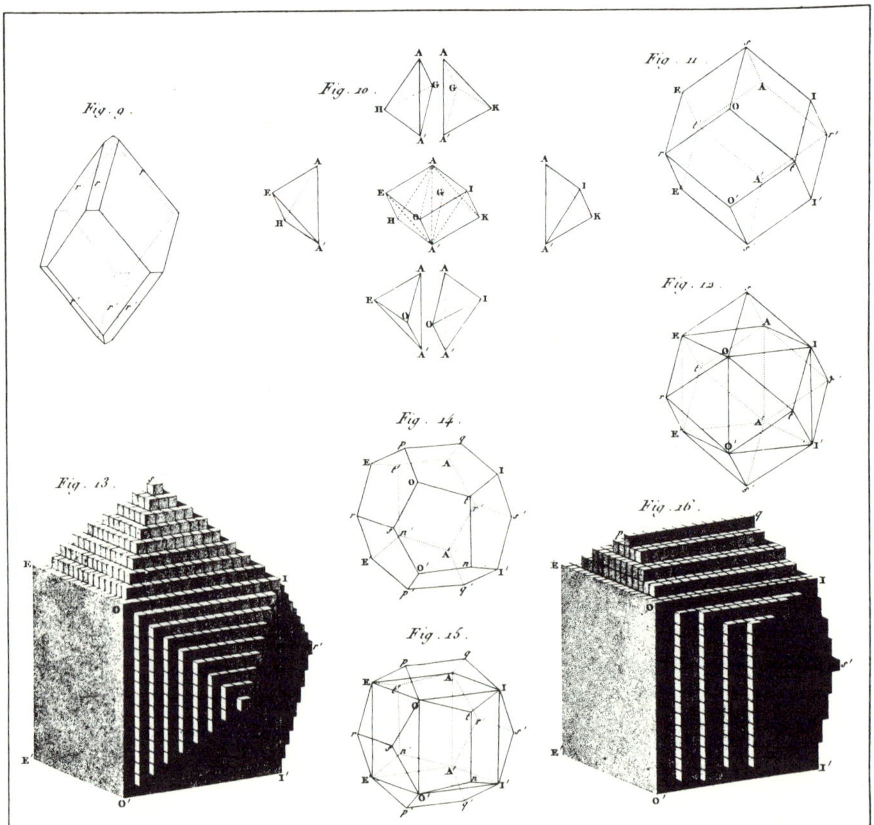

3

THE CRYSTAL CONSIDERED AS A SIMPLE ASSEMBLAGE OF POINTS

Auguste Bravais

This abridged translation is by Francette Bonnabaud, pp. 165–170, 193 of Études cristallographiques, Part I: Du Cristal considéré comme un simple assemblage de points, *Paris, 1849, pp. 101–194*

[*Editor's Note:* In the original, material precedes this excerpt.]

X. *The determination of the crystalline mode* [*lattice*] *and the primitive form of a mineral species.*

Once the crystalline system to which a mineral species belongs is known, it remains to determine the particular mode to which it belongs and the linear parameters that fix the final form of the solid [unit cell] generating its structure.

One may still call this solid a primitive form, its faces always being parallel to those of one or several crystal forms of the substance. This polyhedron is supposed to have one molecule on each vertex, and sometimes in the center of its figure or even on each center of its faces.

Two primitive forms, or, what amounts to the same thing, two solids generating the structure, are equivalent once they give birth to the same structure. One then has only one solution to the problem of the determination of the primitive form, but presented in two different ways.

Regarding crystals belonging to the regular or terquaternary system [isometric], one can consider the question solved as soon as one knows whether the formation of the terquaternary structure is based on the hexahedral, octahedral, or dodecahedral mode.

In the senary [hexagonal], quaternary [tetragonal], and ternary systems, there exist only two distinct parameters, of which one can be taken to be unity; the other parameter then makes up the only unknown magnitude of the question.

In the terbinary [orthorhombic] system, the mineralogist will first have to make his choice from one of the four modes that belong to this system. Then he will have to determine two of the three linear parameters $\bar{a}$, $\bar{b}$, and $\bar{c}$; the third can always be assumed to be equal to unity.

The binary [monoclinic] system presents four linear parameters, $\bar{a}$, $\bar{b}$, $\bar{c}$, and $\bar{d}$; if one assumes $\bar{d} = 1$, $\bar{a}$, $\bar{b}$, and $\bar{c}$ remain to be calculated, i.e., the three sides of the main triangle of the network normal to the binary axis of the system. One will then have to choose in the same manner between the two crystalline modes which pertain to this system; if one chooses the octahedral mode, according to what I pointed out in my *Mémoire sur les systèmes de points* (fn. p. 97), one will have to determine on which one of the three sides of the main triangle of the network normal to the axis the vertex, the center of the generating prism, is projected.

Finally, in the asymmetric [anorthic or triclinic] system, the number of clearly defined undetermined values is equal to five: it is, for example, the five edges of the main tetrahedron, the sixth one being considered the unit of length.

To obtain the value of the parameters, we have at our disposal knowledge of the slopes of the faces that limit the crystal and the indices derived from the arrangement of the cleavage planes. As a result, after determining a system of three linear parameters *a*, *b*,, and *c*, satisfying the conditions derived from these indices, we can, nevertheless without failing to take into consideration the external structure, replace *a*, *b*, and *d* by three other parameters *ma*, *nb*, and *pd* of the same relative direction, but with different ratios of magnitude; here *m*, *n*, and *p* are integers that can be any number, except in cases where symmetry would require that two or three of these parameters be equal.

To remove as much as possible the indeterminism of the numerical factors *m*, *n*, and *p*, Haüy and his successors have been following up to now this rule. Among the parameters, they adopted ratios of magnitude that made the symbolic notations of the most common faces of the crystal as simple as possible. For example, Haüy, speaking about his research on "euclase," says

> I was looking for that ratio which the three dimensions of the prism must have so that the ensemble of the laws of decrement, which would have been derived from it, would depart as little as possible from the simplicity of the ordinary laws[a]

Now, the application of such a rule obviously is to a large extent arbitrary, since crystallographers do not even agree with each other on the symbols of the plane faces, and the same face with simple notation for one author is often given a much more complex notation in another author's system. Consequently, mineralogists usually give up the matter once it has been pursued up to the point where one, two, or three systems of values for *m*, *n*, and *p* give a sufficiently simple account of the external structure. However, out of two or three solutions so obtained, only one is natural.

To remove this indeterminism as much as possible, I shall use a hypothesis that will not offer, it is true, the degree of rigor of the results to which we have been led so far, but which will have the advantage over the old methods of having a theoretical foundation, of leaving only a slight role to the calculator's discretion and, finally, of [being susceptible] to the most diversified verifications.

I will assume that in a crystalline structure the reticular planes that are easiest to separate from their neighbors by using cleavages are in general the ones which serve as limits to the thicker layers.

These are the theoretical reflections that support this point of view.

[a] Haüy, *Cristallographie*, 2nd ed., Vol. II, p.155.

When the cleavage separates two adjacent reticular planes, one from the other, one has to consider, on the one hand, the force with which the molecules on one plane are bound together, called *the cohesion tangential to the plane*, or the resistance to rupture along lines drawn in this plane; one must consider, on the other hand, the force with which the molecules on a plane maintain a certain distance from the molecules of the adjacent plane, which can be called the *cohesion normal to the plane*, or resistance to rupture in the direction parallel to this plane. We know in general that the cohesion of bodies increases as the average intervals of the molecules decrease. From that, one can deduce that, all else being equal, the tangential cohesion will be the greater as the network of the plane is denser, and that the normal cohesion will be the lesser as the spacing between two adjacent planes increases.

When the various systems of reticular planes of the crystal are examined one after the other, the density of the network and the spacing of the two adjacent planes for each system are two quantities that increase and decrease together, always remaining proportional to one another. One has indeed, according to equation (51) of the *Mémoire sur les systèmes de points*,

$$\Delta S(ghk) = \Omega = \text{constant}$$

and, consequently (equation 68 of the present *Mémoire*),

$$\Delta = \Omega \rho$$

Thus, as we deal with faces whose reticular density increases more and more, the tangential cohesion will increase and the normal cohesion will decrease; consequently, the tendency to cleavage along these faces will become more and more obvious, since tangential cohesion favors it, whereas the normal cohesion tends to prevent it. Therefore, the easiest cleavage will have to be parallel to the reticular plane of maximum density; and if the crystal can be split parallel to two or three crystalline forms, the order of ease of these cleavages will have to correspond to the order of decreasing densities of the corresponding reticular nets.

In almost the same way, we will show that the faces most likely to occur in the process of the crystallization, or, in other words, the reticular planes most likely to limit the crystal, are also, *in general*, the ones with the most considerable reticular density. For the internal movements that act incessantly at the surface of the crystalline mass during its formation play, up to a certain point, the role of external forces that would cleave a crystal. Crystals must thus respond to the molecular ranks, which are arranged in reticular planes whose networks have the most closely knit textures. From this point of view, the faculty of effective limitation, which each possible face of the crystal possesses, must be, at least in part, proportional to the density of its texture. However, this criterion seems to be less certain than the one inferred from the consideration of cleavage, for the formation of natural faces has been subject, during crystallization, to many external forces that are difficult to analyze, and which may have favored the growth of one or another crystalline form or may not even have acted equally on the various faces of the crystal. On the contrary, when the crystal, definitively constituted, is later subjected to the processes of cleavage, it is removed from all traces of the action of the forces which originally modified its exterior morphology, and may provide the observer with independent indices of its initial state.[b]

We continue by stating clearly the rule that we substitute for the empirical law of the least complex notation, and from now on we will use this rule to determine the crystalline mode of a mineral species, together with the ratios of its parameters. We must, in every particular case,

> choose, from the infinite number of solutions which the determination of the primary form entails, the one which assigns to the various possible faces of the crystal reticular densities whose decreasing series represents as accurately as possible the decreasing series of ease of production, either naturally (through crystallization) or artificially (through cleavage) of these same faces.

It is understood that the readiness of natural production of the faces will be ascertained after a careful examination of a great number of samples of the same species, gathered in circumstances as varied as possible.

In cases where the indices deduced from the cleavage would not be completely in agreement with those inferred from the examination of dominant external forms, our rule will allow for arbitrary interpretation until a close application of the method enables us to judge the degree of validity that these various indices deserve.

This correspondence of a certain kind of cleavage with the ratios of length of the linear parameters has already been noticed by Mr. Frankenheim,[c] who especially points out that in the quaternary crystals of apophyllite and uranite, where $a < d$, the cleavage is parallel to the base.

Given the reticular density of the face (ghk) equals $1/S(ghk)$, instead of arranging the faces according to the decreasing values of this expression, we will be able to classify them according to increasing values of the elementary area $S(ghk)$, or rather of the square of this area, in order to avoid second-degree radicals.

If there exists in $S(ghk)$ or in $S^2(ghk)$ factors independent of g, h, and k, we will be able to omit them, without changing the order of succession of the crystalline forms. We will then designate as $\Sigma(ghk)$ and $\Sigma^2(ghk)$ the values assumed by $S(ghk)$ and $S^2(ghk)$ after being rid of such factors; and after having replaced, in equations (69) to (102),[d] S and S^2 by Σ and Σ^2, we will be able to omit or introduce on the right side any factor as long as it is not a function of the indices.

[b] This proposition must not be taken in too rigorous a sense. The same mineral species does not always have the same equally easy cleavages; but such variations are far less in extent than those which occur in the degree of predominance of the same faces among various specimens of the same species.

[c] Frankenheim, *Die Lehre von Cohaesion*, p. 306.

[d] *Editor's Note*: Equations 69 through 102 are for the calculation of reticular densities of faces (hkl) for each lattice type.

[*Editor's Note:* Material omitted at this point.]

It could even happen that these actions favor one whole series of forms to the detriment of another series, which would alter, in an even more serious manner the results deduced from our fundamental hypothesis. Despite these restrictions, I have no doubt that this hypothesis can be really useful to crystallography and molecular physics. I will have achieved my aim if I succeed in encouraging mineralogists to submit a certain number of species properly studied to the comparative calculation of the reticular density of their faces with the tendency of these same faces to occur in crystallization.

[*Editor's Note:* Material omitted at this point.]

4

THE CRYSTAL CONSIDERED AS AN ASSEMBLAGE OF POLYATOMIC MOLECULES[a]

August Bravais

This abridged translation is by Francette Bonnabaud, from pp. 194–195, 197–202 of Études cristallographiques, Part II: Du Cristal considéré comme un assemblage de molécules polyatomiques, *Paris, 1849, pp. 194–247*

[*Editor's Note:* In the original, material precedes this excerpt.]

1. SYMMETRY OF THE MOLECULES OF CRYSTALLINE BODIES

In the preceding research, we dealt exclusively with the relative arrangement of the centers of gravity of the molecules of crystalline bodies, an arrangement that constitutes the *crystalline structure*, to designate it properly. We will now deal with a more delicate problem, that of the molecular structure; by that we mean the geometric disposition of the elements which constitute the molecule around its center of gravity.

The external signs capable of revealing the structure of the molecule, if it is compound, or at least its form, if it is simple, can only be the results of attractive or repulsive forces exerted by the molecule in various directions. If these forces spread spherically from the center, if they were the same for a given distance in all directions, if in short all was even about the center, the molecule would act as if it were spherical, or even better, as a simple point, the pole of forces; its own structure would not manifest itself through any perceptible phenomenon and could only be deduced from theoretical considerations, for example the atomic laws of chemistry.

But things do not happen this way. The forces emanating from the molecule are not the same in every sense; the reactions which engage it do not always pass through its center of gravity, and the effects of these irregularities permit us, when they are properly studied, to penetrate its inner organization.

We shall try to show that a molecule possesses various centers or poles of forces. These poles are probably the constituent atoms of the molecule; but this last point of view, purely chemical, is to a certain extent foreign to crystallography, and I will only establish the existence of *molecular polarity* independently of further consequences that could be deduced from it.

[a] Presented to the Academy of Sciences 6 Aug. 1849. Feb. 24, 1851, a committee of MM. Dufrénoy, Regnault, Lamé, and Cauchy, speaker, reported favorably on this memoir to the Academy.

[*Editor's Note:* Material omitted at this point.]

All molecules of the same substance must be identical and superposable. In crystallization, the homologous straight lines of these molecules become parallel, as we have indicated at the beginning of the first part of this Mémoire; thus the molecular orientation is the same from one end of the crystal to the other. The molecular polarity and the variation of forces with direction are consequences of the polyhedral figure of the molecule and of the law according to which the forces emanating from each vertex vary with distance. When the intervals between two molecules become sufficiently important, the influence of the interior dimensions of each decreases, and they tend more and more to act as simple points; but the distances at which they occur in crystallization are necessarily less than the limit we have just indicated. One sees thus why their interactions do not generally pass through the centers of gravity, and one understands how the parallel orientation of homologous lines of two adjacent molecules, an effect connected with all crystallization, must result from the rotation that these actions tend to produce about their centers of gravity.

After having established the existence of molecular polyhedrons, we will prove that these polyhedrons are symmetrical and that they possess, as does the crystalline assemblage at whose vertices they settle, appropriate axes, center, or planes of symmetry.

I call the center of symmetry of a polyhedron a central point such that, by extending [back] the line joining it to any vertex of this polyhedron by a quantity equal to itself, the end of this straight line is also a vertex of the polyhedron. It is obvious that, in a limited polyhedron, there can be only one center of symmetry and that it necessarily coincides with the center of gravity.

The axes of symmetry will be straight lines of necessity passing through the center of gravity and such that, by rotating the polyhedron around them through an angle equal to an aliquot part of 360°, the apparent location of the vertices is not disturbed by this rotation. These axes can be binary, ternary, quaternary, etc., according to whether the rotation that restores the places of the vertices is equal to ½, ⅓, ¼, . . . revolution.

Finally, the planes of symmetry divide the polyhedron into two geometrically symmetric halves and must also contain its center of gravity.

This granted, and to fix our ideas, consider the molecule of a crystal belonging to the terquaternary [isometric] system, whose most general crystalline form consists of 48 faces of the same species. Two different cases may occur: either the molecular polyhedron will possess all elements of symmetry of the corresponding assemblage, or it will possess only part of it.

In the first case, the molecular polyhedron will have three quaternary axes like the quaternary axes of the assemblage, four ternary axes, and six binary axes coinciding with the corresponding axes of the assemblage; it will possess nine characteristic planes of symmetry at its center of gravity.

Let F and F' be two directly similar faces of the same form, and consider the molecule M belonging to the reticular plane F, and also the molecule M' belonging to the reticular plane F'.

We translate the plane F parallel to itself, so as to bring M' into M; then rotate the mobile assemblage around the center of gravity common to the coincident molecules so that the mobile network of F' coincides with the fixed network of F. We will produce this superposition through an angular rotation of (p/q) 360° around a particular axis of symmetry L^q of the assemblage, q being the order number of this axis and p being an integer smaller than q. Now, by hypothesis, this axis also belongs to the molecule M'; if, then, this molecule takes part in the motion of F' and the mobile assemblage, it will resume, after the rotation, the same apparent position and will again coincide with M.

From this we see that, in this case, everything is the same for the molecules of plane F, on the one hand, and the molecules of plane F', on the other hand; thus the two faces F and F' are in identical conditions of occurrence as limiting faces of the crystal.

It would be the same if the face F' were inversely similar to the face F with respect to a plane of symmetry P. If the plane drawn parallel to P through the center of gravity of the molecule is a plane of symmetry of the molecular polyhedron, F and F' would still have to coexist.

Finally, in the case where the faces F and F' are inversely similar with respect to a center of symmetry of the assemblage, i.e., when these faces are parallel but with internal and external sides turned in an inverse sense, let C (Figure 8) be the center of symmetry of the system formed by the two faces, and p that pole of the molecule M with center of gravity at M; then all will be the same between the upper surface of F and the lower surface of F', if p_1', the homolog of p with respect to C, is also a pole of the molecule M'. Now, if one draws $M'p'$ equal and parallel to Mp, p' must be a pole of M' by virtue of the parallelism of the molecules of crystals. Thus the coexistence of the faces F and F' depends on that of the poles p' and p_1', i.e., on the condition that M' is a center of symmetry for the molecule M'. If then the molecular polyhedron possesses such a center, the presence of F as a limiting face of the upper half of the crystal will entail the presence of its parallel F' as a limiting face of the lower half.

From the above, it follows that if one restricts consideration to the conditions of coexistence of the 24 faces of the direct demiform (see the definitions on p. 110) in the terquaternary system, one will recognize the necessity of the presence of the 13 axes of symmetry of the system in the molecular polyhedron, and that the inverse demiform will accompany the direct demiform, if the molecular polyhedron also possesses a center of symmetry.

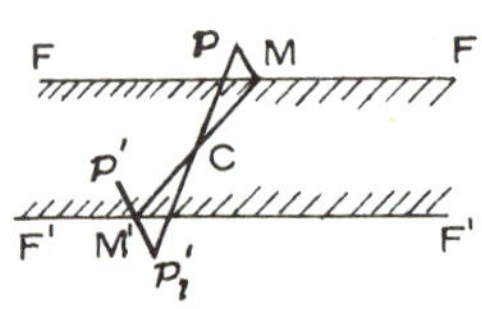

Figure 8

Now we examine the second case and we assume that the axis L^q, which serves to effect the coincidence of F with F', does not exist as an axis of symmetry in the molecular polyhedron.

If a rotation of (p/q) 360° has superposed the two networks of the faces F and F' and the centers of gravity of M and M' coincide, the polyhedron M' is no longer coincident with M, since L^q, by hypothesis, is no longer an axis of symmetry of this polyhedron. Therefore, if one considers F and F' as being faces bounding the crystal at a given moment during its formation in the middle of the fluid mass that produces it, the situation of all poles or centers of forces above and below the plane F will not be rigorously the same as that of the homologous poles above and below the plane F'; the resultant forces that attract a molecule still free and floating at a short distance from the plane F will not be the same as those which attract another molecule similarly placed with respect to the plane F'. Thus crystallization will be able to proceed differently at the two surfaces F and F' that separate the crystal from its generating medium; consequently, part of the faces of the form will not be able to occur.

If the polyhedron, though possessing the 13 axes of the terquaternary system, was deprived of a center of symmetry, the 24 faces of the inverse demiform could disappear and the direct semiform would occur alone.

One designates by the name *holohedral crystals* those with oblique [general] forms that are always complete, possessing 48 faces in the terquaternary system, 24 in the senary [hexagonal] system, 16 in the quaternary [tetragonal] system, etc. One sees therefore that:

> *In all holohedral crystals, the molecular polyhedron possesses the same axes and planes of symmetry as the assemblage and a center of symmetry at the figure's center.*[b]

One has designated *hemihedral, tetartohedral* those crystals for which the number of faces of the general form is reduced to one half or one fourth of its value: still more generally, we will designate these as *merohedral crystals.* We see, therefore, that:

> *For all merohedral crystals, some of the axes, center, or planes of symmetry of the assemblage do not appear in the molecular polyhedra.*

With appropriate inspection of these crystals, one can always recognize the missing axes and planes and break down the elements of the symmetry of the assemblage into two groups, one of which, common to both the assemblage and the molecule, accounts for the simultaneous presence of a certain number of faces of the same form, while the other, by its omission in the molecule, explains the suppression of the missing faces.

Whenever the superposition of the networks (réseaux) of two faces of the same species involves the complete coincidence of the molecular polyhedra, these faces will be called not only faces of the same species, but also *identical*, according to the definitions on page 106; in the opposite case, the faces will be of the same species, but not identical.

[b] The conditions stated in this theorem are redundant, and of the three kinds of elements, axes, planes, and center, two suffice to generate the third.

According to this, the law called "law of symmetry," which rules the mode of distribution of the faces of a crystal, will have to be stated in the following manner:

> *All faces belonging to the same crystalline form, and moreover identical, must coexist.*

We will now consider a molecular polyhedron whose degree of symmetry we will vary. In the beginning, this polyhderon will be supposed to possess the complete symmetry characteristic of its assemblage; then, by removing certain axes or planes of symmetry, we will gradually lower it, and this will give birth to merohedral crystals in which groups of identical faces will become less and less complex, without however their relative situations being altered. It seems, at first, that there is no limitation in nature to this progressive decrease of the symmetry of the molecule for a given crystalline system, and that one may encounter every possible case in the various mineral species. But observation proves that it is not so; molecular polyhedra poor in elements of symmetry do not aggregate in crystalline assemblages of high symmetry, and, one may say as a general thesis and in spite of a very small number of exceptions, that one does not see the forms and combinations of forms of the less symmetrical crystal systems display the angles and slopes characteristic of more symmetrical systems.

From this we may deduce that there exists a limit to the decrease of the symmetry of a molecular polyhedron crystallizing in a given system and that, if a sufficient number of its terms is removed, the symmetry of its crystallization will decrease also to a lesser degree; we can express this by saying that:

> *Polyhedra with complex symmetry crystallize in the systems with complex symmetry and those with simple symmetry in systems with simple symmetry.*

Such a correspondence cannot be mere chance, and since to a given molecular structure there corresponds a given related crystal structure, it seems very likely that molecular symmetry determines that of the crystal assemblage.

[*Editor's Note:* Material omitted at this point.]

Part II

VECTORIAL SURFACE ENERGY

Editor's Comments on Papers 5 Through 7

5 **GIBBS**
Excerpts from *On the Equilibrium of Heterogeneous Substances*

6 **CURIE**
On the Formation of Crystals and on the Capillary Constants of Their Different Faces

7 **WULFF**
Excerpts from *On the Question of the Rate of Growth and of Dissolution of Crystal Faces*

Gibbs's memoir of nearly 300 pages *On the Equilibrium of Heterogeneous Substances* was a theoretical system of thermodynamics embracing 700 consecutive equations derived as an architectural unit from the first and second laws. The rigor and the power of this edifice and the intellectual austerity contribute to the difficulties of approach. In the passages selected (Paper 5), Gibbs describes the differential coefficients of σ, the surface free energy, with respect to the direction cosines of the surface as "discontinuous functions of the latter quantities"—the cusps, in short, of the modern Wulff plots. Pure mathematics lead him to the condition for equilibrium of a crystal in contact with its mother fluid, but in a lengthy footnote he displays physical insight and subtlety of mind not inferior to the analytic powers of the text. Gibbs's footnote is a warning of those difficulties and complexities with which the subject is beset today.

G. Wulff independently discovered the Bragg equation and published it in 1913. Before the turn of the century, he had undertaken an exhaustive series of meticulous experimental tests of Curie's theory. Paper 7, first published in Russian in 1895, reported the results of those experiments and the theoretical implications of Curie's ideas, confirmed by the same experimental results. His principal conclusions, Wulff's theorem, the concepts of the Wulff point, vectors, and the surface energy plot, appear here for the first time.

It does not occur to Wulff, as it did to Friedel, that there may be a discrepancy between the thermodynamic approach of Curie and the molecular approach of Bravais. He saw these in fact as complementary aspects of the same idea. Bravais mechanics are the origin of the surface energies, which in interaction with environmental factors give rise to the observed differences in capillary constants. It was to these differences that his experiments were directed and that he understood his theorem to describe.

5

Reprinted from pp. 315–316, 320–322, 324–326 of J. Willard Gibbs, *The Scientific Papers of J. Willard Gibbs*, Vol. I, *Thermodynamics*, Longmans, Green & Company, London, 1906

ON THE EQUILIBRIUM OF HETEROGENEOUS SUBSTANCES

J. W. Gibbs

[*Editor's Note*: In the original, material precedes this excerpt.]

As in the case of two fluid masses (see page 257), we may regard σ as expressing the work spent in forming a unit of the surface of discontinuity—under certain conditions, which we need not here specify—but it cannot properly be regarded as expressing the tension of the surface. The latter quantity depends upon the work spent in *stretching* the surface, while the quantity σ depends upon the work spent in *forming* the surface. With respect to perfectly fluid masses, these processes are not distinguishable, unless the surface of discontinuity has components which are not found in the contiguous masses, and even in this case (since the surface must be supposed to be formed out of matter supplied at the same potentials which belong to the matter in the surface) the work spent in increasing the surface infinitesimally by stretching is identical with that which must be spent in forming an equal infinitesimal amount of new surface. But when one of the masses is solid, and its states of strain are to be distinguished, there is no such equivalence between the stretching of the surface and the forming of new surface.*

* This will appear more distinctly if we consider a particular case. Let us consider a thin plane sheet of a crystal in a vacuum (which may be regarded as a limiting case of a very attenuated fluid), and let us suppose that the two surfaces of the sheet are alike. By applying the proper forces to the edges of the sheet, we can make all stress vanish in its interior. The *tensions* of the two surfaces are in equilibrium with these forces, and are measured by them. But the tensions of the surfaces, thus determined, may evidently have different values in different directions, and are entirely different from the quantity which we denote by σ, which represents the work required to form a unit of the surface by any reversible process, and is not connected with any idea of direction.

In certain cases, however, it appears probable that the values of σ and of the superficial tension will not greatly differ. This is especially true of the numerous bodies which, although generally (and for many purposes properly) regarded as solids, are really very viscous fluids. Even when a body exhibits no fluid properties at its actual temperature, if its surface has been formed at a higher temperature, at which the body was fluid, and the change from the fluid to the solid state has been by insensible gradations, we may suppose that the value of σ coincided with the superficial tension until the body was decidedly solid, and that they will only differ so far as they may be differently affected by subsequent variations of temperature and of the stresses applied to the solid. Moreover, when an amorphous solid is in a state of equilibrium with a solvent, although it may have no fluid properties in its interior, it seems not improbable that the particles at its surface, which have a greater degree of mobility, may so arrange themselves that the value of σ will coincide with the superficial tension, as in the case of fluids.

With these preliminary notions, we now proceed to discuss the condition of equilibrium which relates to the dissolving of a solid at the surface where it meets a fluid, when the thermal and mechanical conditions of equilibrium are satisfied. It will be necessary for us to consider the case of isotropic and of crystallized bodies separately, since in the former the value of σ is independent of the direction of the surface, except so far as it may be influenced by the state of strain of the solid, while in the latter the value of σ varies greatly with the direction of the surface with respect to the axes of crystallization, and in such a manner as to have a large number of sharply defined minima.* This may be inferred from the phenomena which crystalline bodies present, as will appear more distinctly in the following discussion. Accordingly, while a variation in the direction of an element of the surface may be neglected (with respect to its effect on the value of σ) in the case of isotropic solids, it is quite otherwise with crystals. Also, while the surfaces of equilibrium between fluids and soluble isotropic solids are without discontinuities of direction, being in general curved, a crystal in a state of equilibrium with a fluid in which it can dissolve is bounded in general by a broken surface consisting of sensibly plane portions.

[*Editor's Note*: Material omitted at this point.]

Let us apply the same method to the case in which the solid is a crystal. The surface between the solid and fluid will now consist of plane portions, the directions of which may be regarded as invariable. If the crystal grows on one side a distance δN, without other change, the increment of energy in the vicinity of the surface will be

$$(\epsilon_V' - \epsilon_V'') s\, \delta N + \Sigma' (\epsilon_{S(1)}' l' \operatorname{cosec} \omega' - \epsilon_{S(1)} l' \cot \omega') \delta N,$$

* The differential coefficients of σ with respect to the direction-cosines of the surface appear to be discontinuous functions of the latter quantities.

where ϵ_V' and ϵ_V'' denote the volume-densities of energy in the crystal and fluid respectively, s the area of the side on which the crystal grows, $\epsilon_{S(1)}$ the surface-density of energy on that side, $\epsilon_{S(1)}'$ the surface-density of energy on an adjacent side, ω' the external angle of these two sides, l' their common edge, and the symbol Σ' a summation with respect to the different sides adjacent to the first. The increments of entropy and of the quantities of the several components will be represented by analogous formulæ, and if we deduce as on pages 316, 317 the expression for the increase of energy in the whole system due to the growth of the crystal without change of the total entropy or volume, and set this expression equal to zero, we shall obtain for the condition of equilibrium

$$(\epsilon_V' - t\eta_V' - \mu_1''\gamma_1' + p'')s\,\delta N + \Sigma'(\sigma' l' \operatorname{cosec}\omega' - \sigma l' \cot\omega')\delta N = 0, \quad (664)$$

where σ and σ' relate respectively to the same sides as $\epsilon_{S(1)}$ and $\epsilon_{S(1)}'$ in the preceding formula. This gives

$$\mu_1'' = \frac{\epsilon_V' - t\eta_V' + p''}{\gamma_1'} + \frac{\Sigma'(\sigma' l' \operatorname{cosec}\omega' - \sigma l' \cot\omega')}{s\gamma_1'}. \quad (665)$$

It will be observed that unless the side especially considered is small or narrow, we may neglect the second fraction in this equation, which will then give the same value of μ_1'' as equation (387), or as equation (661) applied to a plane surface.

Since a similar equation must hold true with respect to every other side of the crystal of which the equilibrium is not affected by meeting some other body, the condition of equilibrium for the crystalline form (when unaffected by gravity) is that the expression

$$\frac{\Sigma'(\sigma' l' \operatorname{cosec}\omega' - \sigma l' \cot\omega')}{s} \quad (666)$$

shall have the same value for each side of the crystal. (By the value of this expression for any side of the crystal is meant its value when σ and s are determined by that side and the other quantities by the surrounding sides in succession in connection with the first side.) This condition will not be affected by a change in the size of a crystal while its proportions remain the same. But the tendencies of similar crystals toward the form required by this condition, as measured by the inequalities in the composition or the temperature of the surrounding fluid which would counterbalance them, will be inversely as the linear dimensions of the crystals, as appears from the preceding equation.

If we write v for the volume of a crystal, and $\Sigma(\sigma s)$ for the sum of the areas of all its sides multiplied each by the corresponding value of σ, the numerator and denominator of the fraction (666), multiplied each by δN, may be represented by $\delta\Sigma(\sigma s)$ and δv

respectively. The value of the fraction is therefore equal to that of the differential coefficient

$$\frac{d\Sigma(\sigma s)}{dv}$$

as determined by the displacement of a particular side while the other sides are fixed. The condition of equilibrium for the form of a crystal (when the influence of gravity may be neglected) is that the value of this differential coefficient must be independent of the particular side which is supposed to be displaced. For a constant volume of the crystal, $\Sigma(\sigma s)$ has therefore a minimum value when the condition of equilibrium is satisfied, as may easily be proved more directly.

When there are no foreign substances at the surfaces of the crystal, and the surrounding fluid is indefinitely extended, the quantity $\Sigma(\sigma s)$ represents the work required to form the surfaces of the crystal, and the coefficient of $s\,\delta N$ in (664) with its sign reversed represents the work gained in forming a mass of volume unity like the crystal but regarded as without surfaces. We may denote the work required to form the crystal by

$$W_S - W_V,$$

W_S denoting the work required to form the surfaces {i.e., $\Sigma(\sigma s)$}, and W_V the work gained in forming the mass as distinguished from the surfaces. Equation (664) may then be written

$$-\delta W_V + \Sigma(\sigma\,\delta s) = 0. \tag{667}$$

Now (664) would evidently continue to hold true if the crystal were diminished in size, remaining similar to itself in form and in nature, if the values of σ in all the sides were supposed to diminish in the same ratio as the linear dimensions of the crystal. The variation of W_S would then be determined by the relation

$$d W_S = d\Sigma(\sigma s) = \tfrac{3}{2}\Sigma(\sigma\, ds),$$

and that of W_V by (667). Hence,

$$dW_S = \tfrac{3}{2} dW_V,$$

and, since W_S and W_V vanish together,

$$W_S = \tfrac{3}{2} W_V,$$

$$W_S - W_V = \tfrac{1}{3} W_S = \tfrac{1}{2} W_V, \tag{668}$$

—the same relation which we have before seen to subsist with respect to a spherical mass of fluid as well as in other cases. (See pages 257, 261, 298.)

The equilibrium of the crystal is unstable with respect to variations in size when the surrounding fluid is indefinitely extended, but it may be made stable by limiting the quantity of the fluid.

[*Editor's Note*: Material omitted at this point.]

The manner in which crystals actually grow or dissolve is often principally determined by other differences of phase in the surrounding fluid than those which have been considered in the preceding paragraph. This is especially the case when the crystal is growing or dissolving rapidly. When the great mass of the fluid is considerably supersaturated, the action of the crystal keeps the part immediately contiguous to it nearer the state of exact saturation. The farthest projecting parts of the crystal will therefore be most exposed to the action of the supersaturated fluid, and will grow most rapidly. The same parts of a crystal will dissolve most rapidly in a fluid considerably below saturation.*

But even when the fluid is supersaturated only so much as is necessary in order that the crystal shall grow at all, it is not to be expected that the form in which $\Sigma(\sigma s)$ has a minimum value (or such a modification of that form as may be due to gravity or to the influence of the body supporting the crystal) will always be the ultimate result. For we cannot imagine a body of the internal structure and external form of a crystal to grow or dissolve by an entirely continuous process, or by a process in the same sense continuous as condensation or evaporation between a liquid and gas, or the corresponding processes between an amorphous solid and a fluid. The process is rather to be regarded as periodic, and the formula (664)

*See O. Lehmann, "Ueber das Wachsthum der Krystalle," *Zeitschrift für Krystallographie und Mineralogie*, Bd. i, S. 453; or the review of the paper in Wiedemann's *Beiblätter*, Bd. ii, S. 1.

cannot properly represent the true value of the quantities intended unless δN is equal to the distance between two successive layers of molecules in the crystal, or a multiple of that distance. Since this can hardly be treated as an infinitesimal, we can only conclude with certainty that sensible changes cannot take place for which the expression (664) would have a positive value.*

[*Editor's Note*: Material omitted at this point.]

---

* That it is necessary that certain relations shall be precisely satisfied in order that equilibrium may subsist between a liquid and gas with respect to evaporation, is explained (see Clausius, "Ueber die Art der Bewegung, welche wir Wärme nennen," *Pogg. Ann.*, Bd. c, S. 353 ; or *Abhand. über die mechan. Wärmetheorie*, XIV) by supposing that a passage of individual molecules from the one mass to the other is continually taking place, so that the slightest circumstance may give the preponderance to the passage of matter in either direction. The same supposition may be applied, at least in many cases, to the equilibrium between amorphous solids and fluids. Also in the case of crystals in equilibrium with fluids, there may be a passage of individual molecules from one mass to the other, so as to cause insensible fluctuations in the mass of the solid. If these fluctuations are such as to cause the occasional deposit or removal of a whole layer of particles, the least cause would be sufficient to make the probability of one kind of change prevail over that of the other, and it would be necessary for equilibrium that the theoretical conditions deduced above should be precisely satisfied. But this supposition seems quite improbable, except with respect to a very small side.

The following view of the molecular state of a crystal when in equilibrium with respect to growth or dissolution appears as probable as any. Since the molecules at the corners and edges of a perfect crystal would be less firmly held in their places than those in the middle of a side, we may suppose that when the condition of theoretical equilibrium (665) is satisfied several of the outermost layers of molecules on each side of the crystal are incomplete toward the edges. The boundaries of these imperfect layers probably fluctuate, as individual molecules attach themselves to the crystal or detach themselves, but not so that a layer is entirely removed (on any side of considerable size), to be restored again simply by the irregularities of the motions of the individual molecules. Single molecules or small groups of molecules may indeed attach themselves to the side of the crystal but they will speedily be dislodged, and if any molecules are thrown out from the middle of a surface, these deficiencies will also soon be made good ; nor will the frequency of these occurrences be such as greatly to affect the general smoothness of the surfaces, except near the edges where the surfaces fall off somewhat, as before described. Now a continued growth on any side of a crystal is impossible unless new layers can be formed. This will require a value of μ_1'' which may exceed that given by equation (665) by a finite quantity. Since the difficulty in the formation of a new layer is at or near the commencement of the formation, the necessary value of μ_1'' may be independent of the area of the side, except when the side is very small. The value of μ_1'' which is necessary for the growth of the crystal will however be different for different kinds of surfaces, and probably will generally be greatest for the surfaces for which σ is least.

On the whole, it seems not improbable that the form of very minute crystals in equilibrium with solvents is principally determined by equation (665), (i.e., by the condition that $\Sigma(\sigma s)$ shall be a minimum for the volume of the crystal except so far as the case is modified by gravity or the contact of other bodies), but as they grow larger (in a solvent no more supersaturated than is necessary to make them grow at all), the deposition of new matter on the different surfaces will be determined more by the nature (orientation) of the surfaces and less by their size and relations to the surrounding surfaces. As a final result, a large crystal, thus formed, will generally be bounded by those surfaces alone on which the deposit of new matter takes place least readily, with small, perhaps insensible truncations. If one kind of surfaces satisfying this condition cannot form a closed figure, the crystal will be bounded by two or three kinds of surfaces determined by the same condition. The kinds of surface thus determined will probably generally be those for which σ has the least values. But the relative development of the different kinds of sides, even if unmodified by gravity or the contact of other bodies, will not be such as to make $\Sigma(\sigma s)$ a minimum. The growth of the crystal will finally be confined to sides of a single kind.

It does not appear that any part of the operation of removing a layer of molecules presents any especial difficulty so marked as that of commencing a new layer; yet the values of μ_1'' which will just allow the different stages of the process to go on must be slightly different, and therefore, for the continued dissolving of the crystal the value of μ_1'' must be less (by a finite quantity) than that given by equation (665). It seems probable that this would be especially true of those sides for which σ has the least values. The effect of dissolving a crystal (even when it is done as slowly as possible) is therefore to produce a form which probably differs from that of theoretical equilibrium in a direction opposite to that of a growing crystal.

6

Reprinted from *J. Chem. Educ.*, **47**, 636–637 (Sept. 1970)

Translated by
Cecil J. Schneer
Department of Geology
The University of New Hampshire
Durham, New Hampshire 03824

Pierre Curie: On the Formation of Crystals and on the Capillary Constants of their Different Faces

In the theory that he has given for capillary phenomena of liquids, Gauss takes as his point of departure, the hypothesis of central forces for intermolecular actions; he asserts that molecules can only affect each other over very small distances. Finally he makes use of the principle of virtual velocity.

He considers the virtual work done by the capillary forces as given by the derivative of a certain function, and he shows that a part of the terms of this function depend only on the volume of the liquid while the other part is proportional to the surface area.

The part which depends on the volume is invariable if one supposes that the liquid is incompressible; the other part shows that the virtual work done by the capillary forces is proportional to the variation of the surface.

Now it is possible to bring forth considerations which are both more elementary and more general without being obliged to make use of the hypothesis of central forces. What we are going to say could apply not only to liquids but to any body whatsoever so long as it is deformable without changing its identity or its volume.

Given such a body and without considering any external forces other than capillary forces, the internal energy is the same for all elements of the same volume sufficiently far from the surface; by contrast at the surface there is an extremely thin bed of transition. The elements of volume of this bed have an average energy differing sensibly from that of the interior elements. From this we consider that within the total energy, one part is proportional to the volume, the other part is proportional to the transition bed which is to say, to the surface.

When the body is deformed the bulk energy is constant and the total energy varies proportionately to the variation of the surface.

The capillary constant A characteristic of the surface of separation of the two *milieux* is the energy that it is necessary to expend to increase by a single unit the surface of separation.

If the body is isolated from all forces except capillary forces, then since the system tends to have a minimum energy, the surface of separation tends to be the smallest possible and the body takes the form of a sphere.

If several surfaces of separation S, S_1, S_2, of capillary constants A, A_1, A_2, limit the body, the stable form will be that which will give a minimum for the quantity $AS + A_1S_1 + A_2S_2$.

Let us now consider a crystal in its saturated mother liquor and let us suppose that certain parts dissolve and come to be deposited on other parts. The crystal is thus deformable without either it or its mother liquor experiencing any change in composition or volume. If one neglects the essentially minimal effects due to the weight, the energy for the surface of separation of the crystal and its mother liquor will be the only variable and the most stable form will be that for which the sum of the surface energies is the smallest possible.

To each kind of face there must correspond a distinct capillary constant, for if not, the crystal in its mother liquor would tend to take the form of a sphere.

The dominant form ought to be that given by the faces for which the capillary constant is the weakest.

Granted these principles we can now treat a certain number of problems.

Problem: Let us suppose that a crystal takes the form of a quadratic prism in its mother liquor, what will be the relationship of the side of the base x to the height [of the prism] y which will characterize the most stable form?

We designate the capillary constant of the lateral faces A and that of the base B. The energy of the lateral face will be xyA, that of the base, x^2B; for the stable form the quantity

$$E = 4xyA + 2x^2B$$

ought to be a minimum.

For the rest

$$x^2y = V$$

V being the constant volume of the crystal, where

$$\frac{E}{2} = \frac{2AV}{x} + Bx^2$$

and

$$1/4\, \frac{dE}{dx} = -\frac{AV}{x^2} + Bx$$

For the minimum one should have [setting $dE/dx = 0$]

$$\frac{AV}{x^2} = Bx$$

TRANSLATOR'S NOTE: J. W. Gibbs in 1878 expressed the conditions for thermodynamic equilibrium for a crystal surface. This article by Pierre Curie, which was originally published in the *Bull. Soc. Fr. Mineral.*, **8**, 145 (1885), derives Gibbs' relation from Gauss' principle of virtual velocities. It was the point of departure for G. Wulff's (1901) classical experiments on crystal growth measuring relative capillary constants. Wulff's theorem in turn was proven in 1943 by von Laue.

The paper is a model of clarity and elegance. Mineralogists should note Curie's warning in the third from final paragraph; —the theory "seems probable, but experiment alone can tell." The second problem for which Curie gives only the result is not obvious. It should make for an interesting evening.

from which

$$Ay = Bx$$

from which

$$\frac{x}{y} = \frac{A}{B}$$

This has to be a minimum because $d^2E/dx^2 = 12B$, a positive quantity.

Problem: Given a crystal taking the form of a cubo-octahedron; what will be the stable form, A and B being the capillary constants of the cubic and octahedral faces?

Let x be the intercept made by the octahedral face on a cube edge (the distance taken from a trigonal corner of the cube reconstructed by the prolongation of the edges b [to the intercept]) we find for the stable form

$$x = \left(\frac{3}{2} - \frac{B}{A}\frac{\sqrt{3}}{2}\right)b$$

This formula shows that the crystal will be entirely cubic when one has

$$\frac{A}{B} < \frac{1}{\sqrt{3}}$$

and that it will be entirely octahedral when one has

$$\frac{A}{B} > \sqrt{3}$$

that it will take the form of a cubo-octahedron absolutely defined for every value A/B taken between $1/\sqrt{3}$ and $\sqrt{3}$.

Problem: Let us now consider several crystals possessing the stable form situated in the same mother liquor. What will happen? The ensemble of crystals will have a minimum energy when their total surface is the smallest possible, that is to say when all the crystals are combined in a single one.

Actually there will be another case of equilibrium; it is when all the crystals are exactly equal in size but it corresponds to an energy maximum and the equilibrium is unstable.

Thus the answer is in perfect agreement with experience; one knows, in effect, that in a saturated mother liquor *the largest crystal eats the little ones.*

The preceding theory therefore permits the resolution of an entire series of questions; it is uniquely based on general scientific principles insofar as it depends upon no particular hypothesis, it does not present any specific notion of the mechanism of formation of crystals. Every specific physical theory which sets out to explain such a mechanism must first not be in contradiction with this more general theory.

The theory predicts that in a saturated mother liquor all the crystals must unite themselves in a single one; it provides the dimensions of various faces of a crystal; it shows the possibility of existence of the smallest facets in a perfectly stable form.

In conclusion we will apply some restrictions on the manner of applying the preceding theory.

When a crystal is bathed in its saturated mother liquor at constant temperature, it is obvious that the crystal will not change form if it already possesses the most stable form, but if it possesses another will it deform itself spontaneously? This seems probable, but it is experiment alone which can reply.

It is certain that the crystal tends to take the form which corresponds to the minimum of energy, and that it will profit by every occasion that presents itself to accomplish this; but nothing in the theory says that it can do this spontaneously in its mother liquor.

Every time that one applies the principle of maximum work to a system, the mechanism of which one knows only imperfectly, one must still ask if a driving force, if a little preliminary work is not necessary to permit the calculated greater effect to occur. It is thus that an object placed on the edge of a table does not necessarily fall and that in chemical phenomena, a very large number of exothermic combinations do not occur spontaneously.

7

ON THE QUESTION OF THE RATE OF GROWTH AND OF DISSOLUTION OF CRYSTAL FACES

G. Wulff

This abridged translation was prepared expressly for this Benchmark volume by Deborah Watson, from pp. 450, 520–530 of "Zur Frage der Geschwindigheit des Wachstums und der Auflosung der Kristallflachen," Z. Krist., **34**, 449–530 (1901).

[*Editor's Note:* In the original, material precedes this excerpt.]

During the growth of the crystal, one usually notices that its diverse faces withdraw unequally from the initial point of origin; i.e., different faces of the same crystal advance with unequal speed. The diversity of external crystal forms (tabular, prismatic, acicular crystals, etc.) can be explained in this way.

Such variation in the rate of growth of the faces on the same crystal proves to be a natural consequence of the anisotropy of the crystalline media, and demonstrates that the energy with which the crystallized particles bond with others from the solution changes with direction.

Experience shows us that this difference in the rate of growth of crystal faces depends not only on the distribution of molecular forces around each individual molecule, but also on the reciprocal effect of these molecules and those which the solution surrounding the crystal contains. The latter forces likewise vary with direction, but they are also dependent on the characteristics of the solution. From this it follows that the same substance, crystallizing from various solutions, is covered with different crystal faces, so that it takes on various external forms, while its inner characteristics remain unchanged.

At this point it is appropriate to define more closely the concept of rate of growth of a single crystal in various directions.

[*Editor's Note:* Material omitted at this point.]

From these numerous examples, the consequent relationship between size of the capillary constant of a crystal face and the distance of the latter from a point common to all faces, which represents the condition of a minimum of surface energy, must follow a general law. This may be easily demonstrated through the following consideration.

The Rate of Growth and of Dissolution of Crystal Faces

Let us assume that we have found a polyhedron with a given orientation of faces determined by the normals $n_1, n_2, n_3, \ldots$, which satisfies the condition of a minimum surface energy at given volumes. It is evident that all polyhedrons, which are similar to the latter and are differentiated from each other only by their volumes, would likewise satisfy this condition. All these polyhedrons will represent a crystal in various stages of growth, from which it is clear that the beginning of growth must coincide with the center of similarity of the whole series of polyhedrons obtained. Thus the area of one of these polyhedrons can be expressed in the following manner:

$$p(n_1^2 + n_2^2 + n_3^2 + \cdots)$$

where p is a constant, and the surface energy E for capillary constants of $k_1, k_2, k_3, \ldots$, for the various faces would be

$$E = p(k_1n_1^2 + k_2n_2^2 + k_3n_3^2 + \cdots)$$

The volume of this polyhedron can be written in the following form:

$$V = q(n_1^3 + n_2^3 + n_3^3 + \cdots)$$

where q is a constant.

Since the polyhedron has a minimum surface energy at constant volume, the following conditions must be fulfilled:

$$dE = 2p(k_1n_1\,dn_1 + k_2n_2\,dn_2 + k_3n_3\,dn_3 + \cdots) = 0$$

$$dV = 3q(n_1^2\,dn_1 + n_2^2\,dn_2 + n_3^2\,dn_3 + \cdots) = 0$$

This, however, is only possible if

$$k_1 : k_2 : k_3 \ldots = n_1 : n_2 : n_3 \ldots$$

The minimum surface energy for a given volume of a polyhedron will be achieved if the distances of its faces from one and the same point are proportional to their capillary constants.

From this we can draw a conclusion very important for the whole investigation. Let us turn our attention to a polyhedron that has developed from the crystal at a time after the beginning of growth. It is evident that the length of the normals, which from the start of development are drawn to the faces of the polyhedron, indicate the rate of growth of these faces. But since the length of the lines is proportional to the capillary constants, as was demonstrated above, we can therefore conclude that the rate of growth of the surfaces of the crystal is proportional to the capillary constants of these surfaces in relation to the mother liquor.

As we develop further P. Curie's theories, we come to the conclusion that, through the measurement of the relative rate of growth of the crystal faces, we are at the same time measuring the relative size of the capillary constants of these faces in relation to the mother liquor in which the crystal is growing.

We have found that between capillary constants of the surfaces and their distances from the beginning of growth the following relationship exists:

$$k_1 : k_2 : k_3 \ldots = n_1 : n_2 : n_3 \ldots$$

At a particular stage of development, the areas of the faces might be s_1, s_2, s_3, ... It can be said that

$$s_1k_1 : s_2k_2 : s_3k_3 \ldots = s_1n_1 : s_2n_2 : s_3n_3 \ldots$$

But sk is the surface energy in the stage of growth under consideration and sn is proportional to the volume of the pyramid of which the surface s serves as base and whose apex lies at the origin of growth. Therefore, the volume of the material that during crystallization settles on any one face is proportional to the surface energy of this face; moreover, the surface energy of the crystal is proportional to the energy expended in the formation of the complete crystal.

The correctness of this conclusion is so clear that in my opinion one could use it as a basic thesis and from it derive the theorem of least energy of the crystal polyhedron. This would be in complete accordance with the fact that each face grows faster, the greater its energy. Since, however, the fast-growing faces disappear more or less rapidly, only the slowest growing, which have the least energy, remain, so that the total surface energy is driven to attain the minimum value. The hypothesis that the greater the energy of the face, the more particles settle out of solution on the face during a period of time seems to me more concrete and mechanically clearer than the principle of greatest and least work. There is no doubt that the principle of greatest and least underlies many natural phenomena. But that proposition alone, that a phenomenon takes a certain course because somehow nature is attempting to realize the principle of economy, cannot replace an exact explanation of the occurrence; we will only then be satisfied when we have succeeded, through analysis of the whole phenomenon, in demonstrating how this principle of economy is realized in the preceding case.

In conclusion, let us turn our attention to an inference from the theorems derived above. If the capillary constants are equal, the polyhedron will circumscribe a sphere. It seems to me that the ideal combinations in the diagrams in accordance with this deduction should be presented, so that the faces of different forms that comprise a combination touch the same sphere.

9. ON THE MOST COMMON FACES IN CRYSTALS OF LIKE SUBSTANCE

It is known that if the symbols (100), (010), (001), and (111) are assigned to any four faces of a crystal, all the other faces, which occur most often[a] in polymorphous crystals of the same substance, may be designated by

[a] *Editor's note:* Polymorphous here signifies diversity of *habit* of crystals of the same species (chemistry and structure).

symbols with simple indices. The greater or lesser simplicity of the indices depends upon the choice of faces to which are given the symbols (100), (010), (001), and (111).

Bravais[b] sought a reliable criterion for the selection of primitive form. He assumed that in the molecular lattice, which is the basis of crystal structure, the cleavage faces, from which the molecules of the adjoining faces most easily separate, will be those which border the densest layers. In support of this hypothesis is a characteristic of the Bravais lattices, that the distance between two bordering faces occupied by molecules is directly proportional to the density of the molecular distribution in these faces (respectively inversely proportional to the surface content of the elementary parallelogram), so that the cohesive strengths of the molecules in the plane are to some extent inversely proportional to the cohesive strengths of the planes to each other.

From this one can also assume that surfaces which are most suited to form during crystallization, or, stated differently, those which are most suited to serve as the outer form of the crystal, generally have a large reticular density, since local motions that take place continuously at the surface of a developing crystal act to some extent as external forces striving to split the crystal.

Bravais' rule states:

> One must choose from the infinite number of solutions which will give the primitive form, those which will assign such reticular density to the crystal surface that this series, arranged in decreasing size, corresponds exactly with the order of decreasing ease of formation of the faces during crystallization or during cleavage.

A significant flaw in the above-stated ideas of Bravais is the incomplete explanation of the difference between the cohesive forces within the crystal and those which work at the surface of the crystal during its growth. The first depend exclusively on the distance of the molecules from each other and on the distribution of the size of the attractive forces around each molecule, and are therefore always the same for the same substance. The second, on the other hand, depend both on the capillary action of the crystal and the surrounding medium, and therefore change with the characteristics of the mother liquor. This may be explained by an example. In a pure aqueous solution, sodium chloride crystallizes in cubes; in a solution which contains an admixture of urea, in octahedrons. The relative ease of formation of the two forms is in each case identical; therefore, to apply Bravais's rule, one must in this case restrict it with the condition that only those crystal forms which are obtained from pure solutions be grouped in a series, or that cleavage be given precedence over the faces. The cleavage of the sodium chloride crystal is actually in both cases cubic, i.e., it is not dependent upon the characteristics of the solution.

[b] Bravais, *Études cristallographiques*, p. 166.

Sohncke,[c] in conformity with Bravais, presents the hypothesis that in general faces most densely packed with molecules occur most commonly at the crystal surface. He compares this hypothesis of Bravais with Curie's theoretical conceptions of least surface energy and concludes that both theories lead to a mutually concurring result.

According to Sohncke,[d]

> a connection must exist between the surface energy of a crystal face and its surface density. That is, in the most densely packed surface the molecular forces can do no further work because the particles cannot draw closer to one another; the potential energy of such a surface must therefore be a minimum. In the same way, when the surface densities of other crystal faces are smaller, their surface energies (or capillary constants) must be greater. Indeed, it seems necessary that the various possible crystal faces at the actual boundary play a smaller role to the same degree that their surface densities are less.

I will try to examine more closely the relationship of the reticular density to the cleavage and the external crystal boundary. In doing so I will rely on Bravais's ideas.

To destroy the crystal, one must overcome (1) the forces which hold the reticular planes parallel at a fixed distance apart, i.e., the cohesive forces perpendicular to the plane, and (2) the forces which retain molecules at a fixed distance in the each plane, i.e., the cohesive forces parallel[e] to the plane. Thus the process of crystal destruction consists of processes against both these types of cohesive forces. It is obvious that, in the cleavage of the crystal, only those particles parallel to the cleavage planes are separated, and the individual particles of the crystal are not destroyed. Therefore, in cleaving the crystal, the energy expended is against the perpendicular cohesive forces holding the planes at a fixed distance; this distance we term ϵ.

> It is well-known that the cohesion of bodies is greater the less the distance between molecules. It can therefore be assumed that under like conditions the tangential cohesion will be greater the greater the surface density of the plane, and that the perpendicular cohesion will be weaker the more significant the distance between two adjacent planes.

Therefore, if we designate the energy against the perpendicular cohesion as N and a constant of proportionality as A, we may assume that

$$N = \frac{A}{\epsilon} \tag{1}$$

[c] F. Sohncke, "Ueber Spaltungsflächen und natürliche Krystallflächen," *This journal*, **13**, 214–235 (1888).
[d] *Loc. cit.*, p. 221.
[e] Bravais, *loc. cit.*, p. 167.

> As we examine different systems of reticular crystal planes in turn, we find that the density of the net and the distance between two neighboring planes of this system are quantities which increase and decrease together, and therefore always remain proportional to each other.[f]

If the reticular density of the planes is designated ρ and the distance of the planes is ϵ, then $\rho/\epsilon = \Omega =$ constant. We can therefore write

$$N = \frac{A}{\rho} \tag{2}$$

In the crystal, the constant of proportionality A changes with direction, since the molecules affect each other variously in different directions, not only because of the difference in distance, but also because of their polarity. From equation (2) it is evident that the energy N must be small in faces for which ρ is large; but among such faces, the cleavage face will be the one for which the relationship A/ρ is the smallest.

The energy of the crystal faces during growth is the same as we have demonstrated above (Chapter 8), i.e., proportional to the capillary constants of the faces. If we assume with Sohncke that, under the same conditions, the smaller the capillary constant k the greater the reticular density of the face, we can assume in analogy to equation (2) that

$$K = \frac{a}{\rho} \tag{3}$$

We will call the factor a the molecular growth energy; this energy must vary with direction with respect to the molecules and at the same time depends on the nature of the solution. Of course, the minima of energy, cleavage, and growth do not need to be equal to one another.

If the capillary constant of any face is given by the reticular density ρ, the quantity a can be determined for the corresponding direction. It is clear from this that the quantities of a cannot differ very much from each other for the various directions in the crystal, at least not for those faces which form the outer boundary, since it is not possible to assume that the crystal molecules are so asymmetrical that they could evoke a very strong polarity in their bonding. That this is in fact the case can be shown by the crystals of Mohr salts. To calculate the quantity a for the faces of this crystal, it is only necessary to assume as before the cubic arrangement of molecules.

Index		Reticular density	Growth velocity	Molecular growth energy
Early	Normal			
(201)	(001)	1.00	1.00	1.00
(110)	(111)	0.58	1.96	1.16
(001)	(201)	0.45	2.25	1.01
(111)	(310)	0.32	2.50	0.80
$(11\bar{1})$	(112)	0.41	2.64	1.08
(011)	(211)	0.41	2.77	1.14

[f] Bravais, *loc. cit.*, pp. 167, 168.

It can be seen from the table that the product $k\rho$ (the molecular growth energy) deviates very little for all faces examined; at least, the differences between individual values of $k\rho$ are much smaller than the differences in growth rates. From the table one could draw the conclusion that a, within the margin of error of the measurements, is a constant for various surfaces. Against such a conclusion, however, is the simple consideration that not all surfaces with equal reticular density and not even those with the same indexes grow equally fast (e.g., the surfaces of hemihedral forms).

The correlation of rates of growth to the reticular densities stands out as even more surprising if the results of experiments with revolving crystallizing vessels are used for the rates of growth. If we set the exactly measured growth rate of the form (110) equal to $\sqrt{3} = 1.73$, the rates for the other surfaces will be as follows:

(201)	(110)	(001)	(011)
0.94	$\sqrt{3}$	2.08	2.38

The reticular densities of these surfaces will be

(201)	(110)	(001)	(011)
$\sqrt{1} = 1.00$	$\sqrt{3} = 1.73$	$\sqrt{5} = 2.24$	$\sqrt{6} = 2.45$

The negligible difference between these two series of values is so surprising that I decided to investigate whether this was a universal phenomenon. Accordingly, I proposed to Herr Z. Weyberg an experiment on the growth rate of another crystal of the cubic system, for which we could be much more certain in the assignment of reticular densities for the faces. Weyberg's research, which forms the substance of the following article (see p. 534), led to the result that the relationship of the two types of values comes out only qualitatively.

While the molecular energy of growth varies continuously with respect to the molecule, the reticular density on the other hand is a discrete quantity; this circumstance easily explains the polyhedral form of the crystal (Figure 13).

If the surface S represents the distribution of the values a around the molecule σ, the values a within a sufficiently small solid angle α cannot differ much one from another, and have a certain average value a'. Assume that the angle α includes the perpendicular to a face with a large reticular density. For all the other faces with perpendiculars within the angle, the energy

$$k = \frac{a'}{\rho}$$

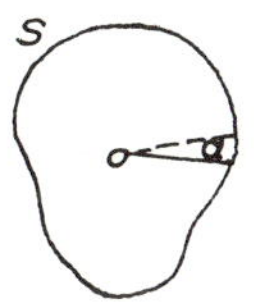

Figure 13

will be very great, because ρ will be very small. These faces will have a very high growth rate in comparison to the face with large reticular density that lies between them. Therefore, the elements bounding the crystal cannot vary continuously with respect to each other; i.e., they can form no curved surfaces, but consist of planes with high reticular density. But if the crystal is rich in faces, among which are some with complex indices, there can be two explanations.

First, this can occur during normal growth, if, in the faces with low reticular density, the molecular growth energy is so small that the product $k\rho$ will be very small. To cause the natural formation of faces with complex indices, the properties of the solution must be changed by adding foreign matter. For example, the addition of borax to a magnesium sulfate solution works this way (according to v. Hauer, 1878). In a pure magnesium sulfate solution, very long prisms with poorly developed ends are formed; the addition of borax retards the growth rate longitudinally and causes a uniform growth of the crystal in all directions. In a solution mixed with an ample amount of borax, crystals often form that are marked by a great number of faces.

The second reason is connected to the interruption of growth by solution. The process of dissolution tends to round off the corners and edges of the crystal. If a period of growth follows such a period of solution, for the reasons just stated, the crystal assumes a polyhedral form, in which it is covered with numerous faces, the greater part of which quickly grow over. If the growth process is interrupted at the proper time, a crystal is obtained with many faces, often with very complex indices. In all probability, natural crystals with many faces are formed in this manner (e.g., calcite, garnet, beryl). This condition is very fully explained in Lehmann's *Molekularphysik*.

At this point, we must discuss the question of the artificial creation of faces on the crystal, and especially the spontaneous formation of faces symmetrical to those made artificially.

K. v. Hauer[9] through his experiments proved beyond doubt that the artificial creation of faces, which normally do not appear in the crystal, is possible in many cases.

With a file he ground various "rational and suitable irrational" faces on chrome alum crystals and grew crystals prepared thus in an iron ammonia–alum solution. Culture in the iron ammonia–alum was a necessary condition for the success of the experiments, because in its own solution the chromium alum crystal grew completely irregularly and the newly formed surfaces did not remain plane. In this way he proved that, of the total amount of salt which had precipitated from solution during a given time period, the greater part was used for the completion of the parts missing in the crystal, and less was used for the general enlargement of the crystal. All artificially formed combinations represent transition stages, and

[9] K. Ritter v. Hauer, "Krystallogenetische Studien: I. Reihe," *Sitzungsber. Wien. Akad. Wiss.*, **39**, 611 (1860) "II. Reihe," *ibid.*, **49**, 539 (1860).

the crystal faces formed thus grow over in time. Von Hauer correctly explained this circumstance by the difference in rates of growth, although he ascribes the difference in the abstract to "changes in the directions in which the molecules settle," while believing the cause of such change unknown.

If a cube is cut from chrome alum, it does not remain as such; rather, in the same time that the smoothing out of the abraded surfaces occurs, all the octahedral and dodecahedral surfaces again appear very minutely outlined. In fact, the dodecahedral surfaces disappear when the crystal is allowed to develop further.

If the edges of an octahedral crystal are beveled by faces of a dodecahedron, the surfaces of the cube appear automatically; crystallization shows a pronounced inclination to destroy dodecahedral surfaces rather than cubic.

All these results of von Hauer's experiments are very simply interpreted by the results I have obtained. It is apparent that the dodecahedral surfaces in the chrome alum crystal with respect to the iron alum solution have the greatest capillary constant; the octahedral, on the other hand, the smallest; with the cubic in the middle. The same applies also to the relative rates of growth.

Cut cubic and dodecahedral faces, which possess high growth energy, produce currents of high concentration in their environment, which not only attract the immediately adjacent parts of the solution, but also those in contact with other crystal faces, so that currents of concentration must form at the surface of the crystal. From this it is clear that the growing crystal faces of lower energy must appear as new faces truncating the corners and edges, in the same proportion as the edges and corners are retarded in growth. These faces are overgrown in turn, as soon as the surface content of the crystal face, which yields the excess of nutrient, decreases sufficiently; concurrently, the surface energy of these faces diminishes.

In view of such an arbitrary appearance of faces with great growth energy, the overall surface energy of growth must increase, which seems to contradict Curie's theory.

The growing crystal should not, however, be observed independently from the solution surrounding it, as was cited earlier. The energy of crystal growth cannot consist solely of the energy of the crystal, but rather also of the energy of the solution, which is expressed, moreover, in the concentration currents. The concentration currents are exerted to take nutrient from some parts of the crystal and to bring it to other parts that are growing more intensively. Therefore, the less nourished parts also seem to possess a smaller energy. The appearance of faces that are symmetrical with respect to randomly formed faces can likewise be explained by the concentration currents. Von Hauer states that, if two opposite ends of a chrome alum octahedron parallel to the cubic faces are filed down until the crystal is tabular, a complete assemblage is obtained with further growth in an iron alum solution.

It is easy to understand that v. Hauer's cited experiments are in close correlation with the phenomenon of the so-called regeneration or healing of the crystal. This phenomenon may be very easily explained by the asymmetrical rate of growth of the crystal in various directions and by the concentration currents which are caused by the rapidly growing parts of the crystal. The subject seems so obvious to me that I will not dwell on it further.

Part III

THE FIRST STRUCTURES

Editor's Comments on Papers 8 and 9

8 BARLOW
Probable Nature of the Internal Symmetry of Crystals

9 BARLOW and POPE
Excerpts from *The Relation Between the Crystalline Form and the Chemical Constitution of Simple Inorganic Substances*

William Barlow began as an amateur crystallographer and finished by deriving independently the 230 space groups. He ". . . united the mathematical theory of symmetry . . . with an anticipation of the new determinations of atomic structure that were to follow after 1910. . . ."[1] From the external symmetry of crystals, Barlow undertook a systematic study of the symmetrical arrangements of close-packed spheres compatible with the external symmetry of crystals, beginning with Paper 8 (1883) in which he promulgated the then radical conception that the separate atoms entering into a crystal compound ". . . occupy distinct portions of space and do not lose their individuality." The major twentieth-century concepts of atomism and symmetry were ignored in the nineteenth century by physicists and chemists alike. It remained for mineralogists and mathematicians to establish their significance.

In Paper 8, Barlow described the body-centered cubic, simple cubic, face-centered cubic, hexagonal close-packed, and extended hexagonal (Zhdanov's 4H) structures and attempted (sometimes erroneously) to assign structures derived from these arrangements to the alkali halides, ice, quartz, calcite, etc. Paper 8 foreshadows the structure determinations of Bragg and the 1927 paper in which Linus Pauling defined the rules of ionic structures.[2] In delimiting a finite number of possibilities consonant with the symmetry of stoichiometry, Barlow here provides the first structural explanations of isomorphism, polymorphism, and twinning. Determination of radius ratios, the idea of minimization of potential energy at distances determined by attractive and repulsive forces, and the determination of atomic grouping by radius ratios were expressly stated in a paper read by Barlow in 1891 at meetings of the British Association.[3]

For many years Barlow collaborated with W. J. Pope, professor of chemistry at Manchester, in the elaboration of their law of valency volumes, according to which ionic volumes were proportional to valence. The success of this scheme in determining the structures of the elements is apparent from their paper of 1907 (Paper 9). Even the failures in unraveling the structures of binary and more complex compounds were suggestive. Bravais had understood the relationship between the symmetry of the molecule represented by the lattice point and merihedry, and Barlow, who had already applied this principle in the derivation of the space groups, used it in proposing tetrahedral groups of close-packed spherical atoms for AgI and ZnS. His groups were each tetrahedra of four of the same kind of atoms rather than the interpenetrant tetrahedra of modern structure theory. Nevertheless, in 1913 this work of Barlow and Pope led W. L. Bragg to the distinction between the rocksalt and the sphalerite structures and the first determinations of the atomic structures of minerals.

In 1883, symmetry had led Barlow correctly, if inelegantly, to propose for the structure of binary compounds an alternation of two kinds of atoms in either simple or body-centered cubic packing. According to the Bravais rule, the lattice of the alkali halides should have been simple cubic.[4] In 1907, the dimorphism of silver iodide and the axial ratios approximating 1.63 of binary hexagonal compounds led Barlow and Pope to the correct lattices (fcc), but an erroneous reading of the symmetry led them away from the correct structures. One sees in Papers 8 and 9 the germs of modern structure theory, and a foreshadowing of labors that brought Nobel prizes to Bragg and later Pauling.

REFERENCES

1. W. T. Holser, *Dictionary of Scientific Biography*, Scribner's, New York, 1970, p. 460.
2. *J. Am. Chem. Soc.*, **51**, 1010 (1929).
3. *British Association Reports*, 1891, p. 581.
4. E. Mallard, *Traité de cristallographie*, V. 1, Paris, 1879, p. 310.

8

Reprinted from *Nature*, **29**, 186–188, 205–207, 404 (1883, 1884)

PROBABLE NATURE OF THE INTERNAL NATURE OF CRYSTALS

William Barlow

SOME studies pursued by the writer as to the nature of molecules have led him to believe that in the atom-groupings which modern chemistry reveals to us the several atoms occupy distinct portions of space and do not lose their individuality. The object of the present paper is to show how far this conclusion is in harmony with, and indeed to some extent explains, the symmetrical forms of crystals, and the argument may therefore in some sort be considered an extension of the argument for a condition of internal symmetry derived from the phenomena of cleavage.

If we are to suppose that crystals are built up of minute masses of different elements symmetrically disposed, it is natural to inquire in the first place what very symmetrical arrangements of points or particles in space are possible.

It would appear that there are but *five*, which will now be described.

If a number of equal cubes are built into a continuous mass (Fig. 1), a system of points occupying the centres and angles of these cubes will furnish an example of one of these symmetrical arrangements. In this system each point is equidistant from the eight nearest points, and if a number of equal-sized spheres be stacked on a base layer arranged so that the sphere centres when joined form a system of equal squares, a side of which bears to the diameter of the spheres the ratio $2 : \sqrt{3}$ (see plan *a*), the sphere centres in such a stack will also furnish an example of this first kind of symmetry (Fig. 2).

A second kind of symmetry will be presented if one-half the points in the first kind be removed so that we have only those at the cube centres, or only those at the cube angles. In this system each point is equidistant from the six nearest points, and if equal-sized spheres be stacked upon a base layer, arranged so that the sphere centres when joined form a system of equilateral triangles, a side of which bears to the diameter of the spheres the ratio $\sqrt{2} : 1$ (see plan *b*); and if the layers be so placed that the sphere centres of the fourth layer are over those of the first, those of the fifth over those of the second, and so on, the sphere centres in such a stack will also furnish an example of this second kind of symmetry (Fig. 3).

A third kind of symmetry will be presented if again one-half the points be removed, *i.e.* so that when cubes of two colours arranged in such a way that each cube is surrounded by cubes of the other colour are used (see Fig. 1), we have only the points at the centres of the cubes *of one colour*. In this system each point is equidistant from the twelve nearest points, and if equal-sized spheres be stacked upon a base layer in which the spheres are in contact, and whether they form a square pattern (see plan *c*), or a triangular one (see plan *d*)—provided that, if triangular-pattern layers be employed, the sphere centres in the fourth layer must be over those in the first, those in the fifth over those in the second, and so on—the sphere centres (the arrangement being the same in either case) will furnish a second example of the third kind of symmetry (Figs. 4 and 4*a*, the latter showing a stack with the angle removed to display the triangular arrangement).

A fourth kind of symmetry, which resembles the third in that each point is equidistant from the twelve nearest points, but which is of a widely different character from the three former kinds, is depicted if layers of spheres in contact arranged in the triangular pattern (plan *d*) are so placed that the sphere centres of the third layer are over those of the first, those of the fourth over those of the second, and so on. The symmetry produced is hexagonal in structure and uniaxal (Figs. 5 and 5*a*).

A fifth kind of symmetry, and this completes the number of very symmetrical arrangements possible, resembles the second kind of symmetry in that each point is equidistant from the six nearest points, and bears the same relation to the fourth kind (Fig. 5) as the second (Fig. 3) bears to the third (Fig. 4); that is to say, it may be regarded as produced by the insertion of additional points in positions midway between points arranged in the fourth kind of symmetry. It is depicted if triangularly constituted layers identical with those depicting the second kind of symmetry (plan *b*) are deposited in the following way (Fig 6): —First place three layers as though to produce the second kind of symmetry; then place the fourth with its sphere centres over those of the second layer; then the fifth so that the third, fourth, and fifth, like the first, second, and third, are in the second kind of symmetry; then the sixth with its sphere centres over those of the fourth and second; and then the seventh, so that the fifth, sixth, and seventh layers are also in the second kind of symmetry; and so on. The symmetry produced is, like the last, hexagonal in structure and uniaxal.

FIG. 1.

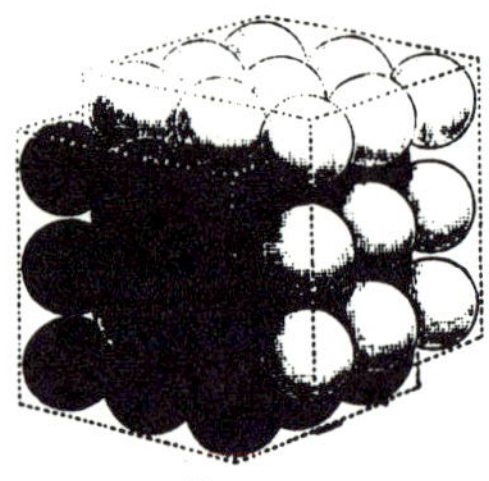

FIG. 2.

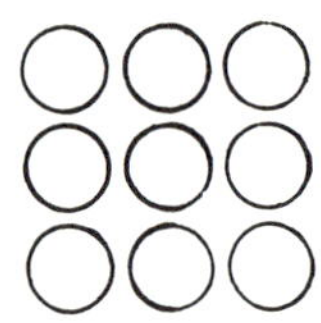

Plan *a*.

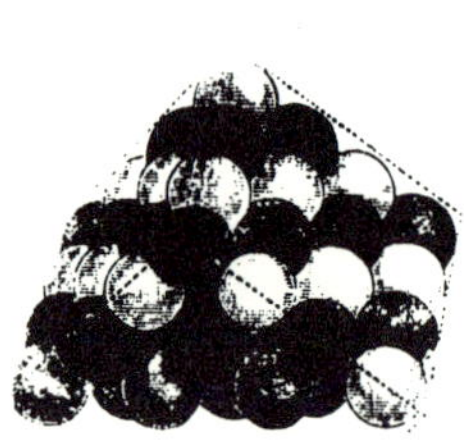

FIG. 3.

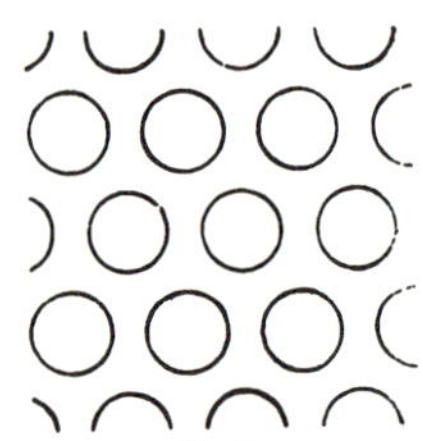

Plan *b*.

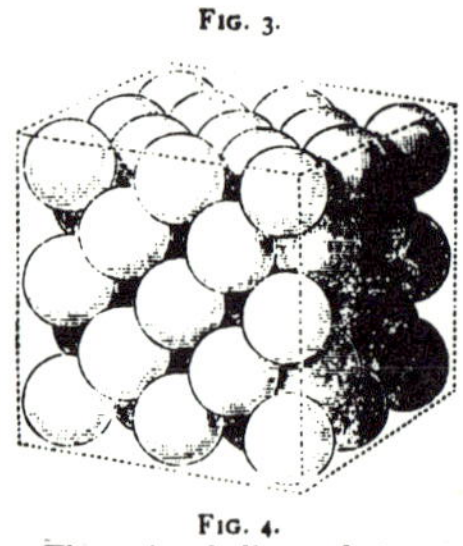

FIG. 4.

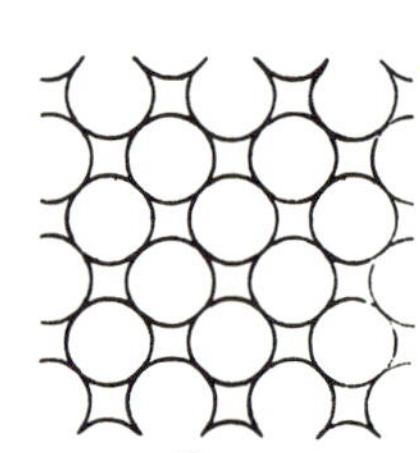

Plan *c*.

The writer believes that every one of the various symmetrical forms presented by crystals can be shown to be consistent with the subsistence of an arrangement of the atoms of the crystallising compound in one or other of these five kinds of symmetry *at the time when crystallisation begins*; and proposes to show that a relation subsists between the atomic composition of very many bodies and their crystal forms in harmony with this conclusion.

To proceed then to facts, we notice first that, as a rule, compounds consisting of an equal number of atoms of two kinds crystallise in cubes.

The following may be mentioned :—
Potassic chloride, KCl.
Potassic bromide, KBr (sometimes elongated into prisms, or extended into planes).
Potassic iodide, KI.
Sodic chloride, NaCl.
Sodic bromide, NaBr.
Sodic iodide, NaI (anhydrous above 40° C.).
Cæsi chloride, CS. Cl.
Plumbic sulphide, PbS.
Argentic chloride, AgCl.

When we have named lithic chloride, crystallising above 15° in octahedra, we have mentioned most of the compounds consisting of two elements in equal proportions known to us in a crystalline state.

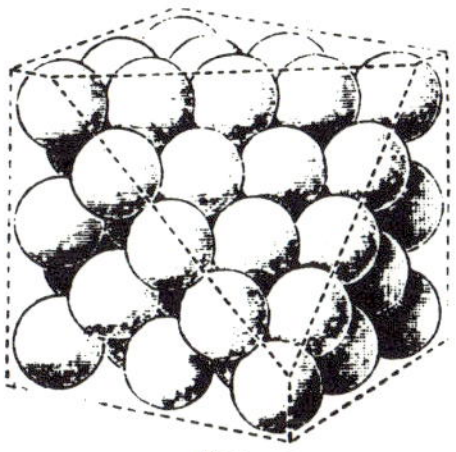

FIG. 4*a*.

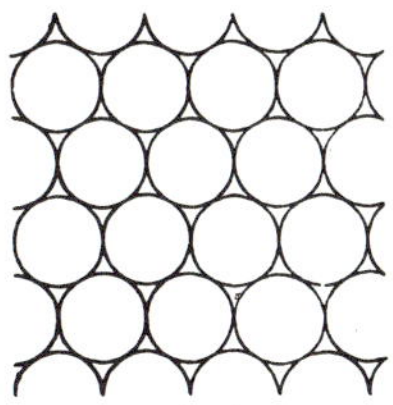

Plan *d*.

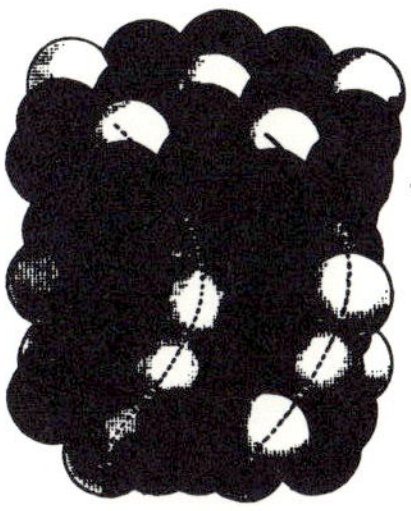

FIG. 5.

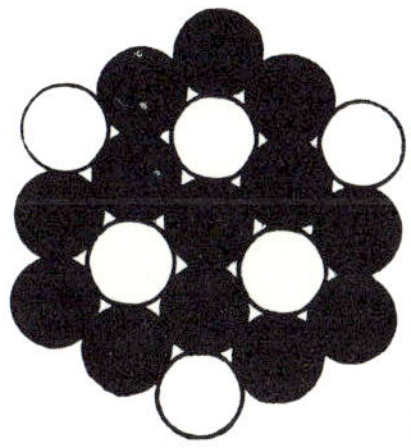

Plan *e*.

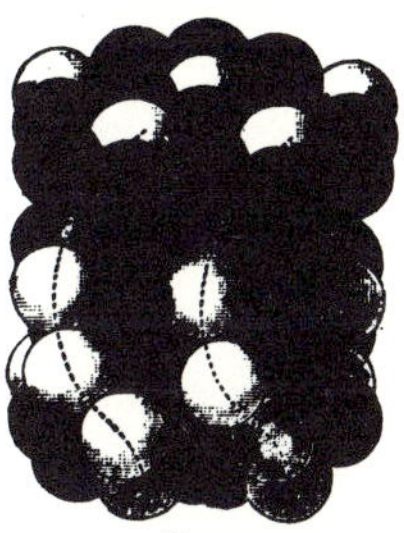

FIG. 5*a*.

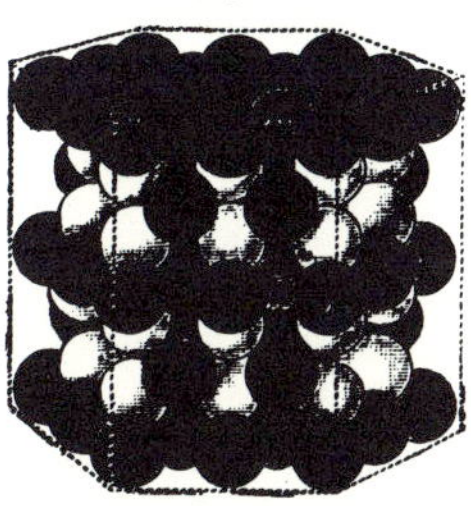

FIG. 6.

Plan *f*.

Mercuric sulphide, Hg. S, which crystallises in six-sided prisms, is an apparent exception, but if we were guided by the gaseous volume of mercury in determining its atomic weight, we should have to write the compound Hg_2S.

Other apparent exceptions are :—
Zincic oxide, ZnO, crystallising in six-sided prisms.
Cadmium sulphide, CdS ; and
Glucina, GO, crystallising in minute six-sided prismatic crystals.

Now three out of our five possible kinds of internal symmetry have three axes or directions at right angles to each other, in reference to which they are disposed in the same symmetrical manner, and two kinds, the first and second, admit of a very symmetrical arrangement of two kinds of particles *in equal numbers* (see Figs. 2 and 3). Surely this coincidence is very significant, and at lea t suggests the probability that when a compound consists of two kinds of chemical atoms in equal numbers, these atoms are symmetrically placed according to either the first or the second kind of internal symmetry.

We observe next that the third and fourth kinds of symmetry (Figs. 4 and 5) readily lend themselves to the symmetrical arrangement of particles of two kinds present in the proportion 1 : 2. For, as already pointed out, these two kinds of symmetry may either of them be produced by piling up layers of spheres placed triangularly in contact (see plan *d*), and spheres of two colours present in the proportion of 2 : 1 can be arranged in a most symmetrical manner in layers of this kind (see plan *e*).

As to what varieties of position of bi-coloured layers of this kind with respect to one another are possible, consistent with great symmetry, we have concluded that, apart from the question of arrangement of colour, there are but two, viz. the third and fourth kinds of symmetry (Figs. 4 and 5); but taking colour into account a greater variety is possible. Thus a little consideration shows us that, while all the possible ways of depositing the second layer produce a practically identical result, a choice of six different equally symmetrical results is presented in depositing the third layer, in all of which the spheres of the less numerous colour form files of spheres in contact running through the layers, and three of which belong to the third kind of symmetry and three to the fourth.

To specify these : We may have the less numerous spheres of the third layer placed with respect to those in the second and first :—

(1) So that the three spheres of each of the files just above alluded to range in line, the lines joining their centres forming a series of parallel straight lines crossing the planes of the layers obliquely. This arrangement belongs to the third kind of symmetry.

(2 and 3) So that the centres of these three spheres, when joined, form a slightly obtuse angle ; a different result being produced as the angle is made to the right or to the left. This pair of arrangements belongs also to the third kind of symmetry.

(4) So that the less numerous spheres in the third layer are vertically over those in the first. This arrangement belongs to the fourth kind of symmetry.

(5 and 6) So that, as in (2) and (3), the triplets of spheres form a system of equal obtuse angles, but the angles now being very obtuse. There are here, as in (2) and (3), a right-handed and a left-handed arrangement. These belong to the fourth kind of symmetry.

The deposition of the third layer, by the necessities of symmetry, determines the deposition of succeeding layers, and it follows therefore from the above that six different equally symmetrical arrangements of spheres of two colours present in the proportion 2 : 1 are possible in the third and fourth kinds of symmetry.

As to (1) the parallel files of the less numerous spheres crossing the first three layers will extend through subsequent layers.

As to (2) and (3) every three continuous layers will display the less numerous sphere centres placed to form the same angles as are presented by the triplets in the first three layers, and consequently these sphere centres lie on spirals which are right-handed or left-handed as the case may be ; the less numerous spheres in the fourth layer being vertically over those in the first, those in the fifth over those in the second, and so on.

As to (4) the less numerous spheres in the fourth layer must lie vertically over those in the second, those in the fifth over those in the third, and so on ; and thus the files of spheres in contact running through successive layers form a series of similar zigzags.

As to (5) and (6) the sphere centres, as in (2) and (3), lie either on right-handed or on left-handed spirals ; in this case the less numerous spheres in the seventh layer being vertically over those in the first, those in the eighth over those in the second, and so on (Figs. 5 and 5*a*).

When we inquire whether the symmetrical arrangements just traced are in harmony with the facts respecting compounds of two kinds of atoms in the proportions 1 : 2, we find some very important evidence.

Thus water, H_2O, crystallises in six-sided prisms or in rhombohedra; forms both of which are compatible with one or other of the above symmetrical arrangements.

And the following most interesting concurrence of facts indicates that the symmetrical arrangements in the fourth kind of symmetry above described (see Figs. 5 and 5*a*) are those of the atoms of quartz.

(*a*) Quartz consists of oxygen two atoms, silicon one atom; just the proportions in these arrangements.

(*b*) It has the property of circular polarisation, from which it has been proved that its molecules must have a *spiral* arrangement, and, since some crystals have the property of rotating in one direction, some in the opposite, that this spiral arrangement is right-handed in some crystals, left-handed in others.

(*c*) It crystallises in six-sided prisms terminated by six-sided pyramids, a form derivable, as we have seen, from the arrangements before us.

As to this last point, just a word of explanation why we must not look for the angles exhibited by our model arrangements to be identical with the angles made by the pyramid faces in quartz.

It is a matter of common observation that the process of crystallisation is generally associated with change of bulk, and if we suppose this change to arise from expansion, or contraction, generally expansion of the different kinds of atoms, and that these different kinds have *different degrees of expansion*, we see that a mass symmetrically arranged in the manner supposed will in crystallising expand or contract more in some directions than in others, and while we should look for a similar change in the direction of each of the three transverse subordinate axes of the crystal, we should look for a different change in the direction of the principal axis. And thus, supposing the mass when liquid immediately before it began to crystallise to have had the internal symmetry which has been depicted, it is evident that the unequal change of dimension in different directions might suffice to bring about such an inclination of the faces of the terminal pyramids to the sides of the prism as is actually found to exist.

In support of this explanation we have the fact that crystals not of the regular system have been found to expand unequally in different directions when subjected to heat.

Further evidence in support of the theories here submitted is found in the fact that, with scarcely any exception, the compounds we are now considering do not crystallise in the regular or cubic system.[1]

(*To be continued.*)

[1] With regard to calcic fluoride (fluor-spar), which appears as an exception, it may be remarked that a different atomic weight for calcium which would enable us to write the compound CaF would enable us to get over a difficulty with regard to another compound of calcium, as we shall see presently.

PROBABLE NATURE OF THE INTERNAL SYMMETRY OF CRYSTALS[1]

THE theory of the modification of crystal angles, just offered in dealing with quartz, is manifestly applicable to all crystals not of the cubic system, and it is submitted that for every such

[1] Continued from p. 188.

crystal there is an *ideal* or *root form* proper to one or other of the five kinds of internal symmetry which have been presented, from which root form the actual form can be derived by a proper proportionate increase of dimension in one or more directions.

It is evident that, while our path must become more and more intricate as we endeavour to establish in the cases of more complex compounds relations similar to those above traced, the reference of whole classes of analogous forms, differing only in their angles, to one root form, removes a very important difficulty, and the wide applicability which it confers on the five kinds of internal symmetry with which we started appears in the fact that there is no crystal form which cannot be thus referred to an appropriate root form in harmony with one or other of these five kinds of internal symmetry.[1]

One more case may be mentioned in which a probable internal symmetry can be assigned to a compound in harmony with its actual crystal form; it is a more difficult one.

The molecule of *Iceland* or *calc-spar* is usually believed to consist of one atom of calcium, one atom of carbon, and three of oxygen. We shall, however, take a liberty, and suppose that the atoms of calcium or the atoms of carbon have but half the mass attributed to them; that in the formula of this compound we should write either two atoms of calcium or two atoms of carbon in place of one.[2]

Making this supposition, we observe that if the calcium and carbon atoms were alike we should have six atoms, three of one kind, three of another; in other words, we should have equal proportions of two kinds of atoms, from which, since the form of Iceland spar is but little removed from a cube, we naturally argue that just before crystallisation its atoms were arranged according to the first or second kind of internal symmetry; these two kinds being, it will be remembered, those in harmony with the cubic form which admit of very symmetrical arrangement of particles of two kinds present in equal numbers.

Since Iceland spar is a uniaxal crystal, the arrangement of the three kinds of atoms, whatever it is, must be symmetrical about one axis only; and we shall now endeavour to show that the atoms can be thus arranged in either the first or second kind of symmetry.

We will show first that they can be hus arranged in the second kind.

Where there are but two kinds of particles present in equal numbers, symmetry requires that the alternate layers of this kind of symmetry (see Fig. 3) shall consist entirely of similar kinds, and therefore in the case before us, one set of alternate layers will represent oxygen atoms; the other, atoms of calcium and carbon. Now particles present, as we suppose the calcium and carbon atoms to be, in the proportion 1 : 2 can be quite symmetrically arranged in these layers (plan *f*), as the sphere centres were in the layers depicting the fourth kind of symmetry (plan *e*), and therefore the only question remaining is the relative disposition of the layers of calcium and carbon atoms with respect to one another.

Now the spheres in alternate layers of the second kind of symmetry considered alone have the relative arrangement of the third kind of symmetry (Fig. 4), and in determining the relative disposition of the calcium and carbon atoms, we may therefore neglect the oxygen atoms, and treat the case as belonging to the third kind of symmetry. The two spiral arrangements in this kind of symmetry, in which the less numerous spheres in the fourth layer are vertically over those in the first (see *ante*), have the necessary symmetry about a single axis, and if the calcium and carbon atoms have one of these arrangements, the requirements of the case are entirely met.

We will now show that the three kinds of atoms can also be arranged symmetrically about a single axis in the first kind of symmetry.

One half the spheres depicting this kind of symmetry will in this case represent the oxygen atoms, and the remaining half the atoms of calcium and carbon (see Fig. 2), and, as previously noticed, the arrangement of either half will be that of the second kind of symmetry. It follows that the question of the relative disposition of the atoms of calcium and carbon is simply the question of the symmetrical arrangement about a single axis of atoms of two kinds present in the proportion 2 : 1 in the second kind of symmetry (Fig. 3). And since the layers of spheres depicting this kind of symmetry have a triangular arrangement (plan *b*), it is evident that this can be accomplished here just as in the former case.

In either of the two arrangements just described we have only to suppose that when the symmetrically placed atoms change volume at the time of crystallisation the dimensions transversely to the axis of symmetry are increased relatively to those in the direction of this axis, and we have an obtuse rhombohedron where formerly we had a cube. And the significant fact that the angle of a rhombohedron of calc-spar diminishes when the crystal is heated supports this theory of its production. Perhaps the arrangement of the atoms according to the first kind of internal symmetry is the more probable of the two, as this would give the cleavage directions coincident with the directions of layers of similar atoms (oxygen).

An important fact supporting our conclusions is that certain definite relations as to their proportions which are found subsisting between the allied forms taken by crystals of the same substance are found inherent in one or other of the five kinds of internal symmetry.

Thus it is well known that if a particular substance is found crystallised in hexagonal pyramids of various kinds—that is, whose sides have various different degrees of inclination to the base—the number of kinds is strictly limited, and they are strictly related to each other. If x be the side of the hexagonal base of the pyramid and y the height for the same substance, while x remains constant, y has not more than fourteen different values, seven related thus: c, $\frac{3}{4}c$, $\frac{1}{2}c$, $\frac{1}{4}c$, $\frac{1}{8}c$, $\frac{1}{12}c$, $\frac{1}{16}c$; and the other seven similarly related thus: d, $\frac{3}{4}d$, $\frac{1}{2}d$, $\frac{1}{4}d$, $\frac{1}{8}d$, $\frac{1}{12}d$, $\frac{1}{16}d$; and c bearing to d the ratio $2 : \sqrt{3}$.

Now, if we turn to the fourth kind of internal symmetry (Fig. 5) to ascertain the possible varieties of inclination of the sides of hexagonal pyramids which can be depicted, we find that the greatest possible height to which we can build a hexagonal pyramid of equal spheres is exactly double the height of a tetrahedron with the same side as the hexagonal base of the pyramid. Thus, if twenty-five spheres form each side of the hexagonal base, giving twenty-four equal distances between the sphere centres in any one side, we find that the highest possible pyramid has forty-nine layers of spheres giving forty-eight equal spaces between consecutive layers.

If we call this height c, it is evident that pyramids corresponding with the first of the above series of actually observed forms will have respectively—

49	layers of balls,	giving	48	spaces between	consecutive layers.
37	"	"	36	"	"
25	"	"	24	"	"
13	"	"	12	"	"
7	"	"	6	"	"
5	"	"	4	"	"
4	"	"	3	"	"

We find, moreover, that such a series can be readily depicted, and that, upon examination, no additional terms appear admissible.

Again, a further inspection of the stack of spheres shows us that with the same heights—that is, with the respective numbers of layers just enumerated—we may, in place of the base layer which forms a hexagon whose sides have twenty-five spheres each, have a base derived from this in which each of the six spheres at the angles becomes the centre of a side, the outline of the base layer being now a larger hexagon described about the hexagon which bounded the former base layer. The sides of this new base thus bear to the sides of the old the ratio subsisting between the side and the perpendicular of an equilateral triangle, *i.e.* the ratio $2 : \sqrt{3}$. And finally, since the distance between the planes containing the centres in successive layers bears to the distance between centres in the same layer the same ratio which the perpendicular from the angle of a tetrahedron upon its opposite face bears to its edge, that is the ratio $\sqrt{2} : \sqrt{3}$, it follows—

That the two allied series of possible altitudes of hexagonal pyramids thus formed, if we take *the same length of side a* for both, will be—

First Series

$$\frac{2\sqrt{2}}{\sqrt{3}}a;\ \frac{3\sqrt{2}}{2\sqrt{3}}a;\ \frac{\sqrt{2}}{\sqrt{3}}a;\ \frac{\sqrt{2}}{2\sqrt{3}}a;\ \frac{\sqrt{2}}{4\sqrt{3}}a;\ \frac{2\sqrt{\ }}{6\sqrt{3}}a;\ \frac{\sqrt{2}}{8\sqrt{3}}a.$$

Second Series

$$\sqrt{2}.a;\ \tfrac{3}{4}\sqrt{2}a;\ \tfrac{1}{2}\sqrt{2}.a;\ \tfrac{1}{4}\sqrt{2}a;\ \tfrac{1}{8}\sqrt{2}a;\ \tfrac{1}{12}\sqrt{2}a;\ \tfrac{1}{16}\sqrt{2}.a.$$

[1] The very symmetrical form the pentagonal dodecahedron is not in harmony with either of the five kinds of symmetry, nor is it found in crystals.

[2] It has already been remarked that the crystal form of fluor-spar favours the supposition that calcium has half the atomic weight usually attributed to it.

Surely the fact thus established, that each term of a series of relative altitudes of the hexagonal pyramids in which a particular substance crystallises always has to some term of the series thus theoretically derived a particular ratio peculiar to the substance, constrains us to conclude that the above fourteen "root" forms are those to which all crystal forms involving regular six-sided pyramids are referable, and that the actual forms are produced from the "root" forms by difference in the degree of expansion in the direction of the axis of the crystal as compared with other directions at the time of crystallisation.

Other allied forms, as allied octahedra or rhombohedra, can be in the same way connected with some one of the five kinds of internal symmetry.

The peculiarities of *crystal-grouping* displayed in twin crystals can be shown to favour the supposition that we have in crystals symmetrical arrangement rather than symmetrical shape of atoms or small particles. Thus if an octahedron be cut in half by a plane parallel to two opposite faces, and the hexagonal faces of separation, while kept in contact and their centres coincident, are turned one upon the other through 60°, we know that we get a familiar example of a form found in some twin crystals. And a stack can be made of layers of spheres placed triangularly in contact to depict this form as readily as to depict a regular octahedron, the only modification necessary being for the layers above the centre layer to be placed as though turned bodily through 60° from the position necessary to depict an octahedron (compare Figs. 7 and 8). The modification, as we see,

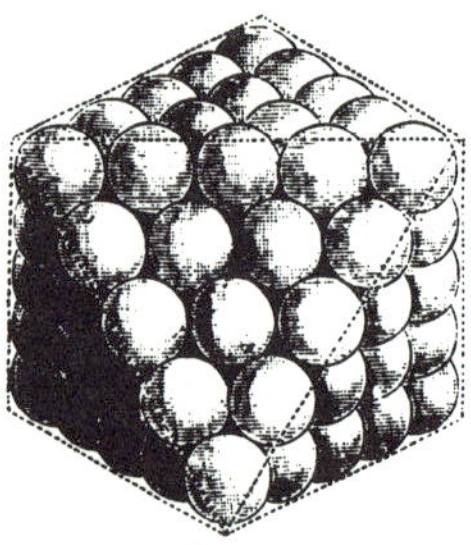

FIG. 7.

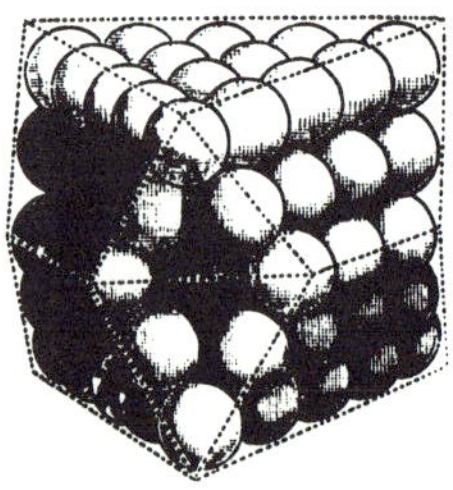

FIG. 8.

involves *no departure from the condition that each particle is equidistant from the twelve nearest particles.*

Before closing, a few words may be said on the bearing of the conclusions of this paper on *isomorphism* and *dimorphism*.

First, as to isomorphism.

The conclusion that there are but five kinds of internal symmetry possible, three of which indicate a cubic form, evidently accords with the fact that not only the simplest combinations—those in which two kinds of atoms are present in equal proportions—but also many very complicated compounds crystallise in cubes.

Out of the regular system we generally find that for the angles of crystals of different compounds to be the same there must be some resemblance in their atom-composition, and the explanation suggested is that the atoms which are common to two isomorphous compounds, *e.g.* the carbon and oxygen atoms in calc-spar and spathic iron ore, have similar situations in the two different crystals, and that the change of bulk which occurs when crystallisation takes place is due to a change in *these atoms only*, the atoms not found in both remaining *passive*.

There are, however, some cases which do not at first seem to be met by this view—cases in which the atom composition of isomorphous compounds has only a very partial similarity. Ammonia compounds may be specially mentioned. Thus, ammonic sulphate, $(NH_3)_2H_2SO_4$, is isomorphous with potassic sulphate K_2SO_4.

The following suggestion would seem to enable us to suppose that in this, as in other cases of isomorphism, the phenomenon is referable to the passivity of some of the atoms in the change of bulk which accompanies crystallisation. Let us write ammonic sulphate thus $(NH_3)_2H_2SO_4$, and let us suppose that the symmetrical arrangement is such that the groups, $(NH_3)_2$ just occupy places which might, without altering the symmetry, be filled by additional groups H_2SO_4; that, in other words, the relative position of the groups H_2SO_4 which are present in the symmetrical arrangement is precisely the same as it would be if the entire mass consisted of these groups instead of consisting partly of NH_3 groups. If now, in addition to supposing that in both compounds the active atoms in the process of crystallisation are the sulphur and oxygen atoms, and these only, we suppose that the expansion of some of the atoms of the active kind checks the expansion of others; that only a certain proportion of these atoms expands, we perceive that we may have both the same amount and kind of atom expansion in the two cases, and, as the natural result, isomorphism.

Next, as to dimorphism.

It is evident that a very small change is requisite to convert one kind of internal symmetry into another. Thus we have already had occasion to notice that the only difference in depicting the third and fourth kinds of symmetry is that for the former the centres of the spheres in the first and fourth layers, those in the second and fifth, and so on, range vertically, while for the latter the centres in the first and third, in the second and fourth, and so on, range in this way.

In the case of a dimorphic compound consisting of two kinds of atoms in the proportion of 2 : 1, *e.g.* water, H_2O, we have only to suppose therefore that the same layers of atoms which under one set of conditions produce hexagonal prisms, are by some alteration in conditions arranged in the slightly different way necessary to produce rhombohedral forms. Other cases of dimorphism are probably to be accounted for much in the same way.

Thus the following interpretation of the fact that calcic carbonate, which we have seen crystallises in obtuse rhombohedra as calc-spar, sometimes crystallises in six sided trimetric prisms as aragonite may be offered.

We have already endeavoured to show that the first or second kind of internal symmetry is that proper to calc-spar. We will now endeavour to show that the fifth kind of internal symmetry (Fig. 6) is proper to aragonite.

Alternate layers of spheres (plan *b*) will represent the oxygen atoms, and the other alternate layers the calcium and carbon atoms; the central layers of the triplets above alluded to, viz. the second, the fourth, the sixth, &c., being the oxygen layers; the calcium and carbon atoms in the remaining layers will be symmetrically arranged (plan *f*). From the fact of the crystals being trimetric, the layers containing the last-named atoms, which, considered apart from the oxygen layers, are in the fourth kind of symmetry, probably have the arrangement above described, in which the less numerous spheres form zigzags, the stack in this case having a different symmetry about three axes at right angles to each other (Fig. 6).

The fact that the dimorphic varieties of the same substance have different densities is in harmony with the supposition that different sets of the atoms are concerned in the different cases; that the active atoms which produce one form are not those, or those only, which produce the other.

It is not always necessary to refer two incompatible crystal forms of the same substance to two different kinds of internal symmetry: for example, from the third kind of internal symmetry we can produce square-based octahedra, and we can also produce right-rhombic prisms, and in accord with this we have the well-known fact that right-rhombic prisms of sulphate of nickel, $N_2SO_47H_2O$, when exposed to sunlight are molecularly transformed, and, though they neither liquefy nor lose their form, when they are broken are found to be made up of square-based octahedra several lines in length. WILLIAM BARLOW

"Probable Nature of the Internal Symmetry of Crystals"

In reply to the important criticisms offered by Herr L. Sohncke on my new theory published in Nature of December 20 and 27, 1883 (pp. 186 and 205)—

Taking first those relating to the geometry of the subject; the following explains why only the five symmetrical arrangements of points in space described in my paper are taken as the basis of the theory.

If it is the case that, prior to the act of crystallisation, the chemical atoms of a body fall into some symmetrical arrangement, it is natural to suppose that they do so through some influence they exert on one another—such, for example, as mutual repulsion—and that a similar influence is exerted by each atom of the same kind on atoms around it. And if this be so, there will be no *stable equilibrium* of the forces thus exerted *until the atoms are very evenly distributed throughout the space allotted to them.*

Now although, as Herr Sohncke has shown, there is a large variety of symmetrical arrangements of points in space in which the points are disposed around every one point of the system in precisely the same manner as around every other, it would appear that only four of these regular systems, the first four described in my paper, signally fulfil the requirement of even distribution, these four systems being distinguished from all the rest by the property that, *if the nearest points grouped around any point of either of these four systems are joined, the solid thus outlined has its edges all equal.*

And further, although the fifth system described in my paper is not one of Herr Sohncke's *regular* systems, its points are more evenly distributed through space than those of any of these systems except the four just referred to. In this system the property is found that either lines joining the nearest points around any point of the system, or lines joining the next nearest, in all cases outline a solid whose edges are all equal.

As the five systems I have in my paper too vaguely distinguished as "very symmetrical" thus stand alone, and moreover, if my views are adopted, they appear to be adequate to all cases of crystallisation, I still incline to think that the chemical atoms of bodies about to crystallise always have one or other of these five kinds of symmetrical arrangement. If I am wrong in this, and some other symmetrical arrangements are admissible, the general lines of the new theory will not however be affected.

Next, as to the bearing of the theory on chemical valency and the usual conception of a chemical molecule, it may be remarked that, while there is no clear knowledge of the nature of the union between the different sorts of atoms in a compound by which to test the new theory, this theory appears to receive support from the phenomenon of electrolysis. For the fact that one *ion* is liberated at one pole, the other at the other, while *no apparent alteration takes place in the fluid between the poles,* goes to show that any particular atom can change its partners without dissolving the chemical ties subsisting between the several atoms of the compound, and thus favours the view that similar atoms equally near to a particular atom are similarly related to it.

As to my supposition that the expansion, or contraction, occurring in the act of crystallisation, is due to the increased or diminished repulsion exerted by *some only* of the atoms of a body on surrounding atoms, it is, perhaps, interesting to notice that if this conception could be extended to the gaseous state, and the expansion to the state of gas of any compound attributed to the agency of certain atoms in each molecule, or ideal unit, to the exclusion of the rest, the simple relations found subsisting between the volumes of compounds and the volumes of their uncombined constituents might in this way be accounted for:—Thus the fact that aqueous vapour has a volume two-thirds that of the added volumes of the hydrogen and oxygen of which it is composed would be explained if all the gaseous expansion of this compound is due to the hydrogen atoms only.

Muswell Hill WM. BARLOW

9

Reprinted from pp. 1150–1152, 1168–1177 of *Trans. Chem. Soc.*, **91**, 1150–1214 (1907)

CXI.—*The Relation between the Crystalline Form and the Chemical Constitution of Simple Inorganic Substances.*

By WILLIAM BARLOW and WILLIAM JACKSON POPE.

IN a previous paper we have shown that the main aspects of the doctrine of valency appear as the expression of certain well-defined geometrical properties of arrangements in space if a very simple conception is adopted as to the nature of crystalline substances (Trans., 1906, **89**, 1675). It has also been shown that the crystalline forms of a large number of chemical substances are derivable from their compositions and constitutions on the basis of this conception. Briefly stated, the concept adopted is that in a crystalline substance the spheres of influence of the component atoms are homogeneously arranged to form a close-packed assemblage; a molecule is therefore a closely arranged group consisting of certain of the spheres of atomic influence. These spheres or domains are conceived to fill space, and are therefore not strictly spherical in form, but are polyhedra enclosed by interfaces drawn at right angles to lines joining adjacent centres of atomic influence. For the purpose of graphically picturing the conditions prevailing in assemblages of atoms, represented by their respective spheres of influence, the mechanical expedient is adopted of squeezing together masses of solid indiarubber balls of appropriate sizes and suitable arrangement so that the interstices become practically eliminated by the interaction of the balls. The balls then exercise mutual repulsions of the nature previously postulated (*loc. cit.*, 1676), and it is suggested that the equilibrium arrangement reached by this conversion of deformable, but incompressible, spheres into space-filling polyhedra accurately represents the allotment among the atoms of a chemical substance of the space which it occupies. Inasmuch as for purposes of demonstration the packing of hard undeformable spheres can be more readily depicted than that of those just described, it is convenient to employ the former in place of the latter, and to apply the term sphere of atomic influence, although the atomic domain is actually polyhedral in shape. The correspondence between the conditions of equilibrium of a close-packed assemblage of hard spheres and of a close-packed assemblage of elastic and deformable, but incompressible, spheres is discussed in connexion with the simplest types of equilibrium assemblages (p. 1157).

In the paper referred to it is shown by numerous instances that homogeneous, close-packed assemblages of spheres, from which the interstices have been eliminated in the manner described, are characterised by the possession of three geometrical properties, two of which may be here again noted.

If from such an assemblage, one homogeneously situated set of spheres, each sphere having the volume *m*, is removed, the vacated spaces can be re-occupied each by several spheres having the total volume *m*, in such a manner that the close-packing may be restored by mere distortion and local adjustment without remarshalling the assemblage. This, the first geometrical property of close-packed homogeneous assemblages, is discussed in more detail on p. 1204; it corresponds entirely with that aspect of valency which states that an atom of valency *m* is replaceable in a compound by several atoms of the total valency *m*.

Again, if in such an assemblage, one homogeneously situated set of spheres, each of volume *m*, is homogeneously replaced by a set of spheres, each of volume $m+1$, or if to each of the original spheres of volume *m* is added one of volume 1 in similar contact with it, the assemblage being homogeneously opened or spread out for the purpose, it is found that, if remarshalling does not take place, close-packing can only be restored by the insertion into gaps produced in the assemblage of additional spheres in the proportion of volume 1 for each original volume *m* concerned. The replacement of a sphere of volume *m* by others of the total volume $m+1$ thus necessitates the simultaneous addition of another sphere of volume 1; in other words, an atom of valency *m* may be replaced by others of total valency $m+2$, $m+4$, &c. This, the second geometrical property, corresponds exactly with the known facts that when an element changes its valency it in general does so by two, or multiples of two, units or minimum valency volumes, and also that a substitution like that of N by C, as in $CH_3 \cdot NH_2$, necessitates the addition of another univalent atom to the added carbon atom, so as to give $CH_3 \cdot CH_3$ in the quoted instance. Many quantitative data were adduced bearing on the applications of the first and second geometrical properties for the purpose of showing that the fundamental valency of an atom is approximately proportional to the volume appropriated by the polyhedron representing it in the crystal structure. It was thus proved that the diagrammatic method of representing molecular constitution which we introduced, and according to which each atom present in a given compound is represented by a sphere of influence approximately proportional in volume to its fundamental valency, is entirely in harmony with accepted chemical theory.

The fact that the methods in question were found to represent adequately the constitutions and inter-relations of a number of aromatic compounds, endows the Frankland doctrine of valency with a real physical signification; the classificatory scheme based on the doctrine of valency and the theory of constitutional formulæ thus acquires a more graphic expression and also a justification. It is, however, desirable that the new method of investigation should be applied to a number of complete classes of simple chemical substances, because by so doing a demonstration of the truth of our former conclusions can be given which is practically independent of the previous one. We proceed to deal in the first place with practically all those substances composed of atoms of the same fundamental valency which have been sufficiently completely examined goniometrically for our purpose; these substances include the elements, most binary compounds, and compounds of the type of the salts KI_3, KI_5, and K_2AgI_3. Some corresponding ammonium salts are also considered.

The validity of the conclusion that the spheres of influence of atoms of the same fundamental valency occurring in the same compound are nearly of the same size is severely tested by the limitation which this imposes on the constitution of the assemblage; the limitation in question is expressed as a fourth geometrical property of close-packed assemblages in the following pages.

The Closest-packed Homogeneous Assemblages of Equal Spheres.

If all the atoms composing a crystalline structure are identical in kind and in environment and are without polarity, the structure may be simulated by a homogeneous, closest-packed assemblage of equal spheres. Equal, deformable, incompressible spheres can, like hard spheres, be closest-packed under a general pressure so as to form a homogeneous structure in two, and only two, ways, or in a homogeneous combination

FIG. 1.

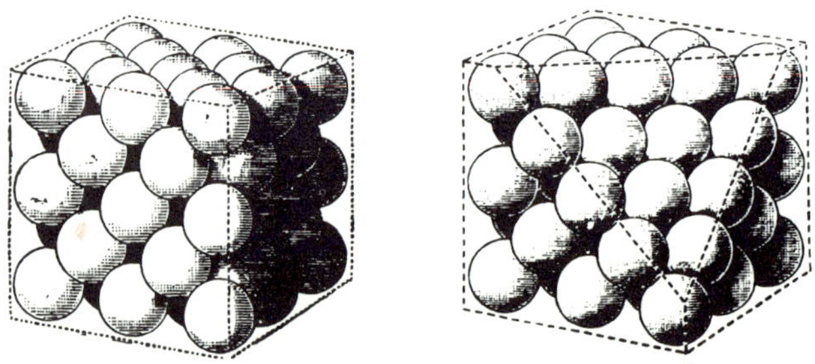

a. *b.*

of them; these two ways are distinguishable as the cubic and the hexagonal closest-packed assemblages of equal spheres (Barlow, *Nature*, 1883, **29**, 186; Kelvin, *Proc. Roy. Soc. Edin.*, 1889, **16**, 693).

[*Editor's Note*: Material omitted at this point.]

The Crystalline Structure of Binary Compounds.

Most binary compounds are composed of two elements of the same valency, and, according to our previous results, the spheres of atomic influence of two such elements possess approximately the same volume. It follows immediately that binary compounds of two elements of the same valency should, in accordance with the fourth geometrical property, in most cases possess crystalline structures in which the marshalling is that of either the cubic or the hexagonal closest-packed homogeneous assemblages of equal spheres. A reason is thus provided for the well known, but hitherto unexplained, fact that some 88 per cent. of the determined crystalline forms of binary compounds are either cubic or trigonal (see Table I). Practically the only crystalline forms of binary compounds known to belong to other than the cubic or trigonal systems are those of the tetragonal mercurous chloride and iodide and silver fluoride, the orthorhombic litharge, PbO, and iron arsenide, FeAs, and the monosymmetric mercuric oxide, HgO, arsenic sulphide, AsS, and cupric oxide, CuO; the crystalline form of silver fluoride is doubtful, and some of the other substances just noted are composed of two elements of different valencies. The known facts are consequently expressed by the statement that practically all binary compounds composed of elements of the same valency crystallise in either the cubic or the hexagonal systems, most, however, belonging to a hemihedral or tetartohedral class of one or other system. The following table (Table III) includes all those binary hexagonal substances which have been crystallographically examined, and which undoubtedly consist of two elements of the same valency.

TABLE III.

	$a : c.$		$a : c.$
GlO *	1 : 1·6305	CdS ¶	1 : 0·8109
ZnO †	1 : 1·6077	AgI ‖	1 : 0·8196
ZnS ‡	1 : 0·8175		

* Mallard, *Ann. des Mines*, 1887, **12**, 427.
† Traube, *Jahrb. Min.*, *Beil.-Bd.*, 1894, **9**, 147.
‡ Friedel, *Compt. rend.*, 1861, **52**, 983.
¶ Mügge, *Jahrb. Min.*, 1882, **2**, 1.
‖ v. Zepharovitch, *Zeitsch. Kryst. Min.*, 1880, **4**, 119.

The measurements from which the above axial ratios are calculated may be regarded as fairly accurate; they differ so slightly from those for the closest-packed assemblage of equal spheres, namely, $a:c = 1:1{\cdot}6330$, or one-half that ratio, $a:c = 0{\cdot}8165$, that they may be considered as practically identical with the value for the model assemblage.

[*Editor's Note*: Material omitted at this point.]

Silver Iodide.

Silver iodide is dimorphous, crystallising at the ordinary temperature in the hemimorphous pyramidal class of the hexagonal system with $a:c = 1:0{\cdot}8196$, whilst at 145—147° it undergoes conversion with almost explosive violence into a cubic modification (Lehmann, *Zeitsch. Kryst. Min.*, 1877, 1, 492). It has already been pointed out (p. 1169) that these two crystalline modifications are almost exactly simulated by the cubic and hexagonal closest-packed assemblages of equal spheres, and in view of the close approximation of the axial ratio for the hexagonal silver iodide to that of the latter assemblage, the spheres of atomic influence of silver and of iodine will naturally be regarded as of almost the same size.

A geometrical distribution of two kinds of nearly equal spheres in equal numbers, representative of the cubic modification of silver iodide, is derivable in the following manner from the closest-packed cubic arrangement of equal spheres.

Partition space into equal cubes distinguished as of two kinds, black and white, present in equal numbers, the cells of one kind being in contact along edges only, as indicated in Fig. 16. Describe small equal spheres about the centres of the cube cells of one kind, say the black, and cause all these spheres to expand uniformly until they touch their next neighbours. A comparison of Figs. 16 and 4 will show that the system of spheres thus obtained is the cubic closest-packed arrangement, for the centres of the spheres in Fig. 4 lie at the centres of one-half of the cube cells of a certain cubic partitioning of space, chosen so that the selected cubes are in contact along edges only.

Next effect an auxiliary partitioning of the system of Fig. 16 into larger cubes, each containing eight of the small cubes, four of each colour tetrahedrally arranged, and treat these large cells of the partitioning in the same way as before, namely, distinguish them as of two colours, say red and blue (*R* and *B*, Fig. 17), those of one colour

FIG. 16.

FIG. 17.

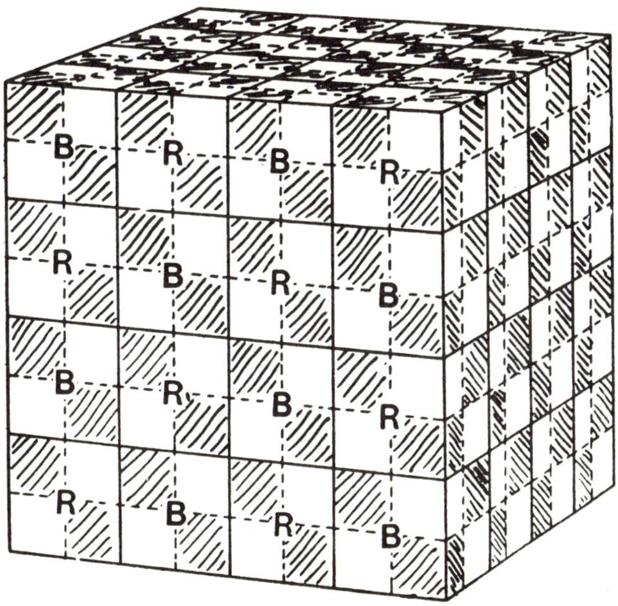

being in contact along edges only; distinguish the tetrahedral groups of spheres having their centres in the large cells of the partitioning in the same manner, so that corresponding to the large red and blue cells, the contained tetrahedral groups of four spheres are coloured red and blue respectively. The two kinds of spheres which are thus distinguished as having their centres within the red or the blue cells of the partitioning are, after modification as described below, to represent the spheres of atomic influence of silver and iodine respectively. The large cell partitioning indicates a division of the assemblage into tetrahedral groups of spheres, because each large cell contains four centres of spheres arranged tetrahedrally; all the tetrahedra thus derived are similarly orientated.

[*Editor's Note*: Material omitted at this point.]

As a result of the foregoing discussion, it follows that from the cubic closest-packed assemblage of equal spheres may be derived an assemblage possessing cubic symmetry, consisting of two kinds of equal spheres in equal numbers; each sphere is so situated with respect to nearest spheres of the same kind that it can, in conjunction with some of them, form a tetrahedral group in two ways, giving oppositely orientated tetrahedra. Every sphere has the same situation with respect to the rest of the system and its centre is a centre of symmetry and lies both on trigonal and digonal axes. The symmetry is consequently very high, being that of type $9a_1$ of Barlow's list (*Zeitsch. Kryst. Min*, 1894, **23**, 45), and the crystallographic symmetry is that of the holohedral cubic system.

[*Editor's Note*: Material omitted at this point.]

An examination of Fig. 19 shows that any two consecutive layers of the cubic assemblage which lie perpendicular to an axis of threefold symmetry can be regarded as consisting collectively of two kinds of oppositely orientated tetrahedral groups of four similar spheres, vertices of the one kind of group pointing downwards, and those of the other pointing upwards; the two kinds of tetrahedral groups are those having their sphere centres in red or in blue cells of the partitioning respectively, and the one kind represents tetrahedra of silver spheres of atomic influence, the other represents the iodine atoms. Such a composite layer is hemimorphous in a direction perpendicular to its plane, although the cubic assemblage of which it forms a stratum possesses holohedral symmetry.

Through the centre of a tetrahedron of either of the two kinds draw a perpendicular to the plane of such a composite layer; make this perpendicular a hexagonal screw axis of such a nature as to locate a second composite layer identical with the first and fitting closely on to it, the two composite layers differing in orientation by 60°. Succeeding similar operations about the same axis will locate other similar layers and complete the assemblage, which will then contain a flock of hexagonal screw axes passing through the centres of all the tetrahedra of the one kind in any composite layer. The relative arrangement of the spheres in the assemblage thus obtained is a mixture of the cubic and the hexagonal closest-packed arrangements of equal

FIG. 19.

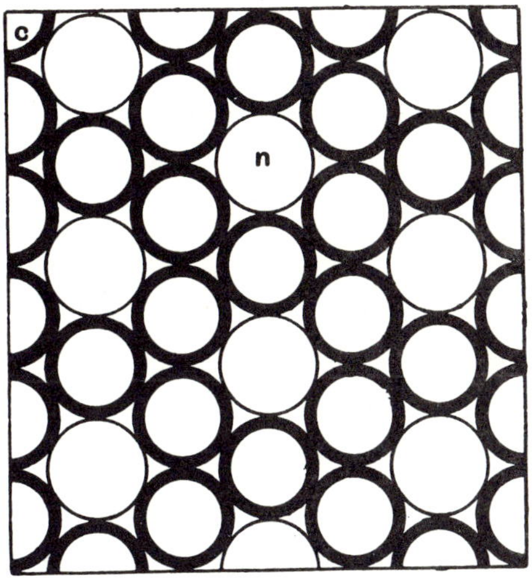

c.

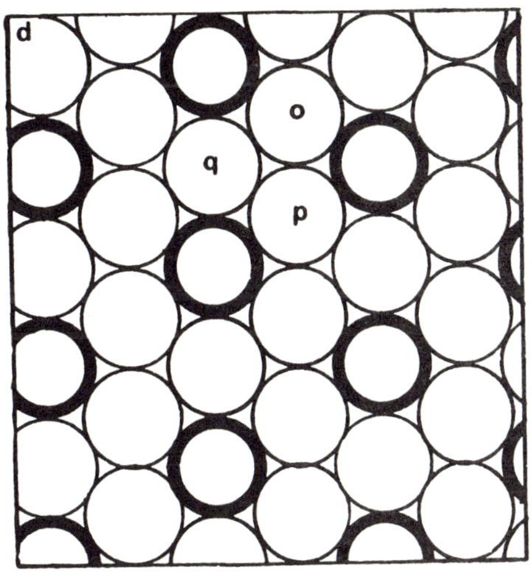

d.

spheres already described. Thus a stratum consisting of three consecutive single layers, the middle layer of which has its less numerous spheres transfixed by the hexagonal screw axes, will present the cubic form of arrangement, whilst, where these axes pass midway between the more numerous spheres of the middle layer, the stratum will have the arrangement characterising the hexagonal form of closest-packing. The packing has the same maximum degree of closeness as in the case of either of the two simple modes of closest-packing provided that the spheres are all equal in size. The completed hexagonal assemblage, like the composite layer from which it is generated, is hemimorphous, but possesses planes of symmetry.

FIG. 19.

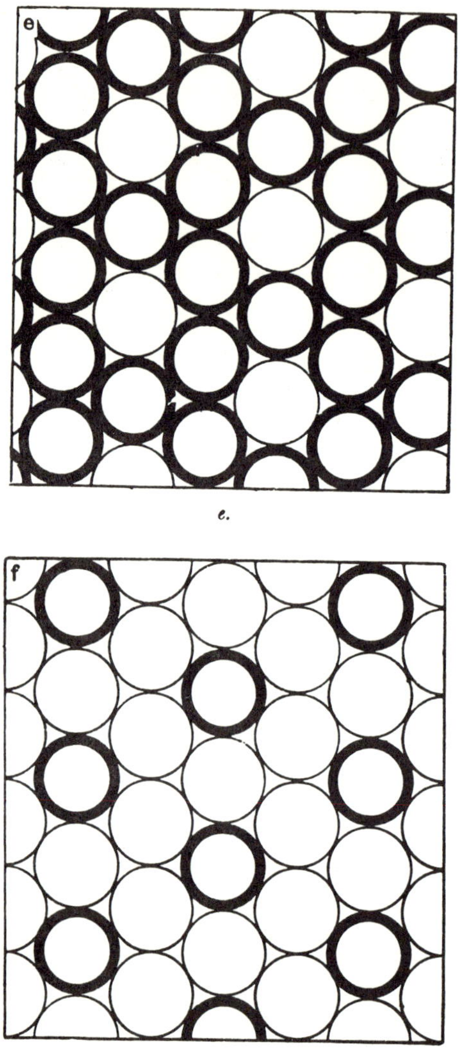

e.

f.

The practical identity of the axial ratio for the hexagonal silver iodide, namely, $a:c = 1:0{\cdot}8196$, with that of the closest-packed hexagonal arrangement of equal spheres, namely,

$$a:c = 1:2 \times 0{\cdot}8165,$$

would seem to indicate the practical identity in volume of the spheres of atomic influence of silver and iodine. It does not, however, absolutely prove this, because it really means that the spacing of the consecutive layers in the composite assemblages of the two dimorphous forms is the same; this similarity of spacing might conceivably accompany a slight difference in volume between the two kinds of sphere of influence. The sudden passage of hexagonal silver iodide to cubic symmetry at 145° may be attributed to a change, probably very slight, in the relative volumes of the spheres of atomic influence of the two kinds produced by change of temperature, as a result of which stable equilibrium occurs in the cubic arrangement at above 145° and in the hexagonal arrangement below that temperature.

It will be seen from the foregoing argument that the conversion of the cubic into the hexagonal modification necessitates a simple and small shear or translation of pairs of the consecutive layers of Fig. 19 over each other, and that the dimorphism of silver iodide consists merely in the existence of alternative methods of close-packing structural units which are dimensionally identical at the transition temperature. This result is of importance as affording a generally applicable explanation of polymorphism, a subject which has hitherto remained almost entirely unelucidated. Polymorphism may, in fact, be defined as the existence of alternative methods of homogeneously close-packing spheres of atomic influence under the condition that the homogeneous partitioning shall yield units or molecules identical in composition, constitution, and configuration. It differs from isomerism in that the latter consists in the existence of alternative methods of close-packing spheres of atomic influence under the condition that the homogeneous partitioning shall yield units or molecules identical in composition but not in constitution and configuration (compare *Sci. Proc. Roy. Dubl. Soc.*, 1897, **8**, 575, 590, 623).

[*Editor's Note*: Material omitted at this point.]

Part IV

BRAVAIS'S RULE AS NATURAL LAW

Editor's Comments on Papers 10 Through 13

10 FRIEDEL
Excerpts from *On the Experimental Bases of the Reticular Hypothesis*

11 FRIEDEL
Excerpts from *Studies on Bravais's Law*

12 GOLDSCHMIDT
Barite (Plate)

13 FEDOROV
The Crystal Empire (Editor's Review)

Georges Friedel (1865-1933), a student of Mallard, gave definitive expression to the law of simple, rational intercepts (Haüy's law) and the lattice theory of Bravais. Logic and concentration on precision and epistemological integrity mark his work. He conceived of his task as constructing principles of mineralogy drawn directly and firmly from nature, starting with the definition of the crystalline state (see Paper 10). Nature alone is sufficient to establish Haüy's law and the lattices of Bravais, but these lattices are a "geometric intermediary suitable to express the law and nothing more."

To establish the lattice theory of Bravais as an empirical law of mineralogy, Friedel undertook statistical studies of hundreds of species. In the second edition of his *Traité de cristallographie*, after developing the "Curie-Wulff concept" in full mathematical rigor, he rejected it as inapplicable to solids. Here is perhaps one weakness in the classical austerity of his thought. "Nevertheless it remains perfectly possible and even reasonable that growth rates are a function of the capillary constants; nothing prevents even that they are proportional according to Wulff's law. But this law of Wulff, based on the hypothesis of Curie, has not been demonstrated."[1]

In 1907, Friedel was able to determine absolute cell dimensions from the combination of morphological data, mineral density, and chemical formula. His results were one half the modern structure values, as he assumed two rather than eight formula units per unit cell.

Friedel in his "Études sur la loi de Bravais" (p. 332) (Paper 11) criticized the studies of Goldschmidt, Fedorov, and Baumhauer as presented in principally arithmetic form and largely lacking in physical significance. Victor Goldschmidt believed that his now forgotten "law of complication"[2] included all the fundamental laws of mineralogy, and with his "law of harmony" applied universally to crystals, the arts, music, and the human spirit. Starting from Haüy's law of simple, rational intercepts, the law of complication was a numerical series expressing the periodicity of the faces in a zone (and the zones of a crystal), which would be the same for all zones of every species. The corresponding indices of two higher-rank faces in a zone would be added to obtain the indices of the face between them of the next lower order of importance, a result that Friedel dismissed as a minor part of the law of Bravais.

Goldschmidt's *Atlas* (see Paper 12) was a complete map of the mineral kingdom known to science, a recapitulation of the entire literature. Each of the nine volumes of illustrations was accompanied by its own volume of tables of morphological data. The *Atlas* occupies a position with respect to morphology corresponding to the series *Strukturbericht* for structure. Ironically, the first volume of the *Atlas* appeared in 1913, one year after the discovery of X-ray diffraction. Its purpose was to provide the data for the determination of the distribution of molecular forces within the crystal from the crystal form.[3]

Fedorov (see Paper 13) also considered this determination the ultimate goal of crystal chemistry. He turned to exhaustive laboratory preparations for statistical confirmation of the fundamental law of Bravais. When quantitative measures of agreement of a number of faces with prediction were considered rather than the sequence of morphological importance, agreement was excellent. Discrepancies of one or of a few forms of a sequence were quantitatively minor compared to the general agreement for the first nine or ten principal forms. Thus, Fedorov showed that the Bravais rule established the lattice. As for the more detailed geometry of the polarities of the molecules of the lattice, these established the individuality of the polarities of an isomorphous series and the chemistry of the species.

REFERENCES

1. *Leçons de cristallographie*, 1926, p. 262.
2. Z. Krist., **28**, (1897); *Ueber Harmonie and Complication*, Springer, Berlin, 1901.
3. G. C. Amstutz, in Victor Goldschmidt, *Dictionary of Scientific Biography*, Scribner's, New York, 1972, Vol. V, 456a.

10

ON THE EXPERIMENTAL BASIS OF THE RETICULAR HYPOTHESIS

Georges Friedel

This article was translated by Francette Bonnabaud, from pp. 104–106, 108–112 of "Sur les bases expérimentales de l'hypothèse réticulaire," Bull. Soc. Franc. Mineral, **18**, *95–149 (1905).*

[*Editor's Note:* In the original, material precedes this excerpt.]

Amorphous matter possesses only continuous vectorial properties. By these properties, when it is anisotropic, it is not essentially distinguished from crystallized matter. A homogeneous portion of a compressed glass, of an animal or plant fiber, for example, has all the optical properties of a uniaxial or biaxial crystal.

On the other hand, amorphous matter totally lacks the discontinuous properties always present in crystallized matter. Consequently, these alone are the ones that can be used to define in a precise manner the limit between the two states.

A crystalline substance could be without any plane faces. In general, then, it has cleavages, which also characterize it. If it has no plane faces, no cleavages, no glide planes, no twin planes or axes, then one has no way of distinguishing it from amorphous matter. If an anisotropic substance, which habitually has plane faces and which has no cleavages, twins, or glides, is accidentally found in the shape of a fragment devoid of plane faces, it is considered crystalline solely by analogy with other known specimens, all whose properties, chemical or physical, scalar or continuous vectorial, are alike. But if a substance was always found in the shape of fragments without plane forms or cleavages, or generally without discontinuous properties, there would be no basis on which to decide whether it belongs to the crystalline or the amorphous class of matter. Perhaps, in certain very special cases, one could still find a way of regarding it as most likely crystalline on the analogy of its continuous properties with those of some clearly characterized crystals (the only example I can think of is the case of pyroelectricity). But one thing is certain, at least; if all crystalline substances were devoid of plane faces, cleavages, or other discontinuous properties, one would never have thought of establishing an essential difference between them and anisotropic amorphous substances and of giving them a special structure. By virtue of what characteristics should an organic fiber be classified as amorphous? Solely because it does not possess these discontinous properties. "Termierite," a kind of clay, is a compact mineral without sides or cleavages, is birefringent, and is therefore anisotropic through large more or less homogeneous areas; is it crystalline or, is it amorphous? Shall it be given a reticular structure or not? No criterion enables us to answer this question as long as no discontinuous vectorial property, faces or cleavages, or even twins or glide planes are found in "termierite."

Consequently, it is clear that the real definition of the crystalline state lies only in the existence of discontinuous vectorial properties. Any theory not based upon these properties is not specific to crystals but is applied or must also be applied to amorphous matter. When, for example, Mallard thinks he can infer the reticular structure from the notions of anisotropy and homogeneity alone, he then commits himself to attribute the reticular structure to every homogeneous portion of a deformed or tempered glass. Actually, this is not his intention. And if he excludes such amorphous solids from the results of his theory, when does he do it? It is at the precise moment when, accepting the reticular hypothesis, he conceives a medium in which the distribution of matter in various directions is an essentially discontinuous function of these directions. Along with other observations that we will try to bring out later, he thus introduces this first essential fact, the real definition of crystalline matter, which is the existence of discontinuous vectorial properties.

[*Editor's Note:* Material omitted at this point.]

HAÜY'S LAW

From the single fact that there exist in crystals discontinuous properties, evidenced by planes and straight lines in definite directions, with properties that are not possessed to any close degree by neighboring planes and straight lines, can we draw interesting deductions? Can we, in particular, reach conclusions about the possible modes of symmetry in crystals? Actually, very little. The discontinuity (i.e., the fact that the discontinuity can be physically ascertained) implies that there is a finite number of these planes and straight lines. This excludes, as impossible in crystals, axes of symmetry of infinite order, whether they are rotary axes or not. Without this restriction, since every polyhedron with a finite number of faces is the representation of an essentially discontinuous ensemble of directions of planes and straight lines, a crystal, if it was defined as possessing discontinous properties without further details, could have the symmetry of any polyhedron whatsoever with a finite number of faces.

It is then necessary from now on to add to the existence of plane faces and rectilinear edges another fact of observation. This fact is the law that rules the directions of these planes and edges: Haüy's law.

The absolute positions of the faces of a crystal (external faces, cleavages, glide planes, or twin planes) are not fixed. This is the law of the constancy of angles, or law of Romé de l'Isle, which is only one form of the notion of homogeneity. If these positions were fixed, the body under consideration would not be homogeneous any more; it would not be a *crystal* any more. The fact that *crystals* exist, i.e., masses of homogeneous crystalline matter, implies, because of the existence of plane faces, both the indeterminacy of the position and the determination of the direction of these faces.

Moreover, let AOB, BOC, COA, and ABC be four faces of a crystal such that none of them is parallel to the intersection of two or the others. Let also $A'B'C'$ be a fifth face not parallel to the first ones. When these faces are plane enough to permit measurements of angles, and within the limits of precision possible for measurements, considering the evenness of the faces and the accuracy of the best goniometers, one always has

$$\frac{OA'}{OA} : \frac{OB'}{OB} : \frac{OC'}{OC} = \lambda : \mu : \nu \qquad (1)$$

λ, μ, and ν are simple rational numbers or, at least, relatively simple if the faces AOB, BOC, COA, and ABC are appropriately chosen. This is the law of rational truncations, or Haüy's law. As a matter of observation, it consists in this: the numbers λ, μ, and ν differ from simple rational numbers only by quantities less than the errors of measurement. By saying that they *are* rational, we mean no more than this. We do not make the useless hypothesis that they are rigorously rational. When we deal with these numbers as if they were really rational, our results will apply to crystals insofar as the real numbers λ, μ, and ν will approach rational numbers, nothing more.

To say that the numbers λ, μ, and ν are simply rational is also to say that they can be taken, they or their reciprocals, as integers and less than a certain limit. In its most general form, Haüy's law leaves this limit undetermined. But it affirms its existence. If, indeed, this limit did not exist, the phenomenon would cease being physically discontinuous and Haüy's law would not make sense any longer. We will agree to take as integers the numbers $l = 1/\lambda$, $m = 1/\mu$ and $n = 1/\nu$. From here on, we will express Haüy's law in the following manner: the numbers l, m, and n can be taken as integers less than a maximum L. These are the *characteristics* of the face $A'B'C'$.

This is the most general form of the law that connects the directions of

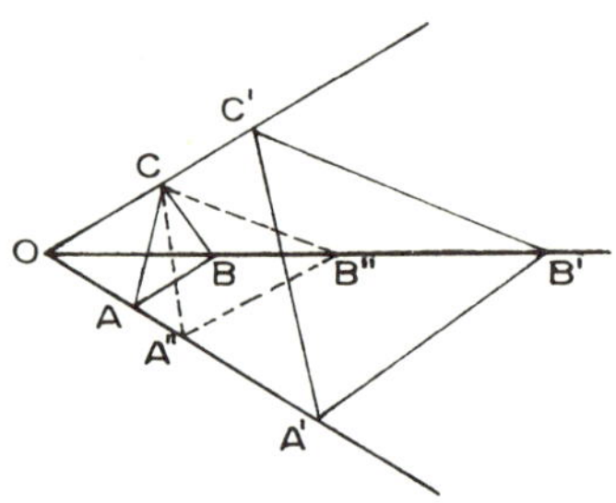

Figure 1

the faces among themselves. Actually, observation allows us to go further and specify, among the planes for which l, m, and n are integers and inferior to L, those which play the most important part in the discontinuous properties. For the moment, we set aside this specific form of Haüy's law. We intend to look for what can be inferred from the law in its most general and vague form, which is the one we have indicated above.

Another expression, more geometric, can be given to this law. The position of the face $A'B'C'$ being undefined, but only its direction being specified, we displace this face until it reaches the point C. The relations (1) become

$$\frac{OA''}{OA} = \frac{n}{l} \quad \text{and} \quad \frac{OB''}{OB} = \frac{n}{m}$$

n/l and n/m are simple integers. Consequently, if we build up a lattice of contiguous and identical parallelepipeds whose edges are in length and direction OA, OB, and OC, each face of the crystal, such as $A'B'C'$, is parallel to a reticular plane of this lattice. Moreover, if S is the area of the mesh of this reticular plane, one has

$$S^2 = Al^2 + Bm^2 + Cn^2 - 2Dlm - 2Emn - 2Fnl$$

A, B, C, D, E, and F being constants that depend on the choice of the four initial faces AOB, BOC, COA, and ABC. So that l, m, and n being inferior to a maximum L, the area S is, for all the possible faces of the crystal, similarly less than a maximum. Haüy's law can be expressed in this way, which is exactly equivalent to the first: the directions of the faces of a crystal are among those of the reticular planes of a lattice and among those whose reticular area is less than a certain maximum.

The lattice we are considering here has nothing to do with physical hypothesis. It is a convenient formula to express Haüy's law, nothing else.

This geometric expression of Haüy's law leads to a great temptation for the synthetic method—to imagine, on a pretext easy to conceive as an approximation but impossible to justify with precision, that the lattice in question really exists in one or another form in the crystal. This would be done not by starting from Haüy's law in order to relate it to something else, which would be legitimate, but by starting a priori from approximate reasoning, which would allow such a system of points, straight lines, and planes in the crystal.

Once this is done, it would be an easy game to infer a great many results, among them Haüy's law and all its consequences, especially those which relate to possible symmetries in crystals. Everything would seem to come out from almost nothing. But, fundamentally, everything would be inferred from Haüy's law, the implicit basis for the hypothesis. The result of this method is that the logical consequences of the law, those which can be deduced from it without hypothesis, would be mingled with the consequences of the hypothesis, with which they should not be confused. And we will see specifically that the reticular hypothesis implicitly assumes other experimental facts than the law of rational truncations, even as regards simply its most immediate consequences relating to the modes of symmetry.

Moreover, one can remark that we give too much importance to the synthetic method in granting that, from the reticular structure acknowledged as fundamental hypothesis, the existence of plane faces and cleavages and the law of rational truncations can be deduced. It is in reality completely incapable of being done through pure logical deductions unless we add to the reticular hypothesis all kinds of other hypotheses relating to material particles, to their reciprocal actions, etc. One would more accurately say that, if these laws of observation are hidden behind the reticular hypothesis, it is in order not to find them again ever afterward. Let us apply the inverse method. It consists, in short, in going back to the wholesome method that led Hessel, as early as 1830, to set up 32 modes of symmetry.

[*Editor's Note:* Material omitted at this point.]

11

STUDIES ON BRAVAIS'S LAW

Georges Friedel

This article was translated by Francette Bonnabaud, from pp. 347–349, 355–357, 454–455 of "Études sur la loi de Bravais," Bull. Soc. Franc. Mineral., **20**, *326–455 (1907).*

[*Editor's Note:* In the original, material precedes this excerpt.]

Our purpose is to show, from the observation of facts only without any kind of theory, that to express a large number of facts that are not explained by Haüy's law, considered in its usual meaning of law of simple rational truncations, it is best to replace this law by the following much more complete law of Bravais:

> The importance of the directions of planes which are revealed in a crystal through discontinuous vectorial properties is first of all in relation to their reticular area in a certain "réseau" [lattice] of parallelepipeds. It is all the greater as this area is smaller.

Other factors also have an influence, but a secondary one, and up to now no generalization is known about them.

Preliminary Remarks

1. The lattice we are talking about has none of the character of a physical concept. (The lattice we used before to state Haüy's law has none either.) It is a geometric intermediary suitable to express the law, and nothing more. If observation confirms the law, we will be able to conceive them, as an hypothesis and if this may be helpful to relate the law to others, that crystalline matter is periodic, and that this lattice represents its period. But the law, if it exists, cannot depend on a theory. We need not suppose that the lattice really exists in the crystal. I know quite well that today all scientists consulted on this matter would answer, without hesitation, that such an assumption is actually useless and wanders from the domain of science. So it is for the atom in chemistry. But many scientists also, as for the chemical atom, behave as if they actually believed in its real existence, and, depending on the present inspiration, determine and try to reach this real lattice through a great number of various means, often contradictory and without any relation to the facts that have led to the notion of lattice.

2. The importance of a plane is not something that can be measured with precision. When exterior forms are concerned, the importance of a plane is expressed by the *frequency* [of occurrence] and the *size* of the face parallel to this plane; it is clearly understood that this appraisal is

based on as many crystals as possible, formed under the most various conditions. We know that for most species the forms often vary according to the conditions of crystallization. We also know that symmetries are found in the forms of crystals, which cannot be those of any lattice. They are called "merohedral." The lattice set forth in Bravais's law has first necessarily all the symmetry of the discontinuous properties that it expresses: two flat sides that cannot be physically distinguished from one another should not differ in anything, by definition, as reticular planes of this lattice. But, moreover, owing to the fact that it is a lattice of parallelepipeds, and therefore subject to certain geometric conditions (such as always having a center), it has necessarily, if the symmetry of the crystal is "merohedral," extra elements of symmetry that do not belong to the crystal. There are, therefore, in a crystal with merohedral symmetry, planes that have the same lattice plane, the same reticular area, but which, however, may not have the same physical importance. If the importance of forms depended on the reticular area only, there could well be "merohedries," but the discontinuous properties would not reveal them. There could be "merohedral" crystals but no "merohedral" exterior forms.

These two facts are sufficient to make useless every attempt at a precise measure of the comparative importance of faces, and, consequently, to forbid us to hope to find in Bravais's law a mathematical rigor that it cannot have. If it were wrongly given this characteristic of precision, the law would mean this: the exterior forms of a crystal depend solely on the reticular areas. Give me the parameters of a crystal and I will tell which are its forms and their relative development. I will draw the crystal for you. We will see how, in a great many cases, this is almost true. But since the conditions of crystallization have an influence on the forms, and since symmetries that belong to no lattice appear in these forms, the forms do not depend solely on the lattice. Thus, Bravais's law would be misunderstood if it was taken to mean that the importance of forms depends on the reticular area only. The law only states that the importance of planes depends distinctly on the reticular area and depends *above all* on this area, and this is so obvious that in many cases other influences are hardly felt. That this fact may stand out and be obvious, we need not have an absolute criterion concerning the order of importance of the planes such as forms, cleavages, etc. It is sufficient that the law classifies the forms in an order which agrees with the order of importance insofar as it can be appraised by examination of crystals. We will see that, for the law to appear with perfect clarity, the mere *occurrence* of forms is already a sufficient criterion of their importance.

[*Editor's Note:* Material has been omitted at this point.]

BRAVAIS'S LAW IN SIMPLE CASES

I will examine in detail a few crystalline species in order to show clearly how the parameters determined by Bravais's law express many more facts than those which result from the law of simple rational truncations. The data I will use are in general Dana's (edition 1903).

Tourmaline: an example of the slight influence of merohedrism upon the classification of forms according to importance, in a species which can however be considered one of the clearest cases of crystals with merohedral exterior shapes. Despite the merohedrism, the lattice intervenes almost alone in the determination of the importance of forms. On the other hand, this is all the more striking an example of the essential difference between crystals with a ternary lattice and those which, having a hexagonal lattice, are ternary from merohedrism.

According to Bravais's law, the lattice is rhombohedral with the parameter $c : a = 0.4477$. With ternary notation, which alone can be used in such a case, one has

$$S^2 = (p - q)^2 + (q - r)^2 + (r - p)^2 + (p + q + r)^2 \times 7.485^{a}$$

Let us first classify the reticular planes of this lattice according to decreasing reticular densities.

$S^2 =$	$\underline{(10\bar{1})}$	$\underline{(100)}$	$\underline{(11\bar{1})}$	$\underline{(11\bar{2})}$	$\underline{(20\bar{1})}$	$\underline{(110)}$	$\underline{(21\bar{2})}$	$\underline{(11\bar{3})}$
	6	9,48	15,48	18	21,48	31,94	33,48	39,48
	$\underline{(21\bar{3})}$	$\underline{(21\bar{1})}$	$\underline{(30\bar{2})}$	$\underline{(30\bar{1})}$	$\underline{(22\bar{3})}$	$\underline{(31\bar{3})}$	$\underline{(111)}$	$\underline{(31\bar{2})}$
	42	43,94	45,48	55,94	57,48	63,48	67,36	67,94
	$\underline{(21\bar{4})}$	$\underline{(210)}$	$\underline{(31\bar{4})}$	$\underline{(11\bar{4})}$	$(40\bar{3})$	$\underline{(22\bar{1})}$	$(31\bar{1})$	(323)
	69,48	73,36	78	79,94	81,48	85,36	91,36	91,94
	$(32\bar{4})$	. . .						
	93,5							

The first 20 forms are all known in tourmaline (the known forms are underscored). The first two, ($\underline{10\bar{1}}$) and ($\underline{100}$), are the imperfect cleavages of the species. The first one is the hexagonal prism, which is absolutely persistent in all crystals; they are almost always elongated parallel to this prism. The second, ($\underline{100}$), is persistent at both ends of the crystals in spite of antihemihedrism. It is not perceptibly affected by the hemihedrism. Then follow ($\underline{11\bar{1}}$), which is constant at the antilogous[b] pole and also very frequent at the analogous pole; then the prism ($\underline{112}$), equally constant, often triangular, but also very often presenting its two half-forms combined, although unequally developed. ($\underline{20\bar{1}}$) and ($\underline{110}$) are especially common at the analogous pole but also exist at the other end. These first six forms are definitely the most important of the species, the first two also

[a] E. Mallard, *Traité de Cristallographie*, Dunod, Paris, 1879, Vol. I, pp. 22, 116, 305. *Editor's Note:* p, q, and r are Miller indices referred to rhombohedral coordinates, used throughout the discussion. To transform to Bravais indices hkl, $h = p - q$, $k = q - r$, and $l = p + q + r$.

[b] *Editor's Note:* Pyroelectric negative; "analogous" is pyroelectric positive.

distinctly prevalent as often as cleavages as faces; and as for the following four, the first two of these are constant, the other two only occasional. The following six forms, $(\underline{21\bar{2}})$, $(\underline{11\bar{3}})$, $(\underline{21\bar{3}})$, $(\underline{21\bar{1}})$, $(\underline{30\bar{2}})$, and $(\underline{30\bar{1}})$, are still common but distinctly accessory and almost always affected by the merohedrism. $(\underline{11\bar{3}})$, $(\underline{21\bar{1}})$, and $(\underline{30\bar{2}})$ exist almost exclusively at the antilogous pole, and $(\underline{30\bar{1}})$ at the analogous pole. All the following forms are even less important. It is not until the twenty-first that one can find an unknown form in the list, $(\underline{40\bar{3}})$, whereas the next, which actually is rare, has been found. After this, the known forms, all rare, become fewer and farther between, mixing with nonrecorded faces, as must be expected.

Here then is a species in which merohedrism very clearly affects the forms, where, consequently, the perturbing influence of the crystalline motif seems to tend to mask, more than in other species, the role of reticular density. However, this role remains so predominant that the merohedrism does not perceptibly affect the order of importance of the forms. For the most important faces of very large reticular density, the two complementary forms maintain about the same importance in spite of the merohedrism. For the following ones, still very important, one of the two complementary forms is persistent, the other less so. Then come forms that are common at one of the poles and rare at the other. Finally, there are faces that are only known to exist at one of the poles. In other words, the influence of the crystalline pattern does not disturb Bravais's law for forms with great density; it is only felt, and more and more, in forms with smaller and smaller density. The same fact is noticed everywhere in merohedral types, and it is very significant. It shows admirably how small is the importance of the factor "motif" with respect to that of the factor "lattice" in the determination of forms and cleavages. The fundamental law of crystallography, whatever form it is given, signifies nothing else: it is the lattice that plays the dominant role in the determination of plane faces, the motif intervening only as a secondary perturbing influence. Bravais's law only adds the following: in the lattice, the importance of forms increases with their reticular density. For tourmaline, this is remarkably clear.

[*Editor's Note:* Material has been omitted at this point.]

POSSIBLE INTERPRETATIONS OF THE DOUBLE LATTICE

It is well known that Haüy's law and the law of rationality of symmetric parameters, or better Bravais's law which makes them explicit by incorporating them both in a single formula, are interpreted by the physical hypothesis of the periodicity of the homogeneous crystalline medium. One can imagine that the lattice defined by Bravais's law represents the period of this medium, the crystalline lattice.

On the other hand, it is also known that, to take into account paramorphic transformations,[c] we have been led to conceive of the crystalline medium as composed of identical particles diversely oriented, distributed at the nodes of an integer submultiple lattice of the crystalline lattice. I remind you that two lattices are said to be *multiples* or *submultiples* (which comes to the same thing) *of one another* when they have the same reticular planes. They are integer multiples of one another if their corresponding parameters are integer multiples of one another.

This integer submultiple lattice of the crystalline lattice is the material lattice. To answer its purpose, the material lattice must be the same, aside from slight deformations, for two forms of a single compound, susceptible of passing from one to the other through paramorphic transformation.

We have just seen that in some crystalline species (and I could give other examples) Bravais's law reveals not only one, but two distinct lattices, which are simple multiples of one another. This immediately suggests a parallel with the theory of the material lattice. The principal lattice that determines the principal forms and cleavages would be the crystalline lattice, for these forms and principal cleavages change completely in paramorphic transformation. The secondary lattice, which is only manifest in the evidences of accessory forms, could be considered as representing the material lattice. The image, it seems, would be extremely satisfactory. But, for this to be acceptable, we will have to find cases of unquestionable paramorphism in which the two forms exhibit, with two distinct principal lattices, the same secondary lattice. It is unfortunately very difficult to find cases for which a thorough study of the forms of two paramorphous varieties of a single compound would be possible, the two varieties not both being stable in general at the same temperature. So far in this matter, I have only been able to examine cases of dubious paramorphism, such as those of calcite and aragonite. Calcite shows indubitable marks of a hexagonal secondary lattice, which, on the contrary, is not clearly found in aragonite. But this example does not prove anything in either sense, since the paramorphic crossing from one to the other in these two species has never been ascertained. Thus the question remains still open.

It is nonetheless interesting to note that the notion of double periodicity imposed by the facts of polymorphism is found again in the detailed examination of crystalline forms in the light of Bravais's law.

[c] *Editor's Note:* Direct polymorphic transformation.

12

Reprinted from Plate 168 of V. Goldschmidt, *Atlas der Kristallformen*, Vol. I, Universitätsverlag GmbH, Heidelberg, 1913-1923

BARITE

V. Goldschmidt

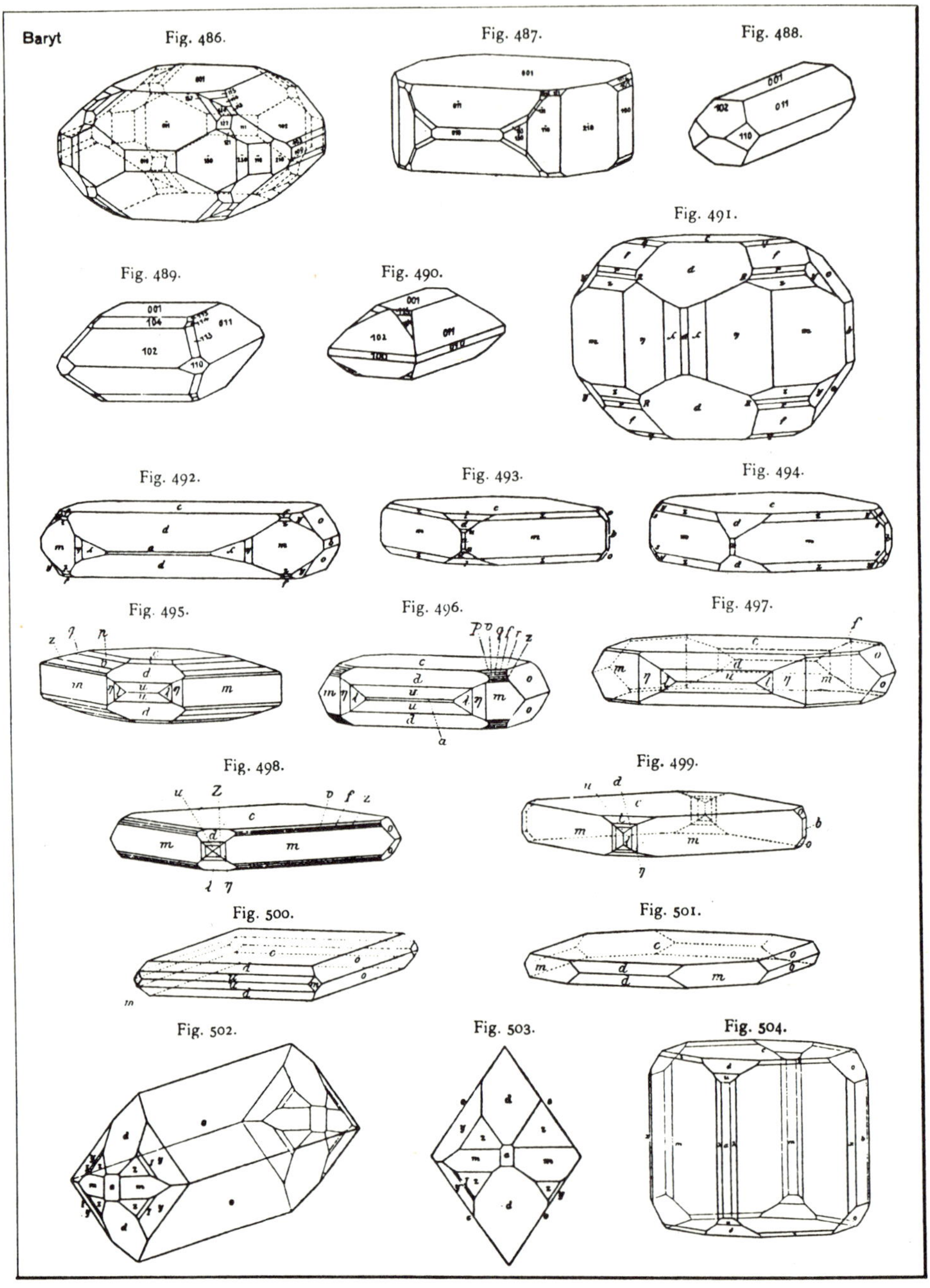

Plate 168

13

THE CRYSTAL EMPIRE (Editor's Review)

E. S. von Fedorov

This review was prepared expressly for this Benchmark volume by Cecil J. Schneer, from Das Krystallreich, Mém. Acad. Sci. Russ. *VIIIe Sér.* **36** *(1920)*

E. S. Fedorov (Fyodorov), 1853–1919, derived the 230 space groups in 1890 and proposed the two circle goniometer and the universal stage. In a major work in 1903 (yet to be translated from the Russian), he developed a unique system of crystal structures derived from morphology. In 1911 he presented these ideas again in a massive survey of the data of crystal chemistry, *Tsarstvo kristallov, The Crystal Empire*, to the St. Petersburg Academy of Sciences. Derived from the morphology, it was a *structural* classification of the entire crystal kingdom (heavily dependent on inorganic chemistry as well as mineralogy).

Fedorov listed the members of an isomorphous series (his most compelling example was the set of 54 double sulfates) with the set of their forms in order of importance, and compared this with the ranking of forms by reticular densities for distinct structures. Even taking into account individual discrepancies, such as the well-known case of the missing basal pinacoid for quartz, he showed that a weighted comparison of a small number of the principal forms of quartz with their reticular densities led to excellent agreement. His structures were essentially lattices, considered as hexahedral, octahedral, and dodecahedral (P, F, I) arrangements, but he recognized what he called the *individuality* of each isomorphous series, evidenced by the occurrence of one or more forms out of its theoretical order of importance. He understood that there was a fine structure, "the distribution of electrical polarity within the crystal-molecule," which remained to be revealed.

Fedorov considered the overwhelming agreement of 54 distinct chemical species among themselves and with theory as irrefutable statistical confirmation of the structure theory, particularly as it was repeated for literally thousands of species in the body of his work. War and revolution prevented publication until 1920, after his death and eight years after X-ray diffraction.[1]

REFERENCE

1. The basic ideas were published in 1912 in "Die Praxis in der krystallochemischen Analyse," *Z. Krist.* **50**.

Part V

THE EXTENSION OF BRAVAIS'S RULE

Editor's Comments on Papers 14 Through 16

14 NIGGLI
The Relation Between the Growth Forms and the Structure of Crystals (Editor's Review)

15 PEACOCK
Calaverite and the Law of Complication

16 DONNAY and HARKER
A New Law of Crystal Morphology Extending the Law of Bravais

Paul Niggli, mineralogist at Zurich, epitomized a generation of mineralogists for whom the discoveries of X-ray diffraction validated the reasoning summed up in Wulff's theorem. Form must be related to the new structure revealed by physical-chemical and not simply numerical and geometric functions. The net reticular density of Bravais—". . . the weights of the planes and their interplanar spacings lose their simple meaning . . . if several species of atom are present." Not only space groups but atomic structures were to enter into the calculations.

Nevertheless, for the actual calculations in Paper 14, dealing with the problem of vicinal faces, Niggli introduced an ad hoc hypothesis. Growth velocity would depend on the depth of an unsaturated surface layer, called the *inner transition layer*, considered as a stepped surface.

In his *Lehrbuch* of 1941, Niggli designed statistical procedures for the quantitative measure of the significance of crystal forms, procedures that were directly applicable to the tables of Goldschmidt's *Atlas*.

Paper 15, a forgotten work by Peacock, is included, first, because it presents the only clear statement of Goldschmidt's law of complication in English, and, second, because it summarizes in English the very remarkable neo-Platonic ideas of Goldschmidt's *Ueber Harmonie und Complication* of 1901. In extending his numerical series of zonal development to music, astronomy, physics (color), and psychology and physiology, Goldschmidt believed that he had uncovered a universal law of nature that was only in part of natural origin (he did explain his law of complication as due to the composition of forces). Small wonder that Friedel dismissed it as equivalent to the observation that the indices of a plane are the sum of the indices of the bracketing planes in the zone! Nevertheless, the observed rank-ordering of the faces in zones was in agreement with predictions made by the various series of the type of Goldschmidt's law of complication, and the law, stripped of metaphysical overtones, was empirically valid and contributed indirectly to later developments.

J. D. H. Donnay writes that he has approached the subject of form and structure through the work of Friedel, Junghann, V. Goldschmidt, Baumhauer, Niggli, Ungemach, Peacock, J. D. H. Donnay, R. Kern, G. Donnay, Evzikova, McLachlan, and Schneer. Beginning with a paper by Gustav Junghann in Poggendorff's *Annalen* for 1874 entitled, "A Simple Law for the Development and Grouping of Crystal Zones," attempts had been made to reconcile apparent discrepancies between axial parameters derived from different zones of crystals of the same species. After 1912, the validity of the several serial approaches to zone development was confirmed as, for example, by Ungemach's paper in 1904, which showed that the revisions of cell parameters that he had derived by such considerations were confirmed by the new method.

The explanation came with a brilliant extension of the Bravais law by Donnay and David Harker (Paper 16), a modification so simple and withal so elegant as to put an end at once to the harmonic, arithmetic, etc., series, and at the same time to connect the pure mathematics of group theory to the tangible, physical forms of crystals. A generation of mineralogists engrossed in the determination of structures through X-ray diffraction saw in this work confirmation from classical forms of the physical significance of the new ideas, a demonstration in short of the unity of science, a link that validated both the old and the new and appeared an additional warranty, if one were required, of the essential interpendendence of form and structure.

14

THE RELATION BETWEEN THE GROWTH FORMS AND THE STRUCTURE OF CRYSTALS (Editor's Review)

Paul Niggli

This review was prepared expressly for this Benchmark volume by Cecil J. Schneer, from "Beziehung zwischen Wachstumformen und Struktur der Kristalle," Z. Anorg. Chem., **110**, *55–81 (1920)*

The reticular densities of planes through the centers of gravity in a crystal are proportional to the interplanar spacings δ. Both are comparable to growth velocities but lose their simple meaning if there are numerous centers of gravity or if several species of atoms are present in which case the space group—and in the general case the structure—is required. The dependence of the growth velocities on δ breaks down when one considers the vicinal faces and their concentration in definite zones. In Figure 1a

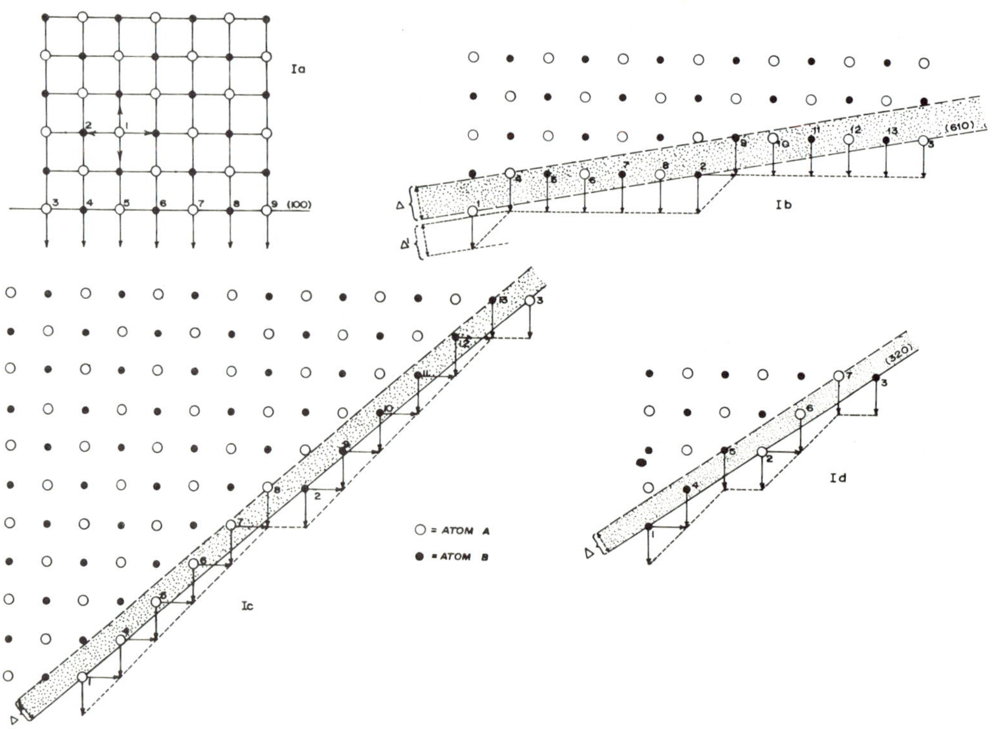

Figure 1 Rock salt type.

the direct bonds define a boundary plane (100). The stepped surface atoms 1 to 13 of (610) Figure 1b are incompletely saturated and comprise an inner transitional layer of width Δ equal to the total depth of the steps. Growth velocity should be proportional to Δ. Figure 1c shows a smaller Δ for (650) relative to the absolutely stable face (110) than the Δ for (610) relative to (100) therefore the expected sequence of faces in the zone should be (100), (650), (610) although δ_{610} is greater than δ_{650}.

For zinc sulfide, positive and negative forms display different Δ-values proportional to the magnitudes of the bonds stemming from zinc relative to those from sulfur for a given form.

Statistical distributions of the forms grown under different conditions for all species considered are in strong agreement with the distributions calculated from the Δ-values. The succession for the absolute growth faces with $\Delta = 0$, namely {100}, {110}, {111}, {211}, and {311} in the examples given, remains undefined. The chemical bond treatment must be applied to these faces, (number, strength and direction of the valences).

15

Reprinted from *Am. Mineralogist*, **17**, 317–337 (1932)

CALAVERITE AND THE LAW OF COMPLICATION

M. A. Peacock, *New Haven, Connecticut*

Few crystallographers would venture to think that the Law of Simple Rational Intercepts and Indices is not a perfectly general truth; yet this is the conclusion reached by Victor Goldschmidt (Heidelberg), Charles Palache and the present writer[1] in a recently published account of an extended study of calaverite, the crystallized di-telluride of gold which has puzzled crystallographers since the end of the last century. Tersely written in German, assuming familiarity with the Goldschmidt method of crystal measurement and discussion, and making use of a new and significant crystallographic conception, this joint account may have presented some difficulties to the English-speaking reader. Encouraged by Professor Palache, the junior author has therefore prepared the present paper which gives the essential theory used in the joint paper and its application to the crystallography of calaverite, followed by a formal statement of the crystallography of this remarkable mineral. The opportunity is also taken to give a brief account and discussion of Professor Goldschmidt's application of the Law of Complication to fields outside of crystallography.

The Classical Laws of Crystallography

The goniometric study of crystals has led to three empirical generalizations which have long been accepted as the laws of systematic crystallography.

1. The Law of Constancy of Angles.
2. The Law of Zones.
3. The Law of Rationality.

The Law of Constancy of Angles states that the angular relations of the forms of crystals of any species are constant and proper to the species.

The Law of Zones states that crystal faces occur in groups in each of which the faces are parallel to a common axis.

By the Law of Rationality, the Law of Simple Rational Intercepts and Indices is generally understood. This law may be stated as follows: With suitably chosen axes and axial lengths, every face of a crystal is parallel to a plane which cuts one axis at the unit

[1] Ueber Calaverit, *N. J. f. Min.*, Beil. Bd. **63A**, pp. 1–58, figs. 1–22, pls. i–x, 1931.

length and cuts the other axes either at infinity or at distances from the axial centre which are simple rational multiples or fractions of the corresponding unit lengths; and therefore the angular relations of every face may be given, with reference to suitable crystal axes, by symbols involving, on any system of notation, only simple rational numbers, zero and infinity.

These laws determine the procedure of crystallographic investigation. A number of crystals of a species are measured, and it is observed that like forms on all the crystals have the same angular relations. The measurements are projected and the faces are found to lie in zones. Suitable axes are chosen and the appropriate simple rational symbols are given to all the observed plane faces. The consistent success of this procedure has consolidated the classical laws of crystallography and placed them in a position of apparently unassailable security.

The study of calaverite was naturally approached in this way; but repeated investigations[2] of excellent material showed that, while the first and second laws held strictly true, the third failed. It is not possible to choose axes for calaverite so that the majority of the forms receive simple symbols; and no admissible assumptions of twinned or heterogeneous structure will serve to bring this crystal species within the Law of Rationality in its generally accepted form. No one seriously thought that the Law of Rationality might be inadequate, and thus the tiny, brilliant, golden crystals of calaverite remained complete enigmas.

Goldschmidt's Law of Crystallography—The Law of Complication

In the gnomonic projection, which forms the graphical foundation of Professor Goldschmidt's system, the Law of Constancy of Angles is proved by the constancy of the co-ordinates of like nodes (projection points of face normals) in crystals of the same species. The Law of Zones is apparent in the fact that the nodes of any one crystal lie on a network of straight lines extending to infinity. And

[2] Of these only two were published: S. L. Penfield and W. E. Ford, *Amer. Jour. Sci.*, **xii**, pp. 225–246, 1901; G. F. H. Smith, *Min. Mag.*, **xiii**, pp. 122–150, 1902. Laborious unpublished studies by Palache, Goldschmidt, Moses, Goldschmidt and Görgey, Goldschmidt and Neff, and Palache and Peacock, representing efforts spread over more than thirty years, confirmed and extended the previous observations, but failed to reconcile them with the accepted laws of crystallography.

the Law of Rationality appears in the fact that the linear distances between the nodes in a zone are simple rational fractions or multiples of a suitably chosen unit length, when the plane of the projection is normal to a zone axis.

In any zone certain nodes, usually those corresponding to the largest and most frequent faces, may be recognized as principal nodes. In zones having one such node at infinity (∞), another being taken as the origin of measurement (0), and a third important node being taken to give the unit of spacing (1), it is found by statistical study[3] that the remaining nodes of the zone segment between the nodes (0) and (∞) occur most frequently at distances $^1/_2$, 2, less frequently at $^1/_3$, $^2/_3$, $^3/_2$, 3, and rarely at other distances from the end-node (0) (fig. 1). Thus, out of the great variety of simple rational spacings which might occur in a zone, there are ev-

FIG. 1. Gnomonic projection of a normal zone having one principal node at infinity. Between the end-nodes 1, ∞, the derived nodes are spaced according to the terms of the normal complication series N_3.

idently certain preferred arrangements which Goldschmidt has recognized as normal series, N_0, N_1, N_2, N_3, · · ·

N_0:	$p=0$								∞
N_1:	$p=0$				1				∞
N_2:	$p=0$		$^1/_2$		1		2		∞
N_3:	$p=0$	$^1/_3$	$^1/_2$	$^2/_3$	1	$^3/_2$	2	3	∞
. . .	. . .	. . .	. . .	. . .	. . .	. . .	. . .	. . .	. . .

Except in the series N_0, in which there is no development between the end-nodes, 0, ∞, the series show a numerical symmetry about the node 1, which is named the dominant,[4] the terms on the one side being the reciprocals of those on the other. From any series N_n the terms of the next higher series N_{n+1}, are obtained by adding 1 to each term of the series N_n, giving the terms 1 · · · ∞ of the series N_{n+1}, and then writing the reciprocals of the terms

[3] V. Goldschmidt, Ueber Entwicklung der Krystallformen: *Z. f. Kryst.*, **xxviii**, pp. 1–35, 414–451, 1896–97. This account is based on these two papers and on conversations with Professor Goldschmidt. The statistical study referred to appears on p. 10.

[4] In the diatonic series, the term 1 corresponds with the dominant of the scale (see below, p. 330).

thus obtained to give the terms 1 · · · 0 of the series N_{n+1}. The number of series may be extended indefinitely, but series higher than N_3 are infrequent in crystal zones.

In the belief that these preferred crystallographic number series are of fundamental significance, Professor Goldschmidt has framed a theory of crystal structure and zonal development which leads naturally to the observed preferred arrangements of crystal faces in a zone. This theory rests on two hypotheses:

1. A crystal is a rigid system of like and similarly oriented particles.

2. Every crystal face is a plane normal to a force of attraction exerted from the centre of the crystal particle.

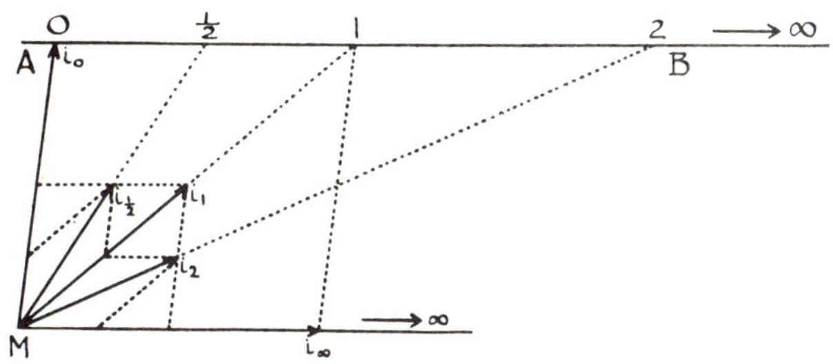

FIG. 2. Section in the plane of a normal zone showing the development of derived forces (and therefore faces) by complication between the principal forces (or faces) i_0, i_∞.

The construction in fig. 2 is a section in the zone-plane of a zone with principal (end) nodes 0, ∞. AB is the trace of the gnomonic plane and M is the crystal centre. Since 0 and ∞ are principal nodes of the zone, i_0, i_∞ are the principal attraction forces, according to hypothesis 2. The forces i_0, i_∞ are fixed in direction by the nodes 0, ∞, but their intensities, given by the lengths of the arrows from M, are arbitrary. If the entire forces i_0, i_∞ or aliquot parts, of them, combine to give a resultant by the parallelogram of forces, a new force is produced giving the dominant node 1 and therefore also the corresponding face. Since the principal forces would be exhausted if they combined completely to give a resultant, Goldschmidt assumes that the principal forces divide into aliquot parts, normally in halves, one half of each principal force continuing to act in its respective principal direction and the remaining halves combining to give the resultant force.

$^1/_2 i_0 + ^1/_2 i_\infty = i_1$, by the parallelogram of forces, and therefore, from the initial series:

$$N_0: p = 0 \ \infty$$

we obtain the next higher series:

$$N_1: p = 0 \ 1 \ \infty$$

By combining one half of the remainder of i_0 with one half of i_1, a resultant $i_{\frac{1}{2}}$ is obtained; and similarly one half of i_1 with one half of the remainder of i_∞ gives i_2.

$$^1/_4 i_0 + ^1/_2 i_1 = i_{\frac{1}{2}}; \quad ^1/_2 i_1 + ^1/_4 i_\infty = i_2$$

Whence we obtain the next higher series:

$$N_2: p = 0 \ ^1/_2 \ 1 \ 2 \ \infty$$

Similarly, the successive combination of halves of the remainders of the principal forces with halves of the newly formed resultants give further resultants corresponding to the new terms of the higher series.

To this process of division and combination of parts Professor Goldschmidt has given the name *Complication*, a process which he believes to be fundamental, not only in crystallography, but also in other widely dissimilar fields. In the realm of crystallography complication means specifically the division of the principal crystal forces (giving the principal faces of a zone) into halves, the combination of these parts to give a resultant force (derived face), and the repeated formation of new resultants between adjacent forces by halving and combination of halves.

The empirically found series:

N_0: $p=0$								∞
N_1: $p=0$				1				∞
N_2: $p=0$		$^1/_2$		1		2		∞
N_3: $p=0$	$^1/_3$	$^1/_2$	$^2/_3$	1	$^3/_2$	2	3	∞
. . .	. .	. .	. .	. .	. .	. .	. .	. .

thus becomes the *normal complication series.*

In many zone segments neither of the principal nodes is at infinity. The number series in such zone segments are not normal complication series. If $p_1 \cdots p \cdots p_2$ be such a series, it must be transformed into the standard form $0 \cdots \infty$ by writing $(p-p_1)/(p_2-p)$[5] for each term of the series. This transformation of series into the standard form plays an important part in the critical

[5] For the proof of this formula and the justification for the transformation, see V. Goldschmidt, *Zeit. f. Kryst.*, **xxviii,** pp. 22–23, 1896–97.

discussion of the number series in crystallography and in other fields.

The complication series and the transformation formula for transforming any number series into the standard form $0 \cdots \infty$, give a valuable criterion for the critical discussion of crystal form systems. Uncertain forms may be retained or rejected according to whether they are likely terms of the complication series or not; and well substantiated forms with unusual complication numbers afford evidence of disturbance in the free development of zones.[6]

We are now in a position to state and understand Goldschmidt's general law of crystallography, which may be briefly expressed as follows:

The form system of a crystal results from complication between the principal forces of the crystal particles.

From what has already been given, it is clear that this law accepts and includes, by implication, the Law of Constancy of Angles and the Law of Zones, and recognizes the principal of rationality in the progressive halving of forces which is inherent in the process of normal complication. Goldschmidt's law is more general, however, as we shall see later; at the same time it is more specific in that it recognizes preferred rational number series whereas any simple rational number series would satisfy the older law.

The Law of Complication Applied to Calaverite

According to the principles developed above, the goniometric investigation of a crystal by the Goldschmidt method involves finding the principal nodes and discussing the complication in each zone. The principal nodes are given the simplest possible symbols by suitable choice of axes and unit form, and thus the best elements of the crystal are determined. Then the remaining nodes, which are derived from the principal nodes by complication, are given rational symbols determined by their spacing with respect to the principal nodes. The practical result is the same as that reached by the older method, except that the choice of elements rests on a securer basis since it is made to give the lowest complication between the principal nodes. The theoretical result of the investigation, however, is to prove the validity of the Law of Complication

[6] Outside of Professor Goldschmidt's laboratory, this method of critical discussion of form systems has been usefully applied by W. T. Schaller, The Mercury Minerals from Terlingua, Texas. *U. S. Geol. Sur., Bull.*, **405**, 1909.

as the principal governing the systematic arrangement of crystal faces.

Since the principal nodes could always be regarded as the simplest forms of the crystal, it was natural that the very first step in the Goldschmidt discussion of a projection was to find axes to which the principal nodes were related in a simple manner. But for calaverite this could not be done, and therefore the Goldschmidt method of approach failed, like the others, in the earlier investigations of this mineral. During the last attempt to solve the calaverite puzzle, which at first seemed as hopeless as ever, it occurred to Professor Goldschmidt that it might be better to disregard the question of symbols at first, and to proceed rigorously by finding the principal nodes and discussing the complication between them. This met with immediate success. The entire form system of ninety-two confirmed forms was found to be developed by complication from a few principal nodes; and thus remarkable confirmation of the Law of Complication was obtained in a case where the Law of Simple Rational Intercepts and Indices had failed.

With the best choice of axes, all the principal nodes, except one pair, received the customary simple symbols; the exceptional pair received the exceedingly complicated symbol ($\bar{3}$.5.29), with only tolerable agreement between the calculated and observed angles. It thus became evident that principal nodes might sometimes be forms with complicated, if not quite irrational symbols. Such principal nodes Professor Goldschmidt has proposed to name *singular nodes.*[7] The singular node pair C was the key to the calaverite puzzle. When it was recognised that the node *C* with its complicated symbol was yet the simplest node, namely the node of origin (0), in each zone containing *C*, and that every zone of calaverite gave a more or less complete N_3 complication series without extra terms, it was apparent that calaverite conformed strictly to the Law of Complication as it was formulated many years ago.

The Systematic Crystallography of Calaverite

The practical problem of formally presenting the crystallography of calaverite still remains. The complicated relations are best shown graphically, but for formal purposes they must be expressed quantitatively.

[7] This Journal, **xvi**, p. 78, 1931.

Calaverite is monoclinic and isomorphous with sylvanite. The orthodome zone is developed prismatically; the measurements are therefore projected and discussed in M_2 position (projection plane

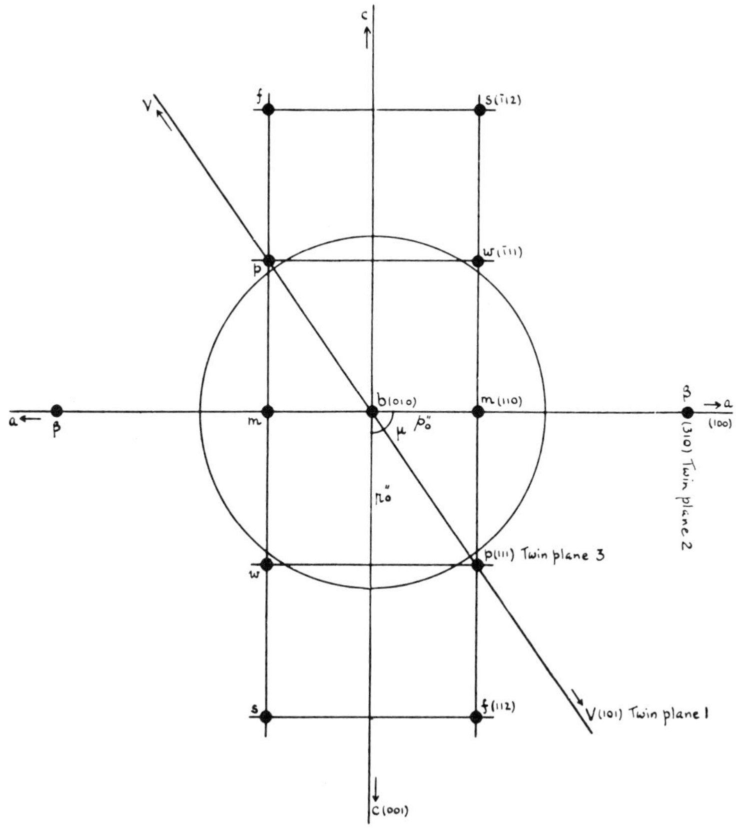

FIG. 3. Calaverite. Gnomonic projection of the S-forms and graphical elements, showing also the three known twinning planes.

normal to the symmetry axis). In this position (fig. 3) the Goldschmidt projection elements are:

$$r_0''=0.8702;\ p_0''=0.6136;\ q_0''=1;\ \mu=89°\ 52'(M_2).$$

The corresponding polar elements in M_1 position (normal position, projection plane parallel to the symmetry axis) are:

$$p_0=0.7051;\ q_0=1.1492;\ r_0=1;\ \mu=89°\ 52'(M_1).$$

These elements are equivalent to the linear elements:

$$a:b:c = 1.6298:1:1.1492;\ \beta = 90°\ 08'.$$

These are the fundamental elements of calaverite; from the isomorphism with sylvanite we have named them the *S*-elements, and the 10 known forms of calaverite which are simply related to these elements we have named the *S*-forms.

CALAVERITE—MONOCLINIC
S-ELEMENTS AND FORMS

$r_0'' = 0.8702$; $p_0'' = 0.6136$; $q_0'' = 1$; $\mu = 89°\ 52'$ (M_2)
or $p_0 = 0.7051$; $q_0 = 1.1492$; $r_0 = 1$; $\mu = 89°\ 52'$ (M_1)
or $a:b:c = 1.6298:1:1.1492$; $\beta = 90°\ 08'$

$c(001)$	$a(100)$	$\beta(310)$	$p(111)$	$f(112)$
$b(010)$	$m(110)$	$V(101)$	$w(\bar{1}11)$	$s(\bar{1}12)$

TWINNING LAWS

1. The twinning and composition plane is $V(101)$; the axes of the prismatically developed zones of the two individuals are parallel.

2. The twinning and composition plane is $\beta(310)$; the axes of the prismatically developed zones of the two individuals intersect at 122° 58′.

3. The twinning and composition plane is $p(111)$; the axes of the prismatically developed zones of the two individuals intersect at 93° 40′.

Thus far the statement of the crystallography of calaverite is perfectly normal; but we have accounted for only 10 of the 92 known forms. On the *S*-elements (M_1) the singular nodes *C* have the exceedingly complicated symbol $\bar{5}.29.3$, and thus it is clear that, since these are the strongest nodes of the form system, all the forms of calaverite, except the 10 simple *S*-forms, will receive symbols of great complexity on the *S*-elements. A table of such symbols would be useless for practical purposes and at variance with the theoretical part of our study.

The only alternative, and it is admittedly a makeshift, is to refer the forms of calaverite to several closely related groups of elements of triclinic character; in this way the great majority of the forms receive simple symbols closely related to their simple complication numbers.

Fig. 4 shows a part of the zone-net of calaverite corresponding to the *S*-elements (dotted), and in full lines sufficient construction

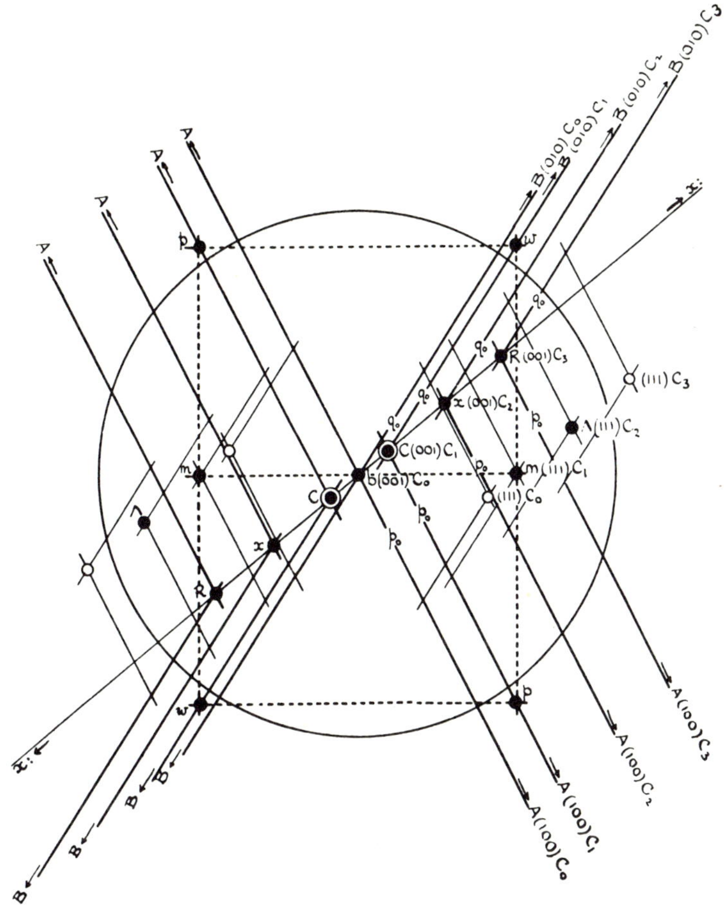

FIG. 4. Calaverite. Gnomonic projection of the C_1, C_2, C_3, C_0 graphical elements (polar axis and unit forms). The principal S-zones are given in broken lines. The blank points are forms not known; the double circles are the singular nodes *C*. *x:* is one of the incongrous CC_2-forms lying in zone with *C* (001) C_1 and a node of the C_2-net.

to illustrate the groups of elements (the calaverite- or *C*-elements) which serve to accommodate most of the forms of calaverite with simple symbols.

C_1-elements: Polar axes ABC; resp. proj. elements $p_0\, q_0\, r_0 = 1$
C_2-elements: Polar axes ABx resp. proj. elements $p_0\, q_0\, r_0 = 1$
C_3-elements: Polar axes ABR; resp. proj. elements $p_0\, q_0\, r_0 = 1$
C_0-elements: Polar axes ABb; resp. proj. elements $p_0\, q_0\, r_0 = 1$

These groups of elements have the same polar axes in the plane of the projection, namely AB, and the same projection elements $p_0\, q_0\, r_0$. The remaining polar axes lie in a zone.

It is of no value to express these groups of elements as triclinic linear elements. The linear constants do not show the close relations which exist between the several groups of polar elements, and they would imply a triclinic interpretation of calaverite which we reject.

If calaverite is measured again, it will be in all probability with the 2-circle goniometer. The most useful practical constants to correlate with future work are therefore the 2-circle angles of the polar axes (or primitive pinacoids) ABC, etc., and the projection elements in the plane of the projection, $p_0\, q_0\, r_0$.

Another reason for not recasting the elements in orthodox triclinic form is the fact that we know that even these four sets of axes, or seven sets, when we include the repetition of three of them by the symmetry axis b, are insufficient to account for all the forms. There remain a number of prismatic forms (CC_2-forms) which lie in zone with the base $C(001)$ of the C_1-elements and with nodes of the incongruous C_2-elements. These forms are therefore incongruous to both groups of elements and cannot be given simple symbols.

The following tabulation[8] is therefore offered as the simplest practical quantitative expression of the remarkable form system of calaverite. It will be observed that all the S-forms appear also in the tables of C-forms.[9]

To the reader who may be tempted to measured calaverite on the two-circle goniometer, the following may be useful. Adjust the prismatically developed zone; the face b then lies in the pole and at the centre of the projection. Find the form $m(111)$ C_1, appearing as a pair of excellent faces with polar distance (ρ) 31° 32′ and with azimuths (ϕ) differing by 180°. With the azimuth

[8] In principle, this mode of presentation is the same as that adopted by G. F. H. Smith, *Min. Mag.*, **xiii**, p. 125, 1902.

[9] In the joint publication, *N. J. f. Min.*, Beil. Bd. **63A**, 1931, will be found the corresponding measured and calculated two-circle angles, tables 6–10, pp. 30–34.

Calaverite

C_1-Graphical Elements and Forms

$A(100)$	$\phi=62°\ 11'$;	$\rho=90°\ 00'$;	$p_0=0.5419$
$B(010)$	$\phi=\bar{5}7°\ 20'$;	$\rho=90°\ 00'$;	$q_0=0.4646$
$C(001)$	$\phi=\bar{3}8°\ 42'$;	$\rho=\ 8°\ 02'$;	$r_0=1$

$C(001)$	$n(302)$	$w(021)$	$\sigma(131)$	$d(211)$
$B(010)$	$p(201)$	$\xi(031)$	$v(1\bar{1}1)$	$\phi(221)$
$A(100)$	$i(301)$	$r(0\bar{1}1)$	$G(1\bar{2}1)$	$\gamma(231)$
$E(110)$	$L(\bar{1}03)$	$W(0\bar{2}1)$	$\kappa(1\bar{3}1)$	$g(2\bar{1}1)$
$c(1\bar{1}0)$	$Q(\bar{1}02)$	$\Xi(0\bar{3}1)$	$M(\bar{1}11)$	$D(\bar{2}\bar{1}1)$
$V(610)$	$y(\bar{1}01)$	$\tau(122)$	$s(\bar{1}31)$	$\Phi(\bar{2}\bar{2}1)$
$l(103)$	$P(\bar{2}01)$	$\iota(\bar{1}22)$	$t(\bar{1}\bar{1}1)$	$\Gamma(\bar{2}\bar{3}1)$
$q(102)$	$I(\bar{3}01)$	$m(111)$	$\Theta(\bar{1}\bar{2}1)$	$\alpha(321)$
$o(101)$	$u(011)$	$\theta(121)$	$\Sigma(\bar{1}\bar{3}1)$	$\chi(331)$
				$f(3\bar{1}1)$

C_2-Graphical Elements and Forms

$A(100)$	$\phi=62°\ 11'$;	$\rho=90°\ 00'$;	$p_0=0.5419$
$B(010)$	$\phi=\bar{5}7°\ 20'$;	$\rho=90°\ 00'$;	$q_0=0.4646$
$x(001)$	$\phi=\bar{3}8°\ 42'$;	$\rho=23°\ 04'$;	$r_0=1$

$x(001)$	$e(101)$	$\psi(0\bar{1}1)$	$\lambda(\bar{1}\bar{1}1)$	$\delta(\bar{2}\bar{1}1)$
$B(010)$	$\omega(201)$	$k(0\bar{2}1)$	$z(\bar{1}\bar{2}1)$	$H(\bar{2}\bar{2}1)$
$A(100)$	$X(\bar{1}01)$	$\Lambda(111)$	$\Delta(211)$	$h(321)$
$E(110)$	$\Omega(\bar{2}01)$	$Z(121)$	$\rho(221)$	$\beta(331)$
$c(1\bar{1}0)$	$\Psi(011)$	$\Pi(1\bar{1}1)$	$j(2\bar{1}1)$	$F(3\bar{1}1)$
$V(610)$	$K(021)$	$O(1\bar{2}1)$	$U(2\bar{2}1)$	

C_3-Graphical Elements and Forms

$A(100)$	$\phi=62°\ 11'$;	$\rho=90°\ 00'$;	$p_0=0.5419$
$B(010)$	$\phi=\bar{5}7°\ 20'$;	$\rho=90°\ 00'$;	$q_0=0.4646$
$R(001)$	$\phi=\bar{3}8°\ 42'$;	$\rho=35°\ 38'$;	$r_0=1$

$R(001)$	$E(110)$	$S(101)$	$\upsilon(\bar{1}\bar{2}1)$	$N(2\bar{1}\bar{1})$
$B(010)$	$c(1\bar{1}0)$	$\pi(201)$	$T(211)$	$\epsilon(\bar{2}\bar{2}1)$
$A(100)$	$V(610)$	$\eta(0\bar{1}1)$	$\nu(2\bar{2}1)$	

C_0-Graphical Elements and Forms

A(100)	$\phi=62° 11'$;	$\rho=90° 00'$;	$p_0=0.5419$
B(010)	$\phi=57° 20'$;	$\rho=90° 00'$;	$q_0=0.4646$
b(001)	$\phi=-$	$\rho=0$	$r_0=1$

b(001)	A(100)	c(1$\bar{1}$0)	J(102)	μ(112)
B(010)	E(110)	V(610)	Y(021)	ζ(212)

CC_2-Forms

a:	Prismatic form in zone with $C(001)C_1$ and $\rho(221)C_2$
e:	$e(101)C_2$
ω:	$\omega(201)C_2$
π:	$\pi(1\bar{1}1)C_2$
z:	$z(\bar{1}21)C_2$
K:	$K(021)C_2$
x:	$x(001)C_2$
Λ:	$\Lambda(111)C_2$

of *m* as zero meridian the remaining forms will fall into the nets given by the above graphical elements or into the same nets repeated by the symmetry axis *b* (fig. 4). This agreement will be obtained only if the measured termination is of the same hand as that which we have chosen. If the termination is left handed with respect to our right handed termination (and right and left handed terminations are equally frequent), the measurements will correspond to nets in mirror image relation to fig. 4 (obtained by plotting azimuths in the negative sense). If the termination is a twin after the first law (prismatic axes in common), the observed forms will fall on fourteen nets, the seven shown in fig. 4 and the seven obtained by constructing a similar figure with negative azimuths.

The Law of Complication in Other Fields

In music. More than thirty years ago Professor Goldschmidt showed some striking similarities between the normal crystallographic number series and the series of numbers obtained by suitably transforming the relative vibration periods of the notes of the diatonic scale.[10]

[10] *Ueber Harmonie und Complication*, Berlin, 1901.

The relative vibration periods (z) of the notes of a major scale, C D E F G A B C′, are very close to the numbers, 1 $^9/_8$ $^5/_4$ $^4/_3$ $^3/_2$ $^5/_3$ $^{15}/_8$ 2, respectively. Transforming this series into the form $0 \cdots \infty$, a series of numbers (p) similar to a normal complication series is obtained.

Like a normal complication series, this series is symmetrical about the term 1, which corresponds with the fifth or dominant of the diatonic scale. The term (3), corresponding to (B♭), is not in the scale of C major, but is closely related to it musically. Notes with complication numbers of equal rank give pleasing combinations; and the passage obtained by combining notes of equal rank in order of increasing complication, over a sustained bass given by the end terms of the series, is musically perfectly satisfactory.

The terms $^1/_7$, 7, are not found in normal crystal series. In the diatonic series the corresponding notes, D, B, make harsh discords against the end terms, C, C′; but at* in the above passage the effect is inoffensive due to the wide intervals between the discordant notes.

On this foundation Professor Goldschmidt has built up an elaborate system of harmonic analysis.

In colour. The most prominent of the Frauenhofer lines of the solar spectrum, *A* (purple), *B* (scarlet), *C* (red), *D* (yellow), *E* (green), *F* (blue), *G* (indigo), *H* (violet), have the measured wave-lengths (λ) in the following table. The lines cover approximately an octave of the spectrum. Their relative wave-lengths are given very closely by the rational number series, 2 · · · 1, and their relative periods (z) by the series 1 · · · 2. Transforming this series into the form 0 · · · ∞, the series of numbers obtained (p) is again essentially a complication series.

Frauenhofer Line	A	B	C	D	E	F	G	H
λ	7608	6870	6563	5893	5270	4861	4308	3969μμ
proportional to	1.935	1.748	1.670	1.500	1.341	1.237	1.096	1.010
or nearly	2.000	1.750	1.667	1.500	1.333	1.250	1.111	1.000
or	2	$^7/_4$	$^5/_3$	$^3/_2$	$^4/_3$	$^5/_4$	$^{10}/_9$	1
$z=2/\lambda$	1	$^8/_7$	$^6/_5$	$^4/_3$	$^3/_2$	$^8/_5$	$^9/_5$	2
$p=2(z-1)/(2-z)$	0	$^1/_3$	$^1/_2$	1	2	3	(8)	∞

The series (p) is remarkably like the normal complication series N_3; the term (8) is additional and it corresponds to the colour indigo, which is a colour difficult to distinguish. Except for this term the series is symmetrical about the term 1, which corresponds to yellow, the dominant (brightest) colour in the spectrum. On this basis Professor Goldschmidt develops an extended analogy between sound and colour, and a method for analysing colour combinations.[11]

In cosmic space. The distances of the planets and the asteroids from the sun, reduced so as to make the distance of the earth equal to 10, have long been known to correspond approximately to a number series due to Titius and advocated by Bode.[12] This number series is obtained by writing the series 0, 1, 2, 4, 8, 16, 32, 64, 128, multiplying each term by 3 and adding 4 to each product. The

[11] *Ueber Harmonie und Complication*, p. 83 *et seq.*

[12] H. Shapley and H. E. Howarth, *A Source Book in Astronomy*, New York, 1929, p. 180.

numbers thus obtained (t) compare with the proportionate mean measured distances (d) as follows:

	Mer.	Ven.	Earth	Mars	Ast.	Jup.	Sat.	Uran.	Nep.
d	3.87	7.23	10.00	15.24	—	52.03	95.39	191.82	300.55
t	4	7	10	16	28	52	100	196	388

Due to the serious discrepancy in the case of Neptune, and the fact that the first term is arbitrary and should really be $(^1/_2 \times 3 + 4) = 5^1/_2$, this number law is now generally regarded as a curiosity without rational explanation.[13]

In a long paper on Harmony in Cosmic Space Professor Goldschmidt[14] has shown that the relative mean distances of the planets from the sun, transformed, when necessary, by the general method already used in the discussion of crystallographic, diatonic and spectral series, conform to the Law of Complication.

In the following table D is the measured distance of each planet from the sun, in millions of kilometers.

	Mercury	Venus	Earth	Mars	Jupiter	Saturn	Uranus	Neptune
D	57.5	107.5	148.7	226.5	773.2	1417.8	2851.3	4467.5
$d=D/733$	0.078	0.147	0.203	0.309	1.055	1.935	3.891	6.096
or nearly..	0.077	0.143	0.200	0.333	1.000	2.000	4.000	6.000
or........	$^1/_{13}$	$^1/_7$	$^1/_5$	$^1/_3$	1	2	4	6

The mean distances are thus very closely represented by the series of rational numbers in the last line of the above table.

The major planets, together with the sun and outer space as end terms, give an incomplete complication series:[15]

	Sun	Jupiter	Saturn	Uranus	Neptune	Outer Space
d	0	1	2	4	6	∞
$p=d/2$	0	$^1/_2$	1	2	3	∞

[13] J. Jeans, *The Universe Around Us*, Cambridge, **1929**, p. 20.

[14] V. Goldshmidt, Ueber Harmonie im Weltraum: *Ann. d. Naturphil.* (Ostwald), **v**, pp. 51–110, 1905.

[15] See Appendix.

Similarly the minor planets, with the sun and Jupiter as end terms, give terms of the complication series:

	Sun	Mercury	Venus	Earth	Mars	Jupiter
d	0	$^1/_{13}$	$^1/_7$	$^1/_5$	$^1/_3$	1
$p=d/(1-d)$	0	$^1/_{12}$	$^1/_6$	$^1/_4$	$^1/_2$	∞
$p'=4p$	0	$^1/_3$	$^2/_3$	1	2	∞
$p''=6p$	0	$^1/_2$	1	$^3/_2$	3	∞

Again, the number series corresponding to the distances of the larger satellites of Jupiter and of Uranus from their respective parent bodies, give incomplete complication series, even without transformation:

	Jupiter	Jo	Europa	Ganymede	Callisto	Outer Space
D	0	419	666	1064	1871	∞
$d=D/970$	0	0.44	0.70	1.11	1.95	∞
p	0	$^1/_2$	$^2/_3$	1	2	∞

	Uranus	Ariel	Umbriel	Titania	Oberon	Outer Space
D	0	205	285	443	583	∞
$d=D/417$	0	0.49	0.68	1.06	1.40	∞
p	0	$^1/_2$	$^2/_3$	1	$^3/_2$	∞

D is the measured mean distance of each satellite in thousands of kilometers.

And in like manner the satellites of Saturn form complication series,[16] and the 700 odd known asteroids are shown to fall into groups concentrated at distances from the sun corresponding to terms of the complication series.[17]

From these facts Professor Goldschmidt develops a cosmogonic theory in which one of the leading thoughts is that groups of bodies forming complication series (or cosmic chords) were each generated in distinct epochs.

[16] V. Goldschmidt, Ueber Harmonie im Weltraum: *Ann. d. Naturphil.* (Ostwald), **v**, p. 77, 1905.

[17] V. Goldschmidt, Harmonie im Reich der Planetoiden: *Ann. d. Naturphil.* (Ostwald), **ix**, pp. 383-392, 1912.

The Significance of the Law of Complication

In the following table we have re-assembled some of the number series found in crystallography, musical harmony, colour and cosmic space. The similarity of the several series is indeed impressive. The examples given, which are the fundamental ones, contain no terms not found in the normal complication series III; the third order of complication seems to be a natural limit which is overstepped only rarely either in crystallography or in the other realms. The collected series show the same order of frequency of terms as that found in the crystallographic series: 0, ∞ (12 times); 1 (10 times); $^1/_2$, 2 (7 to 8 times), $^1/_3$, 3, $^2/_3$, $^3/_2$ (2 to 6 times). In each case the numbers represent lengths or their reciprocals, and where transformation has been necessary to bring the series into the standard form 0 · · · ∞, it has been done by a standard operation whose significance in crystallography is understood, implying a change in the position of the projection plane.

Complication Series

Crystallography									
Normal complication series I	$p=0$	·	·	·	1	·	·	·	∞
Normal complication series II	$p=0$	·	$^1/_2$	·	1	·	2	·	∞
Normal complication series III	$p=0$	$^1/_3$	$^1/_2$	$^2/_3$	1	$^3/_2$	2	3	∞
Music									
Major common chord (root position)[18]	$p=0$	$^1/_3$	·	·	1	·	·	·	∞
Major common chord (second inversion)	$p=0$	·	$^1/_2$	·	·	·	2	·	∞
Minor common chord (first inversion)	$p=0$	$^1/_3$	·	·	·	·	2	·	∞
Chord of the dominant seventh	$p=0$	$^1/_3$	·	·	1	·	·	3	∞
Colour									
Purple, scarlet, red, yellow, green, blue, violet	$p=0$	$^1/_3$	$^1/_2$	·	1	·	2	3	∞
Cosmic Space									
Major planets: Sun, Jupiter, Saturn, Uranus, Neptune, Outer Space	$p=0$	·	$^1/_2$	·	1	·	2	3	∞
Minor planets: Sun, Mercury, Venus, Earth, Mars, Jupiter	$p=0$	$^1/_3$	·	$^2/_3$	1	·	2	·	∞
Satellites of Jupiter: Jupiter, Jo, Europa, Ganymede, Callisto, Outer Spac e	$p=0$	·	$^1/_2$	$^2/_3$	1	·	2	·	∞
Satellites of Uranus: Uranus, Ariel, Umbriel, Titania, Oberon, Outer Space	$p=0$	·	$^1/_3$	$^2/_3$	1	$^3/_2$	·	·	∞
Frequency	12	6	7	4	10	2	8	4	12

Carrying the enquiry forward, more by analogy than by direct comparison, Professor Goldschmidt sees the Law of Complication underlying still further groups of natural appearances; among

[18] The chords are here named according to conventional musical theory.

these are certain processes in organic evolution and even some aspects of human physiology and psychology, especially those concerned with the perception of harmonious things. The general conclusion is that the development of the manifold from the simple, by complication, is a universal process. In the realm of sound complication leads literally to an harmonious relation of parts; and therefore, in a more general sense, the universal action of complication points to an harmoniously constructed universe.

The belief in an harmonious universe is not without precedent. Pythagoras knew exactly the numerical relations of the vibration periods of musical intervals. Fully convinced that "number is the essence of things," he dreamed of a universal harmony which could be expressed in rational numbers. With very imperfect knowledge of the solar system, Ptolemy applied himself to finding harmonious relations in the motions of the planets. With much more exact knowledge, Kepler made an extended search for the "harmony of the spheres." This he found, not in the relations of the radii of the orbits of the planets, but in the ratios of their greatest and least apparent angular velocities as seen from the sun, that is in the excentricities of the orbits. Each planet was thus associated with a definite musical interval. A connection between crystals and the planets appears also in Kepler's fanciful disposition of the five perfect Platonic figures (forms in the cubic system) among the orbits of the planets.[19]

In his discovery of a definite number law underlying several extensive and widely different groups of natural phenomena, Goldschmidt has given new significance and precision to the ancient conception of universal harmony. The close approach of the series discussed to an exact mathematical series cannot possibly be the result of coincidence. Adopting the most conservative attitude, we must concede that the discussion points not only to some pervasive principle of rationality underlying the physical phenomena considered, causing certain measures of these phenomena to have simple rational values, but also to a definite preferred order of frequency in these simple rational measures. In the phenomena considered these measures are functions of length; but it is not unlikely, as Goldschmidt believes to be the case in the crystallographic series, that the measures are really functions of force. A similar principle of rationality pervades chemistry, and there is much in the modern

[19] Ioannis Keppleri, *Harmonices Mundi*, Lincii Austriae, **v**, p. 207, 1619.

outlook in physics to suggest that its fundamentals will eventually be expressed in simple rational numbers.

In crystallography the Law of Complication is evidently a form of the Law of Rationality, and it has proved to be a more general form than the Law of Simple Rational Intercepts and Indices, since it has been found to underlie the hitherto wholly anomalous observations on calaverite. The Law of Complication has furthermore a genetic significance which the older law does not possess; in the hypothesis of crystal development by the interaction of principal crystal forces we have what may be regarded as a first approximation to an explanation of the external form and habit of crystals, an aspect of crystallography—and the one which is after all the most important to the morphologist—which is illuminated neither by the older law nor by röntgenographic studies concerned with the internal arrangement of crystalline matter.

Appendix

Since the above was written, the following publication has been received:

V. Goldschmidt, Der Planet Pluto und die Harmonie der Sphären, *Heidelberger Akten der von Portheim-Stiftung*, **18,** 22 pp., 1932.

In this paper it is shown that the recently discovered planet Pluto, whose orbit has a mean radius of nearly 40 astronomical units, fits squarely into the complication series of the major planets, giving the series:

	Sun	Jupiter	Saturn	Uranus	Neptune	Pluto	Outer Space
$p=$	0	$^1/_2$	1	2	3	4	∞

The new planet thus affords striking proof of the validity of the Law of Complication in the arrangement of the bodies of the solar system.

Professor Goldschmidt briefly reviews his earlier studies of harmonious arrangements in crystallography, music, colour and cosmic space, and gives a generally applicable definition of harmony:

"*Harmony is an arrangement or grouping characterized by the harmonic (complication) series and caused by complication*" (*op. cit.*, p. 17, *trans.*).

The review which we have given is concerned mainly with the numerical basis of Professor Goldschmidt's theory of complication and harmony. To obtain a just conception of the full application and philosophical import of the theory the reader must turn to the original writings. The following list includes those referred to and others not cited:

Victor Goldschmidt (Heidelberg):

1897: Über Entwicklung der Krystallformen, *Zeitschrift für Krystallographie*, **xxviii**, pp. 1–35; 414–451.

1901: *Über Harmonie und Complication*, Berlin.

1904: Über harmonische Analyse von Musikstücken, *Annalen der Naturphilosophie* (Ostwald), **iii**, pp. 449–508.

1905: Beiträge zur Harmonielehre, *Annalen der Naturphilosophie* (Ostwald), **iv**, pp. 417–442.

1906: Über Harmonie im Weltraum, *Annalen der Naturphilosophie* (Ostwald), **v**, pp. 51–110.

1909: Kometen als kosmische Analytiker, *Annalen der Naturphilosophie* (Ostwald), **viii**, pp. 477–482.

1912: Harmonie im Reich der Planetoiden, *Annalen der Naturphilosophie* (Ostwald), **xi**, pp. 383–392.

1919: *Farben in der Kunst*, Heidelberg; second edition, 1929.

1921: Complication und Displication, *Heidelberger Akademie Sitzungsberichte*, A, **12**, pp. 1–90.

1923: *Materialien zur Musiklehre*, vol. **i**, Heidelberg.

1925: *Materialien zur Musiklehre*, vol. **ii**, Heidelberg.

1932: Der Planet Pluto und die Harmonie der Sphären, *Heidelberger Akten der von Portheim-Stiftung*, **18**, 22 pp.

16

Reprinted from *Am. Mineralogist*, **22**(5), 446–467 (1937)

A NEW LAW OF CRYSTAL MORPHOLOGY EXTENDING THE LAW OF BRAVAIS

J. D. H. DONNAY AND DAVID HARKER, *The Johns Hopkins University, Baltimore, Maryland.*

CONTENTS

THE LAW OF BRAVAIS

This law, proposed by Bravais on speculative grounds, only received general acceptance after Georges Friedel (1904, 1905, 1907) proved it to be substantiated by observation. With certain restrictions the law states that, for any crystalline species, a translation lattice can be found such that: (1) The observed crystal faces are parallel to the net planes with the highest reticular densities (or smallest mesh areas); (2) the greater the reticular density (or the smaller the mesh area) the more important the corresponding form.

The importance of a form is estimated from its size, frequency of occurrence, and presence as a cleavage form. The *reticular density* of a net plane *hkl* is defined as the number of lattice points (nodes) per unit surface. It is inversely proportional to the *reticular area*, or area S of the smallest mesh in the net, and directly proportional to the *interplanar distance* d, since $S \cdot d = V$, where V, the volume of the smallest unit cell, is a constant for any lattice.

Morphologists of the French School have advocated the acceptance of the lattice indicated by the Law of Bravais, in preference to any other lattice that would also give the observed forms reasonably simple Miller indices. The lattice chosen on this basis has been called the "Haüy-Bravais lattice" or "morphological lattice," in contradistinction to the "structural lattice" or true translation lattice, determined by x-ray technique. This nomenclature stresses the baffling fact that the two lattices are not always geometrically similar.

Known objections and restrictions to the Law of Bravais include the following:

(1) The forms present on a crystal do not depend only on factors within the crystal, but on external factors as well (conditions of crystallization). A statistical survey of morphological data (locality persistence, in Niggli's sense) usually suffices to detect the influence of external causes, which can then be left out of consideration.

(2) Even in the most favorable instances the law holds for dominant forms only. Minor forms are always irregularly scattered down the list of decreasing theoretical importance.

(3) Frequently the best lattice that can be arrived at leads to an order of importance not wholly satisfactory for the dominant forms and it may even assign small mesh areas to unknown or insignificant forms.

(4) Cleavage forms do not always coincide with the largest and most frequent crystal forms, and conversely.

(5) Complementary merohedral forms have the same mesh area, yet do not always show equal morphological importance.

These objections have given the Law of Bravais the status of a first approximation. G. Friedel, who conceded this point, foresaw an explanation for the encountered anomalies in the "perturbing influence" of the *motif* (the contents of the unit cell).

In this paper we propose to state a law[1] that takes into account one of the important properties of the motif—its symmetry.

Space Group Symmetry

The 230 space groups were derived by combining, in all possible ways, the different symmetry operations: translation, rotation, and reflection (or inversion). This process involved, among others, two composite symmetry elements, which had not been previously recognized (although they are actually implied in some of the 14 Bravais lattices), namely: the screw axis (Sohncke) and the glide plane (Fedorov, Schoenflies, Barlow).

In this paper we shall not concern ourselves with those screw axes and glide planes that are inherent in the translation lattice,[2] since their ef-

[1] A preliminary communication was presented at the January meeting of the *Académie des Sciences* in Paris (1937).

[2] Glide planes and screw axes occur in all the centered modes of all latices, and also in the hexahedral (primitive) mode of the cubic, tetragonal, and hexagonal lattices (we consider the rhombohedral lattice a mode of the hexagonal). The only lattices devoid of

fect is already taken into account by the Law of Bravais. We shall deal only with screw axes and glide planes that can be combined with a translation lattice, in addition to those it may already possess. The distinction between the two kinds of screw axes or glide planes is made in the Hermann-Mauguin notation: those that exist in the lattice itself are implicitly expressed by the Roman capital indicative of the translation group, the others are explicitly symbolized. The effect of the former is to bring a node into coincidence with another node (which could also be reached by applying a translation to, and which is in parallel orientation with, the first). The effect of the latter is to produce additional points within the lattice, points, however, which are not lattice points (nodes) since they cannot be obtained from any node by translation alone, but only by a translation accompanied by either a rotation or a reflection. Additional points produced by a screw axis are congruent to the true lattice points (nodes), although not in parallel orientation with them; additional points produced by a glide plane are enantiomorphous to lattice nodes. For convenience, in case there are additional points resulting from the action of screw axes and glide planes, we shall term "lattice equipoints" the nodes of the Bravais lattice and "space group equipoints" the additional points.

At this juncture one should recall that a node of a Bravais lattice stands for a whole motif (the contents of a complete unit cell, regardless of how matter may be distributed in it) and that the symmetry of the motif cannot exceed that of the parallelopiped outlined by the cell (holohedry), but may be inferior to it (merohedry). When screw axes and glide planes are present the motif (still represented by the "node") may be thought of as split into several fractions or "submotifs" derived from each other by the available screw rotations and glide reflections; any of these submotifs may be represented by the "lattice equipoint," the others are represented by "space group equipoints."

The Law of Bravais defines reticular density in terms of nodes only. It is our purpose to show that, when a node is replaced by a "lattice equipoint" and one or more "space group equipoints,"the reticular density must be defined in terms of "equipoints" for certain faces: the faces perpendicular to a screw axis and some of the faces perpendicular to a glide plane. Inasmuch as the faces affected in this way are determined by the kind and amount of space group symmetry, they become, con-

---

inherent screw axes and glide planes are the triclinic (P), the hexahedral monoclinic (P), and the rectangular hexahedral orthorhombic (P). The figures in the *International Tables for the Determination of Crystal Structures* (Chapter V, Space groups) illustrate the presence of screw axes and glide planes in the lattices mentioned.

versely, criteria indicative of such symmetry. Hence, if the symmetry of the space group is known, its influence on the form development can be predicted and checked; *vice versa* it is possible to ascertain with which space group (or set of space groups) observed morphological features are compatible.

The Effect of a Screw Axis

A screw axis, represented by the symbol p_q in the Hermann-Mauguin notation, is a p-fold axis of symmetry, the operation of which is the product of a rotation through $2\pi/p$ and a displacement by a fraction q/p of the shortest translation parallel to the axis. The displacement is known as the *screw component.* The crystallographically possible screw axes are: 2_1; (3_1, 3_2); (4_1, 4_3); 4_2; (6_1, 6_5); (6_2, 6_4); 6_3. Enantiomorphous screw axes, here collected between parenthesis, may be considered as having the same screw component but opposite screw senses. The fraction q/p, therefore, will always be of the form $1/n$, with $n=2$, 3, 4, or 6.

Consider any two successive lattice planes perpendicular to a screw axis. Let d be their spacing. These two lattice planes will be interleaved by $(n-1)$ other planes brought about by the action of the screw axis, thus resulting in a system of parallel, equidistant planes with the spacing d/n. The lattice planes contain "lattice equipoints" only, the intervening planes "space group equipoints" only. The arrangement of equipoints in any two successive planes of the complete system is the same except for the $2\pi/p$ rotation. All the planes of the system will, therefore, be geometrically equivalent insofar as concerns the crystal form (pinacoid) that parallels them. For such a pinacoid, the faces of which are perpendicular to a screw axis, the "effective" spacing is d/n instead of d. Formally, by virtue of the relations between spacing, reticular density, and mesh area, it is equivalent to consider that the reticular density should be divided by n or the mesh area multiplied by n. Either one of these artifices will adequately express true morphological importance. No other crystal forms, besides the pinacoid just discussed, can parallel equidistant planes containing equipoints of the same sort only; their morphological importance, therefore, cannot be altered. These conclusions are verified by observation.

The reticular area S is given by the formula

$$S^2_{hkl}=h^2S^2_{100}+k^2S^2_{010}+l^2S^2_{001}$$
$$+2(hkS_{100}S_{010}\cos\nu+klS_{010}S_{001}\cos\lambda+lhS_{001}S_{100}\cos\mu),$$

in which $S_{100}=bc\sin\alpha$, $S_{010}=ca\sin\beta$, $S_{001}=ab\sin\gamma$; and $\lambda=(010):(001)$, $\mu=(001):(100)$, $\nu=(100):(010)$. Hence, the value of S will be multiplied by any number n if all three indices are multiplied by n. We propose generally to adopt this convention and to designate by the symbol $(nh \cdot nk \cdot nl)$ any face (hkl) normal to a screw axis, with screw component d/n. The classical symbol (hkl) merely indicates the face direction. The suggested symbol $(nh \cdot nk \cdot nl)$ acquires a further value—that of a quantitative estimate of the form importance, since the mesh area S is directly obtainable from the "multiple indices" (by introducing them as such in the formula giving S or so-called "quadratic form" S^2). The Miller notation, thus but little modified, assumes greatly enhanced morphological significance.[3]

Example of the Screw Axis Effect: Low Quartz. Hexagonal trapezohedral class ($D_3 = 3\ 2$). Axial ratio: $c/a = 1.1000$. As demanded by the Law of Bravais and confirmed by *x*-ray determination,[4] the translation lattice is hexagonal (*not* rhombohedral) and the point group symmetry must be considered holoaxial tetartohedral (*not* holoaxial hemihedral). $a_0 = 4.89$ A.U., $c_0 = 5.38$ A.U. The space group is one of the enantiomorphous pair $D_3^4 = C3_12$ and $D_3^6 = C3_22$.

The 3-fold symmetry axis is a screw axis, whose screw component is $c_0/3$. The basal pinacoid must be written $\{0003\}$ instead of $\{0001\}$. Its mesh area is trebled, or its reticular density divided by 3. In the Bravais sequence of theoretical importance, the base came first ($S_{0001} = 1.13$). In the modified list, it goes down to the eighth place ($S_{0003} = 3.39$), after $10\bar{1}0$, $10\bar{1}1$, $11\bar{2}0$, $10\bar{1}2$, $11\bar{2}1$, $20\bar{2}1$, $11\bar{2}2$ (no distinction is made here between merohedral forms).

Thus is explained the extreme rarity of the basal pinacoid in quartz, one of the most striking morphological characters of the species and a flagrant anomaly to the Law of Bravais.

[3] H. Ungemach (1935), in order to stress the unity of the hexagonal system *sensu vasto*, in the light of recently discovered examples of syntaxy, proposed to use the 4-index Bravais symbols $(hkil)$, slightly modified by the omission of the minus sign over the third index, for crystals having a rhombohedral lattice, with the provision that all four indices should be multiplied by 3 when the sum $h+i+l$ is not a multiple of 3. Examples : (0003), (3030). This amounts to considering the rhombohedral lattice a mode of the hexagonal lattice. M. A. Peacock and M. C. Bandy (1936) followed the suggestion in calling the cleavage of their new mineral ungemachite $\{0003\}$ $\{111\}$.

J. D. H. Donnay (1936) proposed to extend the practice to all systems where the lattice may have more than one mode. Examples.—In the body-centered mode of the cubic lattice, multiply all three indices by 2 if their sum is not even: (110), (200), (211), (222), etc. In the face-centered mode, double all three indices if they are not all odd: (111), (200), (220), (311), etc. Such a procedure expresses the Law of Bravais only.

In the present paper we generalize the symbolization so as to include the effect of screw axes and glide planes.

[4] All the structural data quoted in this paper are taken from *Strukturbericht*, where more specific references may be found.

THE EFFECT OF A GLIDE PLANE

A glide plane is a plane of symmetry, whose operation is the product of a reflection in the plane and a displacement parallel to the plane. The displacement is known as the *glide component.* The possible glide components are (in vectorial notation): $\frac{1}{2}$ **a**; $\frac{1}{2}$ **b**; $\frac{1}{2}$ **c**; $\frac{1}{2}$ (**a**+**b**), $\frac{1}{2}$ (**b**+**c**), $\frac{1}{2}$ (**c**+**a**); $\frac{1}{4}$ (**a**±**b**), $\frac{1}{4}$ (**b**±**c**), $\frac{1}{4}$ (**c**±**a**). The Hermann-Mauguin notation of a glide plane is obtained by replacing, in the point group symbol, the letter *m* indicative of a reflection plane of symmetry by one of the letters *a*, *b*, *c*, *n*, *d*, chosen according to the type of glide component.

Consider a glide plane in an othorhombic primitive lattice (hexahedral mode), parallel to any one of the coordinate planes, for instance (100). Let the glide component be parallel to either one of the two coordinate axes in the plane, for example $\frac{1}{2}$ **b**. The effect of the glide plane will be to bring space group equipoints into the centers of the (001) meshes. (See Fig. 1, where the glide plane is parallel to the plane of the figure, the lattice equipoints are represented by white circlets, the space group equipoints by black dots.) The space group equipoints are in enantiomorphous relation to the lattice equipoints. No new points are introduced between successive (001) planes—the spacing of the latter remains unchanged. In the (001) planes the motif comprises both the lattice equipoint and the space group equipoint derived from it by glide reflection—the translation net of (001) remains what it was, its mesh area has not changed, neither has its reticular density. The same observation applies to all planes that contain both space group equipoints and lattice equipoints, *e.g.*, the system of planes (021).

Among the systems of net planes (0*kl*) perpendicular to the glide plane, there are some that contain one kind of points only, either lattice equipoints or space group equipoints, *e.g.*, the system of planes (011). In any such system of planes, true lattice planes (beset with lattice equipoints only) alternate with other planes (occupied by space group equipoints only). Any two successive planes can be considered geometrically equivalent since they contain identical nets with motifs similar in all respects except for the fact that they are mirror images of each other (the motif of true lattice planes being represented by a lattice equipoint, that of intervening planes being figured by its enantiomorphous "space group equipoint"). For any crystal face that parallels such a system of planes, the "effective" interplanar spacing should be taken as one half of the distance between true lattice planes if it is to express morphological importance; one may as well say that the reticular density should be halved or the mesh area *S* doubled. According to the proposed symbolization the indices should be multiplied by 2, so that (011) is to be written (022). More generally, all the planes ($0kl$) that contain points of the two different kinds are those that fulfill the condition $k=2n$. Those that do not satisfy this condition must have their indices multiplied through by 2. Since $h=0$ for all planes perpendicular to the glide plane, the condition may be written $h+k=2n$, which is the criterion of a lattice centered on (001). We may thus express this result by saying that the lattice acts as a *C*-centered lattice for all faces in the zone [100].

If, in the above example, the glide component had been taken equal to $\frac{1}{2}$ **c**, the lattice would have acted as a *B*-centered lattice for all faces in zone [100].

When the glide plane is parallel to (010) and the glide component is $\frac{1}{2}$ **c** or $\frac{1}{2}$ **a**, the lattice acts as if *A*-centered or *C*-centered, respectively, for all faces in zone [010]. Again, if the glide plane is (001), the effective lattice for the faces of zone [001] will be either *B*-centered or *A*-centered according as the glide component is $\frac{1}{2}$ **a** or $\frac{1}{2}$ **b**.

A glide plane *n*, whose glide component is one half the resultant of the two primitive translations in that plane, *e.g.*, $\frac{1}{2}$ (**a**+**b**), transforms a *P* lattice (primitive) into an *I* lattice (body centered) insofar as faces perpendicular to *n* are concerned.

Similar propositions could be established for all space groups where glide planes occur, invariably leading to the recognition of certain planes (among those perpendicular to the glide plane) whose morphological importance is to be decreased. Such a diminished importance is expressed by a multiplication of the indices so as to comply with the restrictions imposed by the presence of the glide planes. These conditions may be combined with others, resulting from the occurrence of screw axes, or due to the different types of centering of the lattice. It is easily seen that such combined conditions are precisely those that the various orders of *x*-ray reflections must fulfill in order to escape destructive interference (systematic "extinction," *Auslöschung*). These conditions, worked out by structural crystallographers, are directly available in any table of "space group extinctions."

Our conclusions as to the effect of a glide plane are verified by observation.

Example of the Glide Plane Effect: Orthorhombic sulfur. Rhombic dipyramidal or orthorhombic holohedral class ($D_{2h}=2/m\ 2/m\ 2/m$). Axial ratios—$a:b:c=0.8131:1:1.9034$. As demanded by the Law of Bravais and confirmed by *x*-ray determination, the translation lattice is all-face centered (rhombic octahedral mode). $a_0=10.61$ A.U., $b_0=12.87$ A.U., $c_0=24.56$ A. U. The space group is $D_{2h}^{24}=Fddd$.

The space group criteria include the conditions expressing the all-face centered mode of the lattice, namely: $h+k=2n$, $k+l=2n$, $l+h=2n$, or in other words "h, k, l either all odd or all even." For planes whose indices include at least one zero this condition becomes "indices all even." In addition to this first criterion the presence of the glide planes d demands, for those planes whose indices contain at least one zero, that the sum of the indices be a multiple of 4, namely: planes $(0kl)$ with $k+l=4n$, planes $(h0l)$ with $l+h=4n$, planes $(hk0)$ with $h+k=4n$. In accordance with our symbolization by means of "multiple indices" the list of crystal forms according to the Law of Bravais will only contain symbols with indices all odd or all even (indices not all odd are made all even by multiplying through by 2). The first two columns of Table 1 (headed "B–F" and "Com") give the Bravais list in such a form. In order to take the glide plane effect into account, form symbols whose indices include at least one zero, and which are already doubled by virtue of the all-face centered lattice (Law of Bravais), should be multiplied by 2 again if necessary to make the sum of their indices a multiple of 4. The mesh area of

TABLE 1. ORTHORHOMBIC SULFUR

Space group: $D_{2h}^{24}=Fddd$

B-F	Com	D-H	B-F	Com	D-H	B-F	Com	D-H
002	—			135			*424	
	111			315			*139	
020	—		*028	—			155	
	—	004	260	—			—	*048
	113			331			*2.0.10	
	022			*242		*406	—	
200	—		*208	—			?1.1.11	
	202			*137			337	
*024	—			—	400		319	
	115			119			?062	
	220		*402	—			*426	
204	—		046	—			*351	
	131			244			*511	
	133			151			*248	
	026			223			*157	
	224			*317		*064	—	
	311		*420	—			*353	
	206			*153			*513	
	—	040		422			442	
	117			335			260	
	313			*0.2.10			—	408
*042	—			*246		*0.4.10	—	

N.B. Crystal forms that have not been observed on sulfur are preceded by an asterisk (*). Doubtful forms are shown by a question mark (?).

such forms is thus quadrupled by the new law instead of simply doubled as it was by the Law of Bravais. The new sequence is shown by the last two columns (headed "Com" and "D-H") of Table 1. The second column in this table contains the forms common to both sequences of decreasing theoretical importance (Bravais-Friedel, on the one hand; Donnay-Harker, on the other).

The example of sulfur was given by G. Friedel (1926, p. 133) to illustrate the finding of the translation lattice by means of the Law of Bravais. Friedel stresses the presence of certain anomalies in the list, namely the fact that the forms {012} and {102} are unknown[5] for the species, in spite of their relatively small mesh areas. The next two gaps, those of {021} and {014}, occurring among observed forms, are just as anomalous. Moreover the order of the forms does not fit the facts satisfactorily: {001} occupying the first rank implies a tabular habit, which has been observed occasionally, but is by no means representative of the majority of sulfur crystals; the other two pinacoids are also placed too high in the sequence with respect to their minor morphological importance.

The modified law (see Table 1), doubling again the mesh area of the forms in the first column, causes them to recede in the list to ranks more in keeping with their observed importance. The pinacoids become {004}, {040}, and {400}. The first four missing forms become {048}, {408}, {084}, and {0.4.16}, and are relegated far down the list. The rhombic dipyramid {111} takes first place, as one would expect of the habit-controlling form. The first unknown form is the 20th in the new list, against the 8th in the Bravais sequence.

Statement and Discussion of the Generalized Law

The relationship that exists between the systematic "extinctions" of certain orders of x-ray reflection in a given space group and the diminished importance of corresponding crystal forms permits the following statement of the law:

The morphological importance of a crystal face is inversely proportional to its reticular area S if the lattice is of the hexahedral mode (no centering) and the space group symmetry does not contain any screw axis or glide plane. The effect of lattice centering, screw axes, and glide planes is corrected for if the face indices are replaced, in the S formula, by the "multiple indices" of the lowest order of x-ray reflection compatible with the space group symmetry.

Instead of considering the mesh area of a plane, it is largely a matter of personal preference to consider its reticular density or its spacing to the

[5] The form {102}, it is true, has been reported since, but it remains a morphological rarity.

next parallel plane in the system, either one of the last two quantities being inversely proportional to the mesh area.

The law includes the Law of Bravais as a particular case, when, regardless of lattice centering, the space group symmetry is devoid of screw axes or glide planes. It is an extension of the Law of Bravais in the sense that it rests on the same kind of geometrical considerations. The difference is that the reticular constant of a face (its mesh area, reticular density, or interplanar distance) is no longer calculated in *the* translation lattice, but in a variable system of points, changing with the face or zone considered. For instance, in the tetragonal space group $D_{4h}^3 = P\ 4/n\ b\ m$, where the translation lattice actually is primitive tetragonal, the reticular constant is computed as if the lattice were body centered (I) for faces ($hk0$), base centered (C) for faces ($0kl$), simple (P) for faces (hhl), while for faces (hkl)—and this is a general rule—the reticular constant is still to be determined in the true translation lattice.

The new law thus brings out the fundamental importance of zonal development in crystal morphology, a fact which had been anticipated by previous crystallographers (Baumhauer, 1903–1905; Ungemach, 1934 and 1935). It confirms the interest of the investigation of "form series" (Baumhauer's *Zielreihen*, Ungemach's *séries zonaires*, Peacock's *Harmonic-Arithmetic Rule*, 1936).

The law cast in the above statement may, up to a certain point, be said to bridge the gap that existed between structural and morphological crystallographies. It is a further step in the elucidation of the control that the internal pattern of a crystal exercises on its bounding forms. Yet it does not involve any consideration of the special positions that the atoms may occupy, nor does it take into account their nature, their charge, their bonds, or any other physical concept. It only makes use of the "empty frame" of symmetry elements in the cell—it is, in essence, purely geometrical.

Since so many factors are ignored that may well be expected to play a role in the development of crystal forms, the law can claim no other value than that of an improved approximation. Not all the anomalies will be eliminated, as it is easy to foresee that several consequences of the law will not be verified. (1) If space group symmetry were the only influential factor, then all crystalline species belonging to the same space group of the isometric system (where the axial ratios and angles are constant) would have an identical list of forms arranged in the order of decreasing importance, or the same "morphological aspect," as we propose to call such a sequence. (2) Several space groups obey the same general conditions of systematic extinction, thus leading to the same "morphological aspect" for, say, merohedral as well as holohedral species. (3) The generalized law does not discriminate between complementary merohedral forms in all cases. (4) Anomalies may also result from the fact that no provision is made for differences in the conditions of crystallization (rate of cooling, presence of foreign ions in the solution, and the like) in spite of the attempt to eliminate such external factors by a critical examination of available morphological data.

In his masterly work on the subject, Friedel (1907) proved that the influence of the Bravais lattice is so preponderant that it rarely becomes obscured by that of the motif. Now a painstaking survey of a great many crystalline substances for which the space group is known conclusively shows that, although the symmetry of the space group cannot be the only factor either, it so outweighs all the others in influence that its effect, in the large majority of cases, is not obliterated by the action of the disturbing agents. This result, incidentally, will rehabilitate the symmetry (or pseudo-symmetry) of crystals as one of their most significant properties, a notion which of late had lost more and more credence among structural crystallographers.

Apart from any merit that may be conceded or refused to our speculations on the effect of screw axes and glide planes, the outcome of our statistical survey warrants considering the generalized form of the Bravais principle as a law of observation, valid independently of any theory.

This law leads to the recognition of a great many more "morphological aspects" than did the Law of Bravais. We define a "morphological aspect" as a possible list of decreasingly important forms for any set of axial elements (axial ratios and interaxial angles) in any given crystal system. For instance, in the isometric system, the Law of Bravais led to 3 morphological aspects; the present law leads to 17. These are invariable since cubic axial elements are constant. A "morphological aspect" corresponds to a certain set of general extinction conditions, which characterize one space group or may be common to several space groups. There are, therefore, as many "morphological aspects" possible as there are different sets of systematic space group extinctions. They will be presented, for each crystal system, in later publications.

The next sections will illustrate by examples (chosen among isometric substances) the separation of complementary hemihedral forms and the determination of the space group from morphological data. An anomalous case, that of sodium chloride, will also be discussed.

Discrimination between Complementary Merohedral Forms

Since the Law of Bravais is based on the consideration of the lattice only, it cannot make any distinction between complementary merohedral forms. The introduction of space group symmetry enables one to separate (in some cases, at least) complementary hemihedral forms.

Example: Pyrite, FeS_2. Isometric system. Diploidal or parahemihedral class ($T_h = 2/m\,\bar{3}$). $a_0 = 5.40$ A.U. The translation lattice is hexahedral (no centering). The space group is $T_h^6 = Pa3$.

None of the three Bravais lists available in the isometric system, corresponding to the three possible lattice modes, is very satisfactory. The agreement with observed form importance is the least bad in the hexahedral mode (simple cubic lattice), which places {012} at the fourth rank. In spite of many anomalies, the correct translation lattice was thus arrived at by the Law of Bravais.

According to the space group criterion the forms {0*kl*} must satisfy the condition $k = 2n$. If they do not their indices are doubled. This condition separates some complementary pyritohedra; for instance, {021} should have more importance than {012}, which is to be written {024}. The Bravais sequence and the modified list are compared in Table 2. Thefirst column (headed "Bravais-Friedel") gives the Bravais list, which does not discriminate between complementary forms. The second column (headed "Donnay-Harker") gives the new order, in which most of the complementary pyritohedra are separated. The third column (headed "Comparison") presents the data in the same manner as Table 1 for sulfur. In the modified sequence the first three forms are the three most important forms, the first two are the only cleavages (imperfect though they may be). The dominant forms in artificial pyrite crystals are the cube and the octahedron (Groth, 1921, p. 257: "Die nach verschiedenen Methoden erhaltenen Krystalle zeigen meist (100) and (111)").

Two complementary pyritohedra will have equal importance when their symbols contain two odd indices besides the zero; they will be separated if only one of the indices is odd. The data recorded in the standard compendia of mineralogy are insufficient to check the validity of this prediction of the law. Discrimination between positive and negative forms is often uncertain. Hintze (1904) does not attempt to make the distinction. V. Goldschmidt (1920) makes it for a few forms only. Strüver's suggestion that the distinction between positive and negative forms be drawn only if two forms of different sign occur together, in which case the larger and more brilliant one is to be taken as positive, is an arbitrary criterion, at best unsatisfactory. The verification of the law is fortunately made possible by H. Ungemach's valuable observations on form importance in pyrite crystals.[6] It is remarkable that information

[6] Unpublished data of the great French crystallographer and mineralogist (1880–1936), found in the note books he bequeathed to J. D. H. Donnay. The observations are graphically recorded on a work sheet, a gnomonic projection of pyrite showing the

can thus be obtained on the first *eleven* pairs of pyritohedra that are to be separated by the space group criterion. Here are Ungemach's data: {021} overwhelmingly more important than {012=024}; {023} vastly more than {032=064}; {041} much more than {014=028}; {043} probably more than {034=068}; {025} less important than {052=0.10.4} (anomaly); {061} known, {016=0.2.12} unknown; {045} much more important than {054=0.10.8}; {027} unknown, {072=0.14.4} known (anomaly); {065} much more important than {056=0.10.12}; the next four pyritohedra, {081}, {018=0.2.16}, {047}, {074=0.14.8} are known, but are minor forms whose relative importance cannot be estimated from Ungemach's projection (they are all represented by small points). The first three pairs of complementary pyritohedra show perfect agreement between facts and theory. Notice that {023} is a negative form, predominant over its positive complementary.

A strange fact is revealed by an inspection of the pyritohedra with two indices odd, which should have equal morphological importance: the two complementary forms coexist in three cases only, {075}, {0.15.11}, {097}; the negative form is found alone once, {0.7.11}; the positive form occurs alone in fourteen cases, {091}, {071}, {0.17.3}, {051}, {0.11.3}, {031}, {073}, {0.11.5}, {0.15.7{, {0.13.7}, {095}, {053}, {0.13.9}, {0.11.9}. These data conflict with the theory. It should be noticed, however, that the pair {013=026} {031=062} that is encountered first in the sequence of forms occupies the 37th rank! Agreement can hardly be expected to prevail so far down the list. There also remains the looming possibility that forms of uncertain sign may have been labeled positive by Ungemach himself—this would explain the avalanche of positive forms!

The modified list of forms does not make any distinction between complementary diploids, {*hkl*} and {*hlk*}, the effect of the glide plane being restricted to faces perpendicular to it (one index equal to zero).

In conclusion, one should be well satisfied with the verification obtained, which is vastly better than any agreement could be if due to chance alone. A detailed morphological study of pyrite is, nevertheless, badly needed to confirm or ascertain the sign of the hemihedral forms.

The case of pyrite is a first example of space group determination, without the aid of *x*-rays, by means of the scantest morphological information, namely: isometric system (whatever the class); {001}, {111}, {012} dominant (regardless of relative importance). Among the 17 possible "morphological aspects" of the isometric system, only one gives

---

relative importance of the forms by the size of the corresponding poles (the latter vary from tiny pencil points to circlets five millimeters in diameter).

TABLE 2. PYRITE
Space group $T_h^6 = Pa3$

Bravais-Friedel Lattice *P*	Donnay-Harker Aspect *Pa3*	Comparison		
		B-F	Common	D-H
001	111	001	—	
011	002	011	—	
111	— 021		111	
012=021	112		—	002
112	022	012 —	— 021	
122	122		112	
*013=031	113		—	022
113	023 —		122	
023=032	123=132	*013=031	—	
123=132	— 041		113	
014=041	223	— 032	023 —	
223	114		123=132	
114	133	014 —	— 041	
133	024 —		223	
124=142	124=142		114	
233	233		133	
034=043	— 043		—	024 —
*015=051	134=143		124=142	
134=143	115		233	
115	025 —	034 —	— 043	
025=052	234=243	*015=051	—	
234=243	*125=152		134=143	
*125=152	144		115	
144	225	— 052	025 —	
225	334		234=243	
*035=053	*135=153		*125=152	
334	— 061		144	
*135=153	116		225	
*016=061	235=253	*035=053	—	
116	*026=062		334	
235=253	*126=162		*135=153	
*126=162	045 —	*016 —	— 061	
045=054	344		116	
344	*145=154		235=253	
*145=154	335		—	*026=062
335	*245=254		*126=162	
*245=254	136=163	— 054	045 —	
136=163	?236=263		344	
?236=263	345=354		*145=154	
*017=071	117		335	
345=354	* 155		*245=154	
117	— 064		136=163	
*155	*027		?236=263	

TABLE 2. (*Concluded*)

Bravais-Friedel Lattice *P*	Donnay-Harker Aspect *Pa*3	Comparison		
		B-F	Common	D-H
*027=072	*146=164	*017=071	—	
*146=164	*127=172		345=354	
*127=172	255		117	
255	227		*155	
	445		—	— 064

N.B.—Unknown forms marked by asterisk (*), doubtful forms by query (?).

top ranks to the three dominant forms; to that aspect the space group *Pa*3 corresponds uniquely.

Space Group Determination by Morphological Criteria

A "morphological aspect" corresponds to one or more space groups in a crystal system. Within a crystal class, however, a morphological aspect corresponds to one space group only (or, at the most, to two enantiomorphous space groups). Knowledge of the relative importance of the crystal forms usually suffices to limit the possible space groups to a very few, especially if the crystal class (point group) can be determined. Even if the crystal class is not known it is still possible, in certain favorable cases, to determine the space group uniquely. Pyrite is one example. Garnet is another. It will be given here as an illustration of the fact that the dominant form, in an isometric crystal, need not be the cube, the octahedron, or the rhombic dodecahedron, as demanded by the Law of Bravais and implied by the French nomenclature: hexahedral mode, octahedral mode, dodecahedral mode. Derivation of the 17 isometric "aspects" has shown that either one of these three forms can be dominant in all three lattice modes, except that the octahedron can never be dominant in the dodecahedral mode, where, however, the leucitohedron {112} can take first place.

Examples: Garnet, $Al_2Ca_3(SiO_4)_3$. Isomettic system. Hexoctahedral or holohedral class. ($O_h = 4/m\ \bar{3}\ 2/m$). $a_0 = 11.8$ A.U. Translation lattice, body centered (dodecahedral mode). Space group $O_h^{10} = Ia3d$. Here are the most salient morphological data, as given in Dana: the dodecahedron {011} and the trapezohedron {112} are the most common simple forms; the hexoctahedron {123} comes next; the cube {001} and the octahedron {111} are rare.

In order to account for the great importance of the leucitohedron {112} the Law of Bravais can only select the body centered lattice. This choice is confirmed by x-rays. Among the 17 isometric aspects only one gives top ranks to the three most important forms; this aspect uniquely determines the space group *Ia3d*.

Many discrepancies between the facts and the Bravais sequence are corrected by the new list (see Table 3, arranged like Table 1 for sulfur). The Bravais list expresses the condition that the sum of the indices must be even, which is imposed by the mode of the lattice. The new list takes into account the additional restrictions of the space group: $\{0kl\}$ with k and l even, $\{hhl\}$ with $2h+l=4n$. The cube, which came second, is now fourth. The octahedron, which was fifth, goes down to the fourteenth place. The leucitohedron comes first, followed immediately by the dodecahedron and the {123} hexoctahedron.

Analcime, $NaAlSi_2O_6 \cdot H_2O$. This mineral is selected as an example of a pseudo-symmetric species. The dominant "form" is, by far, the trapezohedron {112}, which often occurs alone. Other "forms" that have been observed: cube, dodecahedron, octahedron, tetrahexahedron {012}, and

TABLE 3. GARNET

Space group: $O_h^{10}=Ia3d$

B-F	Com	D-H	B-F	Com	D-H	B-F	Com	D-H
011	—			233			145	
002	—		*015	—		226	—	
	112			134			*136	
	—	022		*125			—	444
013	—		035	—		*017	—	
222	—		334	—			*345	
	123		244	—			046	
	—	004		*116			*127	
*114	—			*235			*255	
	024			—	026	*037	—	

N.B.—Crystal forms that have not been observed on garnet are preceded by an asterisk (*).

trisoctahedron {233}. Of these the octahedron is certainly the rarest. It is hard to assign relative importances to the remaining forms. The only combination given by Hintze that shows five forms (apparently the richest observed) may serve as an indication; its forms are, in order: 112.011.001.012.233. It was found at Friedensdorf, in Hessen.

Incomplete though the data may be, they can be accounted for by one of the morphological aspects only, the same as for garnet, namely: 112.022.123.004.024.233. · · · (octahedron coming 14th in the list). The fact that {123} has not been reported is an anomaly. It does not obscure the otherwise perfect agreement.

The pseudo-isometric character of the species being recognized, simple data on habit and form development permit one to predict that the symmetry of the structure will closely simulate that of the space group $O_h^{10} = Ia3d$. The results of the x-ray investigation have led to the recognition of a "pseudo-structure" ($S6_1$ type of *Strukturbericht*) based on a cubic cell ($a_0 = 13.7$ A.U.) and space group $O_h^{10} = Ia3d$. The confirmation is striking.

Spinel, $MgAl_2O_4$. The most typical feature of this species is the predominance of the octahedron. The dodecahedron and the trapezohedron {113} come next in importance. The cube never is a "habit form."

Out of the 17 isometric aspects four give {111} the first place. One is the aspect of pyrite, corresponding to the space group $T_h^6 = Pa3$, with simple cubic lattice. This aspect is immediately eliminated as it gives the cube second rank, the dodecahedron fifth rank, and {113} seventh rank; not to mention its characteristic hemihedry. It would also demand the presence of {123}, which is unknown, as the first general form, whereas {135}, typical of the F lattice, is observed instead. The other three aspects are: (1) the Bravais list for an all-face centered lattice; (2) the same, slightly modified by symbolizing the cube {004} instead of {002} and making it retreat from second to fourth place; (3) the same list again, but further altered by the "extinction"[7] of all {0*kl*} forms for which the sum of the indices is not a multiple of 4.

The Bravais list is not as good as either one of its two modifications, on account of the rarity of the cube. The third list fits the facts better that the second, since neither {024} nor {046} are known, whereas {026} is observed. Comparison between the three aspects is shown in Table 4. The first aspect (headed "F" because it is the expression of the all-face centered lattice) corresponds to five space groups, one in each point group, namely: $T^2 = F23$, $T_h^3 = Fm3$, $T_d^2 = F\bar{4}3m$, $O^3 = F43$, $O_h^5 = Fm3m$. The second aspect uniquely defines the space group $O^4 = F4_13$. The third aspect is compatible with two space groups: $T_h^4 = Fd3$ and $O_h^7 = Fd3m$. We have seen that the form development points towards the last two space groups, even if the point group is unknown. Since the diploidal class can be ruled out, *Fd3m* is the only possibility.

The x-ray results confirm the morphological conclusions. The röntgenographically possible space groups are given as *Fd3m*, *Fd3*, and $F4_13$, in *Strukturbericht*.

[7] This use of the word "extinction," that recalls the correlated phenomenon in x-ray diffraction, may be found convenient. In the first and second lists, {012}, extinguished as such by the all-face centered lattice, appears as {024}. It is extinguished as {024} and occurs as {048} in the third list, owing to additional restrictions of the space group.

TABLE 4. SPINEL, $MgAl_2O_4$

F	O^4	T_h^5, O_h^7
111	111	111
002	—	—
022	022	022
113	113	113
—	004	004
133	133	133
*024	*024	—
224	—	224
*115	—	*115
135	—	135
244	same	244
026	as	026
*335	list	*335
*117}	F	*117}
*155}	—	*155}
*046	—	—
*246	—	*246

N.B.—Unknown forms are marked by an asterisk (*).

PERSISTING ANOMALIES

Inasmuch as the law presented in this paper is of a strictly geometrical nature only, it cannot be expected to hold in all cases. Some anomalies do persist.

Example: Halite, NaCl. The space group is known to be $O_h^5 = Fm3m$. The translation lattice is all-face centered, the motif is composed of a sodium ion and a chlorine ion, one half of the cube edge apart (which explains why some authors describe the structure as composed of two "interpenetrating lattices"). In this case the Law of Bravais and the generalized law coincide (no screw axis, no glide plane). The new law is, therefore, just as powerless as the Law of Bravais. The anomalies shriek.

As a natural mineral, NaCl usually occurs in cubes. The other forms known are {011}, {111}, {012}, {123}. Artificially produced crystals show various habits according to the conditions of crystallization (the addition of urea to the solution produces octahedra, as is well known) and several new forms, the principal ones being {112} and {122}; the others, {014}, {034}, {035}, {045}, {233}, {345}. The cleavage, which is perfect, parallels the dominant habit form {001}. Only three of the 17 isometric aspects will give the cube the leading rank (see Table 5, first three columns): (1) the Bravais sequence for a simple cubic lattice (Table 5, first column, headed "*P*"); (2) the aspect that corresponds to the space groups T_d^5 and O_h^6, with an all-face centered lattice; (3) the aspect that corresponds to the space group T_h^7, with a body centered lattice. None of these is the one demanded by the law (fourth column, headed "*F*"). The best agreement is found in the "*P*" list. G. Friedel (1926, p. 144) had already stressed the anomaly. He pointed out that, if one did not discriminate between Na ions and Cl ions, the smallest cell would become simple cubic.

It is a curious fact that cadmium oxide CdO, which belongs to the same space group as NaCl and has the same structural type, does obey the law. The data given in Groth permit listing the forms as follows: {111}, dominant; {001} next; {011} third; and last a trapezohedron, which has not been measured, tentatively given in Groth (1906, p. 176) as {112}?. This list is in perfect agreement with the "*F*" sequence (see Table 5), even to the last form if the trapezohedron is {113}, which it may well be. Geometrical considerations are obviously insufficient to account for such cases. The reason why cadmium oxide conforms to the law whereas sodium chloride does not is probably to be sought in the less marked ionic character of the cadmium oxide structure.

Table 5. Halite, NaCl

P	T_d^5, O_h^6	T_h^7	*F*
001	002	002	111
011	022	112	002
111	222	022	022
012	024	222	*113
112	*135	123	*133
—	224	*114	024
122	244	024	224
*013	*026	*233	*115
*113	*226	*134	*135
*023	*046	*125	244
123	246	*334	*026
—	*137	244	*335
014}	028}	*116}	*117}
*223}	*446}	*235}	*155}

N.B.—Unknown forms are marked by an asterisk (*).

It has been shown above that, when the motif is composed of two "submotifs" enantiomorphous to each other or congruent but rotated with respect to each other, the reticular constant (reticular density, mesh area, or interplanar spacing) should be defined in terms of "equipoints" (representing "submotifs") instead of lattice nodes (standing for motifs) for all planes containing one kind of equipoints only (either lattice equipoints or space group equipoints). In the case of NaCl, on account of the ionic character of the structure, the motif may perhaps be thought of as composed of two "submotifs," one Na^+, the other Cl^-. Inasmuch as "submotifs" need not be in parallel orientation, it may perhaps be justified to enlarge the tolerance by supposing that ions of different nature can play the rôle of "submotifs." In the case of the sodium chloride structure, this hypothesis is particularly tempting; if the Na ions are imagined to lie at the nodes of the translation lattice, they can be considered as "lattice equipoints" while the Cl ions will become "space group equipoints." In such an hypothesis all the planes with indices all odd will have their spacing halved (or their mesh area doubled) and hence their indices multiplied by two. All other planes have their indices doubled already by virtue of the all-face centering of the lattice. The final result is thus to have all planes modified in the same manner, namely by doubling the indices, which means that the sequence of forms according to decreasing importance will be identical with that of a simple cubic lattice (the edge of the unit cube halved). All anomalies would then disappear.

We have shown that *the two-dimensional periodicity of a plane is the effective factor determining its morphological importance, regardless of what the motif of the mesh may be;* it need not be the same as the three-dimensional motif of the lattice, but may be a submotif thereof, *i.e.*, a part of the motif obtainable from the other parts by screw rotation or glide reflection. The working hypothesis introduced in interpreting the NaCl morphology is that the submotifs need not even be chemically identical.

Summary

Independently of its theoretical justification the following law can be considered a law of observation, including the Law of Bravais as a particular case:

The morphological importance of a crystal face is inversely proportional to its reticular area S if the lattice is of the hexahedral mode (no centering) and the space group is devoid of screw axes and glide planes. The effect of lattice centering, screw axes, and glide planes is corrected for if the face indices are replaced, in the S formula, by the "multiple indices" of the lowest order of x-ray reflection compatible with the space group symmetry.

Many anomalies of the Law of Bravais are accounted for; some persist, since the law is purely geometrical, involving no physical consideration. The influence of space group symmetry is so preponderant, however, that it seldom becomes obscured by that of other factors. The law introduces a distinction between certain complementary merohedral forms. It leads to the recognition of the various possible "morphological aspects" in any crystal system. The law determines the relative importance of the forms for a given space group; conversely the space group can be predicted from morphological features alone (without the help of x-ray methods).

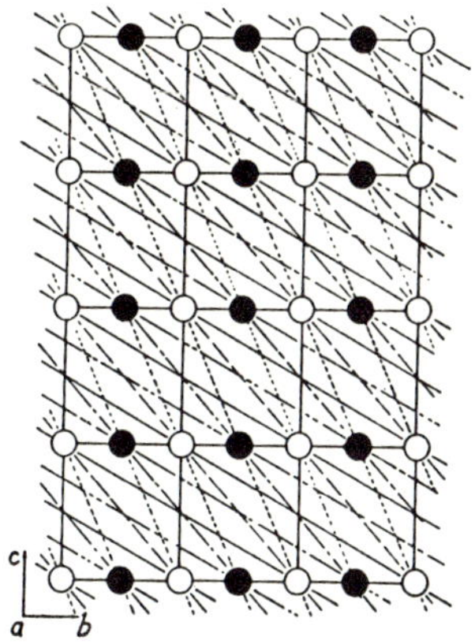

FIG. 1.

Orthorhombic Primitive Lattice.
Glide plane (100) parallel to plane of figure.
Systems of Points:—
Lattice equipoints: Open circles
Space group equipoints: Solid circles
The space group equipoints are mirror images of the lattice equipoints.
Systems of Planes:—
(021) - - - - - - - - - - - - - - -
(011) Through lattice equipoints —— —— —— —— —— ——
(011) Through space group equipoints — - — - — - — - — - — - — -
(012) Through lattice equipoints ————————————————
(012) Through space group equipoints — - — - — - — - — - — - — - -

REFERENCES

BAUMHAUER, H. (1903): *Cbl. Min.*, 665–676.
——— (1904): *Zeits, Krist.*, **38**, 628–655.
——— (1905): *Die neuere Entwickelung der Kristallographie*—Braunschweig.
DONNAY, J. D. H. (1936): *Amer. Min.*, **21** (12, Pt. 2), 4–5.
——— et HARKER, D. (1937): *C.R.*, **204**, 274–276.
FRIEDEL, G. (1904A): *C.R.*, **139**, 221–223.
——— (1904B): *C.R.*, **139**, 314–315.
——— (1904C): *Étude sur les groupements cristallins*—St-Étienne.
——— (1905): *Bull. Soc. franç. Min.*, **28**, 6–12.
——— (1907): *Bull. Soc. franç. Min.*, **30**, 326–455.
——— (1926): *Leçons de Cristallographie*—Paris.

GOLDSCHMIDT, V. (1920): *Atlas der Krystallformen*, **6**—Heidelberg.
GROTH, P. (1906): *Chemische Krystallographie*, **1**—Leipzig.
——— (1921): *Elemente der physikalischen und chemischen Krystallographie*—Leipzig.
HINTZE, C. (1904): *Handbuch der Mineralogie*, **1** (1)—Leipzig.
PEACOCK, M. A. (1936): *Amer. Min.* **21** (12, Pt. 2), 5.
——— and BANDY, M. C. (1936): *Amer. Min.*, **21** (12, Pt. 2), 2.
UNGEMACH, H. (1934): *Bull. Soc. franç. Min.*, **57**, 186–207.
——— (1935A): *Bull. Soc. franç. Min.*, **58**, 97–221.
——— (1935B): *Zeits. Krist.*, (A)A**91**, 1–22.
——— (1935C): *Zeits. Krist.*, (A)**91**, 97–113.

Part VI

APPLICABILITY OF THERMODYNAMICS

Editor's Comments on Papers 17 Through 19

17 **LAUE**
The Wulff Theorem for the Equilibrium Form of Crystals

18 **WELLS**
Crystal Habit and Internal Structure

19 **BUERGER**
The Relative Importance of the Several Faces of a Crystal

In 1914, Max von Laue received the Nobel Prize for what Einstein described as one of the most beautiful discoveries of physics. In 1943, he undertook to vindicate precisely those morphological ideas which his great discovery had cast into shadow (Paper 17). Friedel among others had pointed out that Wulff's theorem (and Curie's hypothesis) was not demonstrated. Donnay did not accept it. Von Laue, a protege of Planck, approached the subject through thermodynamics. The inherent plausibility of Wulff's theorem resides in its thermodynamic foundations. With hindsight, one can read it between the lines of Gibbs (Paper 5). It rests firmly on the second law. The rigorous demonstration of the theorem is not simple. Yet to reject it is to abandon the principal avenue to a *physical* understanding of surface, an avenue moreover which has richly rewarded those who examine surfaces of other than purely mineralogical interest.

Wartime difficulties and the destruction of publisher's stock led to the suspension of publication of the *Zeitschrift für Kristallographie*. An open and outspoken opponent of the Nazis, Laue was not read. There were in any event few students left in Germany. Major Western libraries lack the 1943 volume, and Laue's derivation is rarely mentioned in the literature.

The advantage of the thermodynamic approach is in its generality. It is independent of all assumptions of relationships between surface energies and either internal structure or external conditions. It is valid for vicinal faces and major prisms, for minor facets and dominant forms. It relies upon no calculations of reticular densities or definitions of periodic bond chains, but in fact, any theory of crystal morphology must conform to the requirements of these equations regardless of such contradictory hypotheses as might be conceived, since they stem directly from pure geometry and the foundations of physical science.

One of the few readers of Laue's article was the chemist A. F. Wells, who introduced a wealth of laboratory experience and an extramineralogical point of view. Sharply critical of the restricted crystallographic approach, Wells emphasized thermodynamics and the concept of the equilibrium form (Paper 18). Because of the experimental modification of habit by change of environment, Wells considered the *morphological aspects* (ranking of forms in order of importance) drawn up for minerals as unreliable. He assumed (p. 168) a physical association of atoms with the Donnay-Harker equipoints, and therefore rejected this approach. The same influence of X-ray crystallography, however, which had guided Donnay and Harker, illuminates his discussion, and his complaint is essentially that Donnay and Harker limited their generalization of Bravais's law to a consideration of space-group symmetry without any consideration of the *intensity* distribution as in X-ray structure analysis. Wells stopped just short of suggesting that the X-ray intensities and the geometrical structure factor tables that he drew up are a valid morphological aspect. He used the conditional mode, "*If* we wished to draw up a true M.A. from the X-ray data it *would*. . . . We *could*, for example. . . ." Because of the discrepancy between the sample spaces of mineralogy and chemistry, Wells rejected the crystallographic method to which he had brought such insight and turned instead to thermodynamics. Reviving the hypothesis of Curie for the calculation of surface energies, he demonstrated the agreement between artificially prepared crystals and the known planes of greatest coplanarity of atoms.

In Paper 19, a critical review, the mineralogist and X-ray crystallographer Martin Buerger attempted a physical explanation of the successes of the Bravais-Donnay-Harker rule. As had Wells, in the previous year, he underlined the parallel between form development and X-ray diffraction. "The order of importance of the crystal faces is exactly that of the order of appearance of the corresponding lines on the powder photograph . . .," but the subject was not again raised. Buerger's model of form development assumed growth by molecular blocks, with residual surface energies inversely related to the interplanar spacings of the surfaces of attachment. If this lattice geometry was the major influence on form development, it was nevertheless an oversimplification. "Realized coordination energy or its complement, surface energy, is the feature which governs the preference of sites. . . ." The physical characteristics of the particles arriving at the growing face, physical and not simply geometric, are involved, not only the geometry of the lattice as in the Bravais rule, but also the bonding characteristics of the crystal.

Paper 19 is tantalizing in that it raised issues of major concern in the following years, the parallel with X-ray diffraction, the layers, steps, and kinks of Burton and Cabrera, the calculations of surface energies, the shapes of grains in a polycrystalline aggregate, the form development of superstructures, the effects of nonuniform bond distribution, and hemihedry.

17

THE WULFF THEOREM FOR THE EQUILIBRIUM FORM OF CRYSTALS

Max von Laue

This translation was prepared expressly for this Benchmark volume by Beverly Pfanner and Cecil J. Schneer, from "Der Wulffsche Satz für die Gleichgewichtsform von Kristallen," Z. Krist., **105**, *124–133 (1943)*

§1

For the equilibrium form of a crystal whose N faces F_i are given by the perpendiculars and with specific surface free energies σ_i, J. W. Gibbs in 1878[1] and P. Curie[2] in 1885 derived the principle that the extended sum over all faces bears the least value compatible with the given volume V.

$$\phi = \sum_{i=1}^{N} \sigma_i F_i \tag{1}$$

Since ϕ represents the free energy of the whole surface area, and the bulk free energy is determined by the volume that remains constant, this principle is evidently sufficient; moreover, paragraph 3 presents its derivation.

G. Wulff[3] interpreted this geometrically in 1901: if ϕ has this minimum value, there is in the core of the volume V a point, the Wulff point, such that the perpendiculars h_i from the point to the faces F_i are to the surface energies σ_i as

$$\frac{h_1}{\sigma_1} = \frac{h_2}{\sigma_2} = \cdots = \frac{h_N}{\sigma_N} \tag{2}$$

To construct the equilibrium figure, from any point draw the straight lines $p\sigma_i$ in the prescribed directions, and erect perpendicular planes at their end points. So far as this hypothesis proves that the faces chosen and only these occur in the equilibrium figure, the construction produces a polyhedron bounded by these faces. However, if no closed polyhedron results, or if any of the faces chosen lie completely outside of the polyhedron formed by the others, the hypothesis would have to be changed accordingly. The second of these cases occurs especially if one σ_i considerably exceeds the others. In order to adjust to the prescribed volume, the previously undefined constant p is at one's disposal.[4]

If the faces F_i are arranged in pairs so that, for every pair of equal σ_i, the perpendiculars are opposite, the Wulff point becomes the geometrical center of the polyhedron. Otherwise, there can be no purely geometric designation for it that takes into consideration the values of σ_i. Its definition lies rather in the Wulff point itself; each equation $h_i/\sigma_i = \mu$ fixes, for a given polyhedron, a plane of the Wulff point. That these N (normals to the) planes intersect, by appropriate choice of the common constant μ, in a point is the peculiarity of the equilibrium form.

Here we want to elaborate critically the proof of Wulff's theorem. First, we take the purely geometric proof derived from the requirement for minimum ϕ; next, the direct thermodynamic proof without the detour by way of the minimum principle.

Wulff himself supported his principle only by examples. His attempt at a general proof failed; he surmised that the relations between ϕ and h_i as well as between V and the latter are of the forms

$$\phi = f\sum \sigma_i h_i^2, \qquad V = \sum m_i h_i^3$$

f and m_i are constants; however, the simplest example, in which three pairs of parallel planes are given as boundaries, contradicts these statements.[5] The first correct proof was given by H. Liebmann in 1914, which, however, in spite of its elegance has been largely forgotten. At least it has not gotten into the textbooks.[6]

In 1870, L. Lindelof[7] and, in 1897, H. Minkowski[8] discussed the special case, which is of only mathematical interest, that all σ_i are equal. Both came to the conclusion in agreement with Wulff's theorem, that for minimum ΣF_i, all faces F_i are tangents of the same sphere; thus all h_i are equal. One recognizes especially clearly in this special case the sense of Wulff's principle. A minimum always occurs when a sphere can be inscribed in the polyhedron. The center of the sphere is defined in this way and not in general in any other.

§2

As elegant as Liebmann's proof is, in my opinion there is a still simpler derivation of Wulff's theorem, which is also closer to thermodynamics; one uses the distances h_i as the independent variable instead of, as Liebmann did, the faces F_i. Lindelof's proof for the special case is longer; that of Minkowski is dependent upon a number of other principles not readily separated out. Only we must at the outset define the point from which we calculate the h_i. We establish it as the intersection of [the normals to] three planes,

$$\frac{h_1}{\sigma_1} = \mu, \qquad \frac{h_2}{\sigma_2} = \mu, \qquad \frac{h_3}{\sigma_3} = \mu \tag{3}$$

while we reserve for ourselves the selection of the faces F_1, F_2, F_3, to which these [planes] are parallel, as well as the determination of the constant μ.

The starting point is the stipulation that for the volume V as a function of the h_i,

$$\frac{\partial V}{\partial h_i} = F_i \tag{4a}$$

We derive from this the equation

$$\frac{\partial F_i}{\partial h_\kappa} = \frac{\partial F_\kappa}{\partial h_i} \tag{4b}$$[9]

If we multiply each h_i by the same number p, then the volume V is multiplied by p^3, and each face F_i by p^2. Then, according to Euler's theorem of homogeneous functions,

$$V = \frac{1}{3}\sum_1^N h_i F_i, \quad F_\kappa = \frac{1}{2}\sum_1^N h_i \frac{\partial F_\kappa}{\partial h_i} = \frac{1}{2}\sum_1^N h_i \frac{\partial F_i}{\partial h_\kappa} \tag{5}$$

With the introduction of the Lagrangean factor λ, one can formulate the conditions for the minimum ϕ for a given V as follows:

$$0 = \frac{\partial}{\partial h_\kappa}\left(\sum_1^N \sigma_i F_i - 2\lambda V\right) = \sum_1^N \sigma_i \frac{\partial F_i}{\partial h_\kappa} - 2\lambda F_\kappa \tag{6}$$

or, also, according to (5),

$$\sum_1^N (\sigma_i - \lambda h_i)\frac{\partial F_i}{\partial h_\kappa} = 0 \tag{7}$$

If one combines the N equations (6) with h_κ,

$$\sum_1^N \sum_1^N \sigma_i h_\kappa \frac{\partial F_i}{\partial h_\kappa} = 2\lambda \sum_1^N h_\kappa F_\kappa$$

Therefore, following (5) and writing ϕ_m for the minimum value of ϕ,

$$\lambda = \frac{\phi_m}{3V} \tag{8}$$

The equations (7) however represent N linear homogeneous equations in the N differences $(\sigma_i - \lambda h_i)$. Of course, the determinant of their coefficients $\partial F_i/\partial h_\kappa$ as a functional determinant of the F_i with respect to the h_i is identically zero. There are three independent relations for the F_i since they bound a closed polyhedron. If one projects them on three nonparallel (nearly perpendicular to each other) planes, then the sum of their projections (taking sign into account) is zero in each case.[10] The $(N - 1)$- and $(N - 2)$-rowed subdeterminants also vanish identically. At least one of the $(N - 3)$-rowed is different from zero. One can therefore conclude from (7) that, by choosing three suitable $(\sigma_i - \lambda h_i)$, all the rest can be determined as linear homogeneous functions of these.[11] We denote for these three the indexes 1, 2, and 3.

Now we complete the equations (3) by defining

$$\mu = \lambda^{-1} \tag{9}$$

The three brackets with indexes 1, 2, and 3 disappear and with them, according to (7), all $(\sigma_i - \lambda h_i)$. The Wulff theorem,

$$\frac{h_1}{\sigma_1} = \frac{h_2}{\sigma_2} = \ldots \frac{h_N}{\sigma_N} = \frac{3V}{\phi_m} \qquad (10)^{12}$$

follows.

§3

Let us now derive Wulff's theorem directly from thermodynamics. We observe the equilibrium of the crystal with its vapor and denote by f_c and v_c the free energy and volume per unit mass, respectively, for the crystal, and by f_g and v_g these same quantities for the gas. By the transfer of the quantity dm from the gas phase to the crystal and specifically to its face F_l, so that only h_l of all the h_i varies, the change of the free energy is

$$(f_c - f_g)\,dm + \frac{\partial\phi\, dh_l}{\partial h_l\, dm}\,dm$$

The term

$$\frac{\partial\phi\, dh_l}{\partial h_l\, dm}\,dm = \frac{\partial(\Sigma_i \sigma_i F_i)\, dh_l}{\partial h_l\, dm}\,dm\,,$$

is added because the F_i vary with the transfer, and not only the particular F_l, but, as Gibbs has shown, its neighbors also. Nevertheless, it is unnecesssary to calculate the changes exactly. The equilibrium condition that this increase of free energy is the same as the introduced work $p(v_g - v_c)\,dm$ leads, if we at the same time write ψ for the potential, $f + pv$, to the equation

$$\psi_c - \psi_g = \frac{\partial\phi\, dh_l}{\partial h_l\, dm} = 0$$

Now the volume of the collected transferred matter is, on the one hand, $v_c dm$, and, on the other hand, $F_l\, dh_l$, so that

$$\frac{dh_l}{dm} = \frac{v_c}{F_l}$$

Consequently, the equilibrium condition becomes

$$\psi_c - \psi_g + \frac{v_c\, \partial\phi}{F_l\, \partial h_l} = 0 \qquad (11)$$

Like the other quantities which occur here $(1/F_l)(\partial\phi/\partial h_l)$ must now have a value independent from the index l: we call it 2λ and find again immediately equation (6):

$$\frac{\partial\phi}{\partial h_l}\left(\sum \sigma_i F_i\right) = 2\lambda F_l$$

From this we can now work backward in §2 either to the minimum statement for ϕ or to Wulff's theorem, equation (10). Both principles are equally closely related in this representation of the equilibrium condition (11).

By contrast, K. Jellinek[13] in his textbook of physical chemistry, following the example of J. J. P. Valeton,[14] after he has introduced the volume dV for $F_l dh_l$ in (11) and then $\frac{1}{2}\Sigma_i h_i dF_i$ for dV, concludes then for all kinds of variations dF_i that

$$\frac{d\phi}{dV} = \frac{2\Sigma_i \sigma_i dF_i}{\Sigma_i dh_i dF_i}$$

must have the same value, which would only be possible for a value of the quotient h_i/σ_i independent of subscript. It seems to me that both parts of this conclusion are not pertinent, the first because of the change in the meaning of dV, the second because of the connections between the F_i as discussed in §2; one cannot select the dF_i independently from one another.

According to (8), λ decreases as m^{-1} when we increase the dimensions of the crystal uniformly by the factor m. For very large crystals, the surface area term is omitted in (11); this goes over into the usual equation of the vaporization curve:

$$\psi_c - \psi_g = 0 \tag{12}$$

from which the pressure P_∞ is calculated as a function of T. For a small crystal, nevertheless, according to (11) there is a higher pressure p at a given temperature. We get the difference if we subtract (12) from (11) and make the effect of the pressure equal to ψ_c in both of them, neglecting the actually different ψ_c value because of the negligible volume of the crystal relative to the gas phase. Thus

$$\psi_g - \psi_{g_\infty} = \frac{v_c \partial\phi}{F_l \partial h_l} = 2\lambda v_c \tag{13}$$

and, according to (8),

$$\psi_g - \psi_{g_\infty} = \frac{2\phi_m}{3V} v_c = \frac{2\sigma_l}{h_l} v_c$$

Finally, introducing the ideal gas law,

$$\psi_g - \psi_{g_\infty} = \frac{RT}{M} \ln \frac{p}{p_\infty}$$

M designates the molecular weight of the gas, and the influence of size on the vapor pressure is given by

$$\frac{RT}{v_c M} \ln \frac{p}{p_\infty} = \frac{2\phi_m}{3V} = \frac{2\sigma_1}{h_1} = \frac{2\sigma_2}{h_2} \quad \text{etc.} \tag{14}$$

This is one of the important relations to which I. N. Stranski[15] points in the work immediately preceding the work of M. Volmer[16] on which the above is based.

§4

M. Volmer (*loc. cit.*) in 1939 gave a simple derivation of the Wulff theorem directly from thermodynamics. However, his presentation requires completion.[17]

Take a crystal in Wulff's equilibrium shape bounded by the faces F_i and divided by thin plane partitions, the edges of each of which project from the Wulff point; these partitions extend out into the gas space. Each part of the crystal is a pyramid with a base F_l and height h_l. We assume that no free energy attaches to the surface between partition and crystal and between partition and gas. That this assumption leads to no errors will be proved.

What is the equilibrium between the vapor in that part of space which lies over F_l and the crystal separated from it by F_l relative to the deposition dm, which according to our assumptions changes only F_l and the boundary along the partitions (which is, as far as energy is concerned, of no importance)?

In equation (11), which we keep in other respects, since (in contrast to §2) only F_l changes with h_l, we may now set

$$\frac{\partial \phi}{\partial h_l} = \sigma_l \frac{dF_l}{dh_l}$$

This yields

$$(\psi_c - \psi_g)_l + \frac{v_c \sigma_l dF_l}{F_l dh_l} = 0 \qquad (15)$$

The index l in $(\psi_c - \psi_g)_l$ indicates that this equation refers only to that part of space lying above F_l.

Now, however, let us imagine such accretions on all [the pyramidal] subdivisions and exactly so that the crystal form remains similar to itself. To that end the increases dh_i of the proportion suffice:

$$dh_1 : dh_2 : dh_3 \ldots = h_1 : h_2 : h_3 \ldots$$

and since the h_i according to the hypothesis are proportional to the σ_i, one has to introduce a parameter α which is independent of the index l, so that

$$h_l = \sigma_l \alpha, \qquad dh_l = \sigma_l \, d\alpha \qquad (16)$$

Since F_l then is proportional to α^2, it follows through logarithmic differentiation that

$$\frac{dF_l}{F_l} = 2\frac{d\alpha}{\alpha} = 0$$

and equation (15) goes to:

$$(\psi_c - \psi_g)_l + 2\frac{v_c}{\alpha} = 0 \qquad (17)$$

One sees that the difference $(\psi_c - \psi_g)_l$ becomes independent of the index l, so that, therefore, with the condition that the Wulff equilibrium form is maintained by [the] accumulation, the gas pressure is equal over all faces F_i. Moreover, inserting the value of α from (16) leads directly to the equation

$$\psi_c - \psi_g + 2v_c \frac{\sigma_l}{h_l} = 0 \tag{18}$$

and this is identical to (13) if one [see (12)] bears in mind the equality of ψ_c and Ψ_{g_∞}.

The partitions employed before have now become superfluous with the similar expansion [of the crystal]. For that reason, the earlier assumption about the free energy at their boundaries is unexceptionable.

§5

This whole discussion started out from the assumption that one knows the faces F_i belonging to the equilibrium form with their perpendiculars. If one selects these faces arbitrarily, however, from those which are crystallographically possible, naturally one is not certain if one has made the right choice. And it is doubtful whether the approximate minimum of the surface free energy derived for this choice from the Wulff theorem represents the absolute minimum. Therefore, one must make a selection of different kinds of F_i according to each principle of minimum form, and calculate for them all the minimum value of ϕ, in order to see which of these different relative minima is the lowest. Only then is the problem also physically completely solved.

To enter into the indispensable extensions to molecular theory contained in the investigations of I. N. Stranski and M. Volmer lies out of the realm of our deliberations. What mattered to us was the most feasible clarification possible of the expressions of the equilibrium form based on thermodynamics.

Berlin-Dahlem, Max-Planck-Institut (Kaiser-Wilhelm-Institut für Physik)

NOTES

1. J. W. Gibbs, *Trans. Conn. Acad.*, **3** (1875–1878), reprinted in *Thermodynamische Studien*, Leipzig, 1892, p. 66, esp. p. 376 *passim*.
2. P. Curie, *Bull. Soc. Mineral. France*, **8** 145 *passim* (1885); also *Z. Krist.* **12**, 651 *passim* (1887).
3. G. Wulff, *Z. Krist.*, **34**, 449 (1901) esp. p. 512.
4. The possibility of describing one and the same polyhedron through suitably changing the distances h_i, if one at the same time shifts the origin, yields a relationship between the differential ratios of the volume V to the h_i. If we shift the origin by the vector $J ds$, and taking $\hbar_i$ as unit vector in the direction of the interior perpendicular on F_i, h_i grows by $(J\hbar_i) ds$. Since the polyhedron and its volume remain unchanged, it follows that

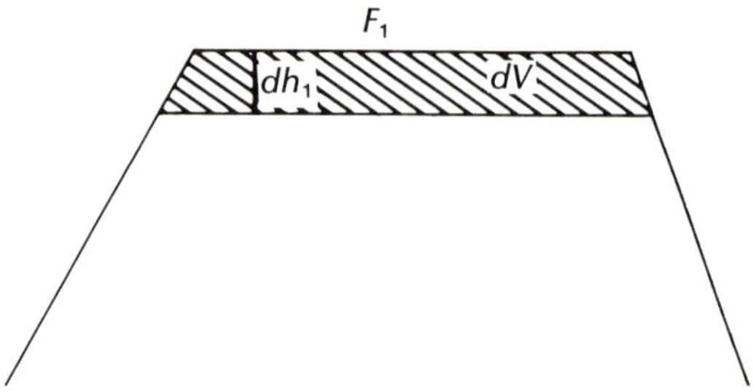

Figure 1

$$\frac{dV}{ds} = \sum_{1}^{N} \frac{\partial V}{\partial h_i}(J\hbar_i) = 0$$

an equation, which because of the completely aribtrary choice of J, has

$$\sum_{1}^{N} \frac{\partial V}{\partial h_i}\hbar_i = 0$$

as consequence. This vector equation contains three scalar functions. The $\alpha V/\alpha h_i$ are accordingly not independent functions of h_i but they make up only an ($N - 3$)-dimensional manifold. Their functional determinant with respect to the h_i's, i.e., the determinant of the $\partial^2 V/\partial h_i \partial h_\kappa$, is not only itself equal to zero, but this is true also for its ($N - 1$)- and ($N - 2$)-rowed subdeterminants as well. Only the ($N - 3$)-rowed are at least in part different from zero.

5. See note 11.
6. H. Liebmann, *Z. Krist.*, **53**, 171 (1914).
7. L. Lindelof, *Math. Ann.*, **2**, 150 (1870).
8. H. Minkowski, *Göttinger Nachrichten*, 1897, p. 198.
9. Figure 1 shows how the V varies with variation of h_l alone; one sees that with consequent decrease of F_l the faces neighboring F_l must increase. The calculation of these changes is found in Gibbs and has passed into the textbooks; repetition is unnecessary.
10. If one represents the angle between the perpendicular on one of the projection planes and the interior perpendicular on the face F_i by $\beta_i^{(\delta)}$, then the equations in question read

$$\sum_{1}^{N} F_i \cos \beta_i^{(\delta)} = 0, \qquad \delta = 1, 2, 3 \tag{a}$$

Thus the cosines are components of the unit vectors $\hbar_i$ of note 4, so that the relationship there for the $\partial V/\partial h_i$ is identical to the three equations above.

11. In the case in which the planes F_1 and F_4, F_2 and F_5, F_3 and F_6 have perpendiculars drawn out from the six faces in opposite directions, and in which these three pairs of perpendiculars are mutually perpendicular, the volume V of the enclosed parallelopipehedron is,

$$V = (h_1 + h_4)(h_2 + h_5)(h_3 + h_6)$$

As a result [see equation (4a)],

$$F_1 = \frac{\partial V}{\partial h_1} = (h_2 + h_5)(h_3 + h_6), \text{ etc.,} \qquad \frac{\partial F_1}{\partial h_2} = (h_3 + h_6) = \frac{\partial F_2}{\partial h_1} = \text{etc.}$$

The three equations for the F_i, with appropriate choice of projection planes, read as follows: $F_1 - F_4 = F_2 - F_5 = F_3 - F_6 = 0$; and the six-rowed determinant of the $(\partial F_i/\partial h_\kappa)$'s, but also each five- or four-rowed subdeterminant, contains the same rows and columns and therefore equals zero. Only the three-rowed are different from zero, i.e., those in which i and k have the values 1, 2, 3.

12. If F_1 and F_2 are parallel, then they enter the three equations (a) of note 10 only in the combination $F_1 - F_2$. As a result one can eliminate this difference and F_3, and the equations between $F_4, F_5, \ldots, F_N$, $\Sigma_4^N \sigma_i F_i = 0$, are left. Then, however, each $(N - 3)$-rowed functional determinant of the latter F_i is zero, according to one or another $(N - 3)$-fold property of h_i. Therefore, one has to avoid parallel faces in selecting the faces F_1, F_2, F_3.

    This necessary condition is also sufficient; for if F_1 and F_2 enter in the three equations (a) of note 10 in the same linear combination, so that such an elimination is possible, then F_1 and F_2 are parallel to each other; only for parallel planes are the direction cosines in coordinates definitely related to one another, i.e., equal except for the initial sign.

    That the definition (3) for the origin of the h_i's fails if, among all the other faces, F_1, F_2, F_3; are parallel is geometrically clear without further comment.

13. K. Jellinek, *Lehrbuch Physik. Chem.*, **2**, 527 *passim* (1928).
14. J. J. P. Valeton, *Sachsische Akad. Ber.*, **67**, 1 (1915), cf. p. 38.
15. I. N. Stranski, *Z. Krist.*, **105**, 91 (1945).
16. M. Volmer, *Kinetik der Phasenbilding*, Dresden and Leipzig, 1939, p. 90.
17. Communicated with Mr. Volmer's concurrence.

18

Reprinted from *Phil. Mag.*, Ser. 7, **37**, 184–199, 217–236 (Mar.-Apr. 1946)

Crystal Habit and Internal Structure.—I.

By A. F. Wells, M.A., Ph.D.

Introduction.

Classical crystallography was restricted to the study of the external forms of crystals and of their physical properties. With the development of X-ray crystallography it has become possible in many cases to determine the atomic arrangement in the interior of the crystal, and in recent years attention has been largely concentrated on this aspect of the subject. It should now be possible to begin to fuse these two aspects of crystallography into one homogeneous whole. Some progress has already been made in relating physical properties to internal structure and, conversely, a knowledge of physical properties such as optical and magnetic anisotropy is often of assistance when determining the structure of a crystal. The problem of accounting for the face development of a crystal is part of the much larger one of interatomic and intermolecular forces, for the external form of a crystal is the expression not only of its internal structure but, if it is grown from solution, also of the interaction between solute and solvent.

The shapes of many crystals are sufficiently characteristic to make it possible to identify them in microchemical analysis, and many minerals are easily recognisable from their face-developments. It is not surprising, therefore, that attempts have been made to account for the shapes of crystals. There have been two lines of approach, the thermodynamical and the crystallographic. The first laid down a general condition for the stability of a crystal in equilibrium with its surroundings in terms of the surface free energies of all the faces. If the correct condition for equilibrium has been stated and if the appropriate property of the different faces could be calculated or measured, this method would obviously lead to a complete specification of the external form of the crystal. The object of the second method was to relate the face development of a crystal to its internal structure. The environment of the crystal is neglected, and only qualitative estimates of the relative stabilities of faces of different types can be made. Both these methods originated long before it was possible to determine the atomic structures of crystals. Now that we have a considerable knowledge of crystal structures it should be possible to make further progress in this field. In the present and later papers we shall review some of the earlier work, record new observations on the growth of crystals, and attempt to interpret these in terms of the structures of the crystals.

Crystals may be grown from the vapour of the substance, from the melt, or from a solution in a suitable solvent. In the first two cases we have only to consider the interactions between molecules of the substance concerned, and we should expect to find simple relations between crystal form and internal structure. The latter case is, however, more complicated, for we are concerned not only with interaction between solute molecules, but also with solute-solvent and solvent-solvent interactions. Apart from their intrinsic interest there are two reasons for dealing with these more complex cases. First, there is the practical point that most solids are purified by recrystallization from solution, and there are accordingly more data about crystals grown from solution than from the vapour or melt. Second, many interesting changes of habit may be brought about either by changing from one solvent to another or by adding a second solute to the solution.

The external form of a crystal is described in terms of (*a*) the types of faces present, as defined by their crystallographic indices, and (*b*) the relative areas of the faces of different types, if faces of more than one kind are present. Since the term "habit" is rather loosely used it will be convenient to distinguish the following types of variation in the shape of the crystals of a particular substance :—

(1) Faces of the same *form* may be developed to different extents, *e. g.*, plates or needles instead of perfect cubes. (A *form* is a set of faces related to one another by the symmetry elements. If one

face of a particular form appears then the others should also be present, developed to the same extent on the ideal crystal.) Changes in habit of this kind are simple cases of distortion from the ideal shape due to accidental causes, *e. g.*, rapid crystallization. Such variations are usual in any batch of crystals unless they are grown carefully, hence it is usual, when illustrating a crystal, to show all faces of a given form developed to the same extent.

(2) On a crystal exhibiting two or more forms it may be possible to vary the relative areas of the faces of different forms. This kind of change of habit is illustrated in fig. 1.

(3) Faces of a new form appear. A simple example is the change in habit of iodoform from a hexagonal bipyramid to a prism, shown in fig. 3.

Factors Influencing the Habits of Crystals.

From the practical standpoint we may distinguish two factors which influence the shapes of crystals of a particular substance : the rate of

Fig. 1.

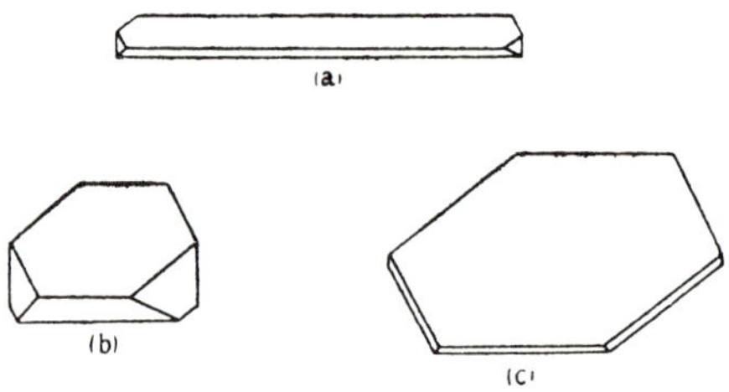

Crystals of naphthalene grown (*a*) rapidly from cyclohexane, (*b*) slowly from cyclohexane or ethyl alcohol, (*c*) rapidly from ethyl alcohol.

growth and external interactions with atoms or molecules of other kinds, *i. e.*, solvent and/or added impurities.

(*a*) *The Effect of Conditions of Growth.*

It is well known that rapid growth often leads to extreme developments. The thin plates or needles formed in rapidly cooled solutions or by sublimation are in many cases far removed from the equilibrium shape of the crystal. The simplest case is that in which the same forms are present but in different relative proportions. Naphthalene may be taken as an example. It is commonly obtained as plates $\{001\}$ bounded by $\{110\}$ and $\{20\bar{1}\}$. Rapid cooling of a solution in many solvents (*e. g.*, the lower alcohols) gives very thin plates. By slow evaporation of such a solution, however, compact crystals may be obtained. Solutions in cyclohexane give needles on rapid cooling, but again compact crystals on slow evaporation (fig. 1).

In some cases the development of extreme shapes would appear to be related to the direction(s) of the strongest interatomic or intermolecular bonds. For example, compounds with layer structures often form thin flakes, and the thin needles of urea, potassium dihydrogen phosphate and α-resorcinol which are formed by rapid cooling of supersaturated solutions represent extreme developments in the directions of the chains or spirals of hydrogen bonds. In cases such as these the different shapes are simply due to different relative developments of the same forms. In other instances faces of new forms appear which are not present on a crystal grown slowly. Two examples may be given.

Pentaerythritol, $C(CH_2OH)_4$. The equilibrium shape obtained from aqueous solution is the tetragonal bipyramid (see below). If a saturated aqueous solution is poured into acetone the crystals produced are thin square plates {001}. As will be shown later this form is very unstable in aqueous solution. Similar plates separate from a solution in β-ethoxy-ethanol on rapid cooling, whereas crystals grown slowly from this solvent are terminated by (101) bipyramid faces.

Fig. 2.

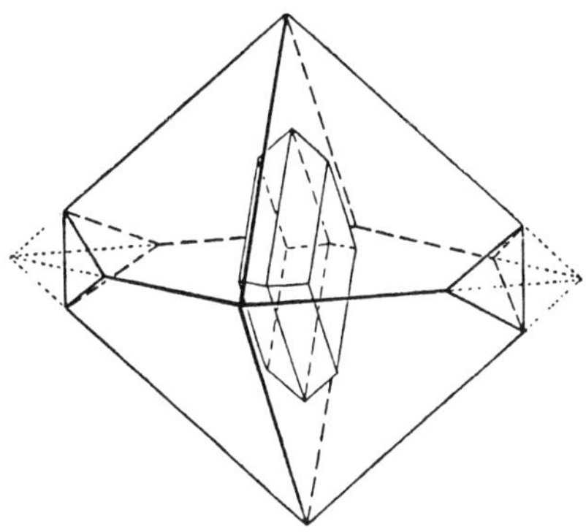

Normal habit of anthranilic acid from ethyl alcohol. Inset is a crystal tabular on (010) obtained by rapid crystallization.

Anthranilic acid, *o*-$(NH_2)C_6H_4(COOH)$, orthorhombic (stable modification). Crystals grown slowly from ethyl alcohol are essentially rhombic bipyramids {211}, often with small {110} (fig. 2). Rapid growth leads to diamond-shaped plates of varying thickness tabular on (010). If a crystal of this type is replaced in saturated alcoholic solution and allowed to grow slowly it tends towards the bipyramidal shape.

(*b*) *Change of Habit with Change of Solvent.*

Most of the studies of variations in crystal habit have been carried out with inorganic salts, and water has necessarily been used as solvent. By using non-ionic solutes, and particularly by choosing compounds which are soluble in a variety of solvents of chemically different types, it is possible to study the habits of crystals grown from different solvents. Many such changes of crystal habit must have been observed but comparatively few appear to have been recorded. Since they are of great importance in the present investigation, two examples are given below.

Iodoform, CHI_3, hexagonal. Forms observed, {0001}, {1010}, {10$\bar{1}$1}. It is normally obtained as plates {0001} bounded by the bipyramid {10$\bar{1}$1}. Simple bipyramids, fig. 3 (*a*), were obtained from a number of amines (furfurylamine, aniline, N-methyl aniline and *m*-toluidine), while the higher halogenated paraffins (*n*-butyl, octyl and dodecyl bromides), nitro-paraffins (1-nitro- and 2-nitro-propanes and 3°-nitrobutane) and cyclohexane and its derivatives (cyclohexanol, cyclohexanone and cyclohexylamine) gave crystals of type (*b*), fig. 3.

Resorcinol, *m*-$C_6H_4(OH)_2$, orthorhombic. From water, compact crystals, {110} and {011}, with faces of the latter form at both ends. This development is interesting since the crystal class is *mm*. The crystals from butyl acetate show the polar development to be expected, squat prisms {110} terminated at one end by two (011) faces and at the other by four (11$\bar{1}$) faces. From carbon tetrachloride and benzene, in both of

Fig. 3.

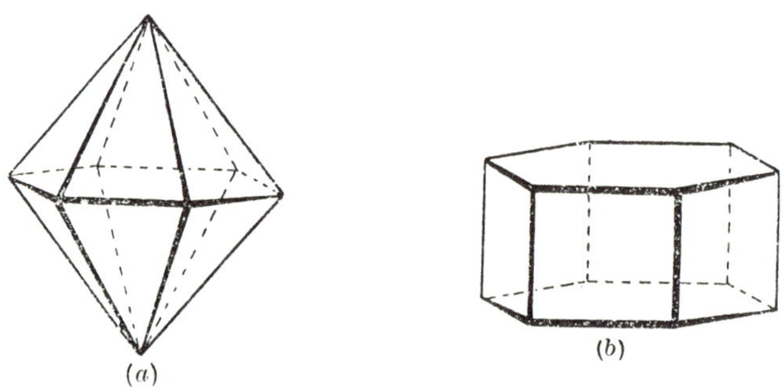

Crystals of iodoform from various solvents.

which it is very sparingly soluble, resorcinol crystallizes in very fine needles {110}. Further observations on the growth of crystals of resorcinol and an analysis of these results in terms of the crystal structure will be given in a later paper.

(*c*) *Changes in Habit caused by the Presence of a Second Solute.*

A considerable amount of work has been carried out on variations in crystal habit caused by the addition of " impurities " to the solution, and attempts have been made to account for these phenomena in a general way. For example, it is reasonable to assume that strong adsorption of a second solute on certain faces of a crystal will result in the development of large faces of that type on the final crystal, since the most prominent faces will be those on which deposition of material takes place most slowly (fig. 4). Following the determinations of the atomic structures of many crystals it became feasible to attempt to correlate the nature and shape of the adsorbed ions or molecules with the atomic arrangement in the different crystal faces (see, *e. g.*, Buckley (1939), France (1935, 1938, 1941), Saylor (1927). An example of a detailed study of this kind is that by Bunn (1933) on the relation between crystal structure, adsorption, oriented overgrowth and mixed-crystal formation, with particular reference to the modification of the habits of alkali and ammonium halides by urea. In all the later work emphasis has been placed on the importance of the structures of the faces of different crystallographic types (*i. e.*, the nature and arrangement of the atoms or ions in the various faces) in relation to the structure of the adsorbed material. Little attention seems to have been devoted to the normal habit of the solute when crystallized from the particular solvent used. It would appear more logical first to attempt to account for this before considering changes in habit induced by adding foreign substances.

It will be apparent that the study of crystal habit is closely related to the study of solubility, for the crystal face is the dividing surface between the crystalline solid and saturated solution. For a solute A in a solvent B the crystal structure of the solute is determined by the interactions A–A, the crystal habit by A–A and A–B, and the solubility by A–B (and also B–B and A–A). Therefore, we shall adopt the view-point that all such

Fig. 4.

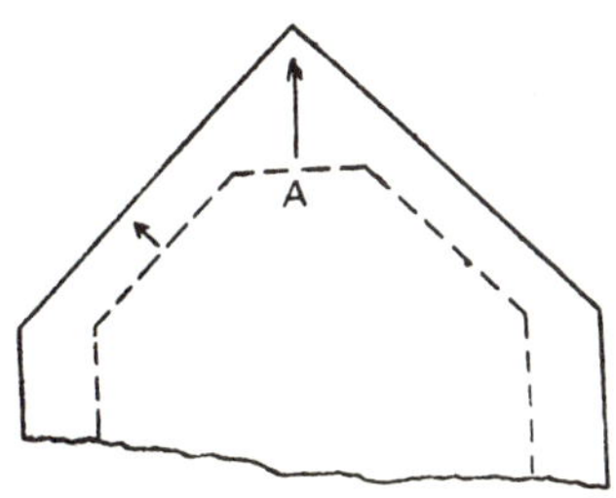

Rapid growth normal to face A leads to disappearance of that face.

phenomena as the dissolution of a crystal, the separation of a solid phase as a colloidal suspension, as compact crystals or as fine needles, must be regarded as manifestations not only of the way in which the primary particles of the solute interact one with another but also of the way in which solute and solvent interact. The very fact that many substances crystallize with different habits from different solvents shows that we must generalize the idea of habit modification by adsorbed impurities and extend it to include interaction with the solvent. If this is very different for different faces it will lead to habit modification just as in the case of the preferential adsorption of impurities. These considerations indicate the restricted validity of any relations between internal structure and crystalline form which do not take account of the solvent. We shall show that the effect of the solvent on crystal habit ranges from one extreme to the other. For ionic crystals (*e. g.*, alkali halides) crystallizing from water, interaction with the solvent is of negligible importance in determining habit, except possibly for the fluorides. Accordingly the general face development is predictable from the crystal structure. On the other hand, the nature of the solvent has a considerable effect in the case of, for example, some hydroxy-compounds, as already mentioned. This is not surprising when we remember that a compound like resorcinol is insoluble in benzene and cyclohexane but extremely soluble in water, cyclohexanol and cyclohexanone. There is a correspondingly large difference between the habits of crystals of resorcinol grown from water and benzene.

TABLE I.

Point-groups in which the general or a single special form is a polyhedron.

Point-group Symbol	General Form (*hkl*)	Special Forms
222	bisphenoid	—
mmm	bipyramid	—
$\bar{4}$	bisphenoid	bisphenoid *hhl* or *h0l*
$\bar{4}2m$	scalenohedron	{bisphenoid *hhl*, bipyramid *h0l*}
42	trapezohedron	bipyramid *hhl* or *h0l*
4/*m*	bipyramid	bipyramid *hhl* or *h0l*
4/*mmm*	{ditetragonal bipyramid}	bipyramid *hhl* or *h0l*
$\bar{3}$	rhombohedron	rhombohedron *hhl* or *h0l*
32	{trigonal trapezohedron}	{bipyramid *hhl*, rhombohedron *h0l*}
3/*m*	bipyramid	bipyramid *hhl* or *h0l*
$\bar{3}m$	{ditrigonal scalenohedron}	{hexag. bipyramid *hhl*, rhombohedron *h0l*}
3/*mmm*	{ditrigonal bipyramid}	{hexag. bipyramid *hhl*, trig. bipyramid *h0l*}
62	trapezohedron	bipyramid *hhl* or *h0l*
6/*m*	{hexagonal bipyramid}	bipyramid *hhl* or *h0l*
6/*mmm*	{dihexagonal bipyramid}	bipyramid *hhl* or *h0l*
and the five cubic classes, 23, 43, *m*3, $\bar{4}3m$, and *m*3*m*		

The Equilibrium Shape of a Crystal.

A crystal is a solid body bounded by a number of plane surfaces forming a closed convex polyhedron. Of the 32 crystal classes 20 may be distinguished from the rest on the grounds that the general form or a special form is a polyhedron, *i. e.*, the set of faces generated by the symmetry elements forms a closed polyhedron. In the other classes faces of at least two forms are required to give a polyhedron. Of course, even in the former (20) classes the general form is not necessarily developed, and the crystal may exhibit two or more forms. In Table I.

are listed those forms which enclose a polyhedron. The cubic classes are omitted since in these all forms are of this type. Judging by the published data, crystals (other than cubic) showing only one form are rare.

We may therefore state our problem as follows :—

(1) If the faces of a crystal are all of one form, we require to know why this and not some other form develops, *e. g.*, why NaCl grows into cubes rather than octahedra or dodecahedra, all of which are consistent with the symmetry of the internal structure of the crystal. In the case of crystals of less than holohedral symmetry it will be necessary also to account for the smaller number of faces in certain forms. Taking urea as an example, we require to show why the form {111} in the class $\bar{4}2m$ consists of only four faces, *i. e.*, to show in terms of the atomic structures of the faces why the face (111) is more stable than ($\bar{1}\bar{1}\bar{1}$), which does not occur.

(2) If faces of more than one form are present we have to show, first, why these particular forms develop and second, what determines the relative areas of the faces of different types.

It is well known that crystals of a given substance differ as regards the relative developments of faces of different forms. Before proceeding further it is essential, therefore, to know whether there is an equilibrium shape for a crystal of a particular substance grown from a certain solvent.

The Concept of Minimum Total Surface Free Energy.

The condition for stability of an isolated drop of a fluid is that its surface free energy, and hence its area, is a minimum. Gibbs suggested that a similar condition should apply to a crystal, viz., that for a crystal in equilibrium with its surroundings at constant temperature and pressure, its total (Gibbs) free energy is a minimum for a given volume. (Curie, 1885, arrived at the same result by a different method.) If we suppose that the volume free energy per unit volume is constant throughout the crystal, then the condition becomes $\Sigma_1^N A_i g_i$=minimum, where g_i is the surface free energy per unit area of the *i*th face of area A_i on a crystal bounded by N faces. In other words, those faces develop which lead to a minimum total surface free energy for a given volume, so that any crystal, if kept for a sufficiently long time in a saturated solution, should attain the equilibrium shape determined by this criterion. If the above condition applies to crystals of all sizes it follows that (1) two or more geometrically similar crystals of different sizes (each equilibrated in itself) brought into the same saturated solution should ultimately become one crystal, and (2) the solubility of a crystalline material is a function of the size of the crystals. We shall shortly consider these points, but first we should point out that Gibbs realized that, owing to the different degrees of internal order in crystals and liquids, the practical realization of an equilibrated crystal might be difficult. In a drop of liquid there is random arrangement of the molecules (or atoms) so that every portion of the surface is similar in structure to every other. In a crystal, however, the structural units are arranged in a regular way in three dimensions so that the whole array possesses certain symmetry. The bounding surfaces of the crystal are plane but not necessarily equivalent, and there is an infinite series of sets of possible faces for any crystal. Gibbs appreciated that the growth of a crystal is a periodic process, in contrast to one such as condensation or evaporation, for it takes place by the deposition of new layers of material on the faces of the crystal. There might, therefore, be " activation energies " associated with the commencement of a new layer, and these might be independent of the area of the face. He concluded that " it seems not improbable that the form of very minute crystals in equilibrium with solvents is principally determined by the condition that $\Sigma_1^N(A_i g_i)$ shall be a minimum for the volume of the crystal, but as they grow larger (in a solvent no more supersaturated than is necessary to make them grow at all), the deposition of new matter on the different surfaces will be determined more by the nature (orientation) of the surfaces and less by their size and relations to the surrounding surfaces. As a final result, a large crystal thus formed will generally be bounded by those surfaces alone on which the deposition of new matter takes place least readily, with small, perhaps insensible, truncations. If one kind of surface satisfying this condition cannot form a closed figure, the crystal will be bounded by two or three kinds of surfaces determined by the same condition. The kinds of surface thus determined will probably generally be those for which g_i has the least values. But the relative development of the different kinds of sides will not be such as to make $\Sigma_1^N(A_i g_i)$ a minimum. The growth of the crystal will finally be confined to sides of a single kind."

Wulff (1901) showed that the Gibbs equilibrium shape of a crystal is very simply related to the relative surface free energies of the faces. The crystal should form a polyhedron such that the perpendicular distances from a point within the crystal to the faces are proportional to the surface free energies of the appropriate faces. von Laue (1943) has recently reviewed, and added to, the various proofs of this theorem of Wulff, and emphasizes that it is not sufficient merely to determine, for any arbitrary combination of faces, the shape corresponding to minimum surface free energy. It would be necessary to consider all possible combinations of faces and to find which of the relative minima is actually the smallest. Now the validity of the minimum surface free energy criterion could only be tested directly if the relative values of these energies of sufficient (ideally, of all the possible) faces of a crystal were known, and unfortunately very little is known of these quantities. Indirect evidence has, however, been adduced to show that the Gibbs condition is not normally applicable to crystals of radius greater than about 10^{-4} cm. The Thomson equation giving the ratio of the vapour pressure of a drop of liquid to that of a plane surface in terms of the surface tension of the liquid and the radius of the drop shows that for a drop with radius 10^{-4} cm. the vapour pressure is only 0·1 per cent. greater than that of a plane surface. The solubility of crystals should vary with their radius and interfacial tension solid-liquid (σ) according to the equation

$$C_r = C_0 e^{\frac{2\sigma M}{RT\rho r}},$$

where C_r is the solubility of a crystal of radius r, C_0 the concentration of a solution which would be in equilibrium with an indefinitely large crystal, M the molecular weight and ρ the density of the solute. Hulett (1901) found, by conductivity measurements, that the solubility of gypsum particles of radius $0{\cdot}3\,\mu$ was 18·2 millimoles/litre at 25°, whereas that of crystals with radius $2\,\mu$ was the normal value, 15·3 millimoles/litre. Although the precise meaning to be attached to such " solubilities " is doubtful, particularly if the crystals have been ground up (see, for example, the discussion by Valeton (1924)), it has been generally concluded that the differences between the surface free energies of different faces are not the only factors operative in determining the shapes of macroscopic crystals (see also Stranski and Kaischew (1934, 1935, 1937), Stranski (1943) and Volmer (1939)).

Our first task is clearly to ascertain whether there is in fact an equilibrium shape for macroscopic crystals of a particular substance grown under specified conditions. Only if this is so will it be possible ultimately to relate the morphology of a crystal to its internal structure and its interactions with the solvent. Evidence of three kinds may be sought. First, large numbers of crystals may be grown slowly to see if they all develop in the same way. Second, crystals of other shapes obtained, for example, by rapid crystallization, may be grown larger under more " ideal " conditions to see if they approach a standard shape. A third type of experiment is possible in cases where crystals grown from different solvents show different habits. Evidence of all three kinds will be presented later. The interpretation of the results of experiments of the second type raises a number of questions.

The Practical Realization of Equilibrium Shape.

As already mentioned, Gibbs foresaw that there might be difficulty in commencing a new layer on a particular face of a crystal. There is, however, a further complication. Let us suppose that there is a definite equilibrium shape for a crystal growing from a certain solution and that a crystal of non-equilibrium shape is placed in the saturated solution, which is not allowed to evaporate. Material should be transferred, at some unspecified rate, from certain parts of the crystal via the solution to other parts until the equilibrium shape is reached. We may imagine that the original crystal has the shape indicated by the broken lines in fig. 5 (*a*), which might represent a section of a tetragonal prism {100}, and that the equilibrium shape is that enclosed by full lines (the prism {110}). Material would have to be transferred as shown by the arrows. The initiation of the process would consist in the movement of molecules from one part of a (100) face to another part of the same face, and moreover, there would have to be a nett overall transfer of molecules to the centre of the face. We should expect that the probability of this process ever taking place would be very small. If, on the other hand, evaporation of the solution is permitted, molecules may be added on all faces of the crystal, which can then grow as shown at (*b*).

Failure to observe any tendency towards an equilibrium shape has already been recorded (*e.g.*, Herzfeld, 1930). We have found that crystals of pentaerythritol of various shapes show no change after shaking for many days in a sealed vessel containing saturated solution at 25° C., though very different results are observed if evaporation is allowed (see below). In view of what we have said it is evident that the failure to change shape spontaneously is not to be interpreted to mean that there is no equilibrium shape. A further point now arises which is of importance in interpreting the growth of crystals of non-equilibrium shape. Is it necessarily possible for a crystal of *any* initial shape to reach the equilibrium shape ? This obviously depends on the relative rates of growth on the various faces. If, for example, the crystal in fig. 5 (*c*)

Fig 5.

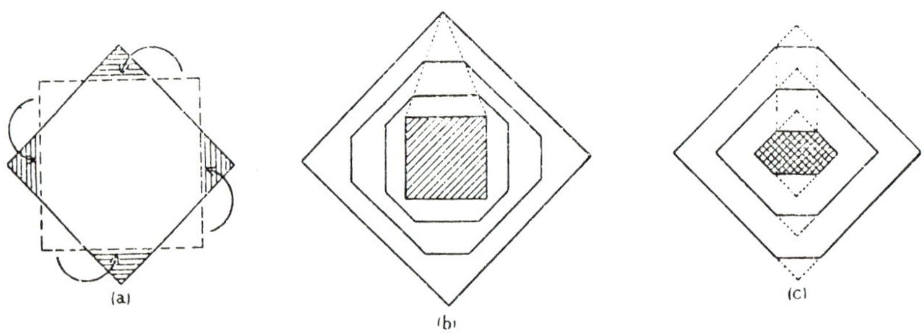

The growth of crystals towards equilibrium shape.

grows as shown, it can never reach the equilibrium shape, though it will approach it.

The following new data on the growth of crystals of certain organic compounds can now be presented. We may summarize under three heads the evidence at present available for the existence of an equilibrium shape for crystals of a particular substance grown from a given solvent :

(i) Constancy of shape of crystals grown from one solvent.

(ii) Growth of crystals of other shapes towards the equilibrium shape.

(iii) Change in equilibrium shape with change of solvent.

(i) *Constancy of Shape of Crystals Grown from one Solvent.*

Measurements have been made of the dimensions of crystals of anthranilic acid grown from aqueous solution to determine whether in fact the same shape is maintained during growth. Large numbers of crystals were grown and batches of ten were removed and measured at intervals. Measurements were made of the overall length and of the breadth and thickness (see fig. 6), and the mean values of the ratios length/breadth and thickness/breadth were calculated.

Owing presumably to the fact that the crystals were grown resting on a (100) face there is some drift in these values (in particular the crystals grew thinner) which cover a volume range of nearly 200 : 1. More constant values for these ratios would be obtained from experiments with

Fig. 6.

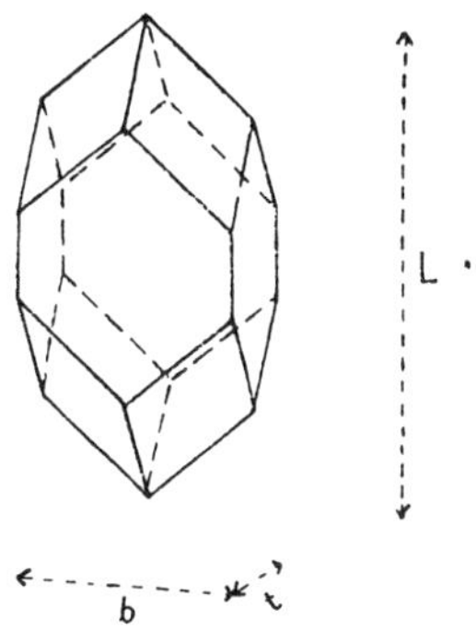

Crystal of anthranilic acid from aqueous solution

crystals suspended in the solution, as we shall show in a later paper, but these rough results are sufficient to show that there is no appreciable alteration in the general shape of these crystals during growth to some 200 times their original volume.

TABLE II.

Dimensions of Crystals of Anthranilic Acid (in mm.)

	L.	*b*.	*t*.	L/*b*.	*t*/*b*.
Set 1	1·17	0·74	0·31	1·58	0·41
,, 2	2·47	1·39	0·55	1·78	0·39
,, 3	8·50	4·66	1·49	1·82	0·32

(ii) *Growth of Crystals of Other Shapes towards Equilibrium Shape.*

Evidence on this point has already been presented for crystals of anthranilic acid growing in ethyl alcohol solution. A more detailed study has been made of pentaerythritol, $C(CH_2OH)_4$.

Crystals of type (*a*), fig. 7, may be readily grown from aqueous solution. Referred to the tetragonal cell with $a=6{\cdot}10$ A and $c=8{\cdot}73$ A, the forms are $\{110\}$, $\{101\}$ and $\{10\bar{1}\}$. The crystals may be twins, since the point-symmetry is only $\bar{4}$, but this is immaterial in the present study. Crystals terminated at one end by (001), shown in elevation at (*b*), may be obtained by cooling a hot saturated solution and allowing crystals to grow downwards from the surface. Alternatively a crystal of type (*a*) may be cleaved, there being a perfect (001) cleavage. Crystals of type (*b*) kept in a saturated solution at constant temperature showed no change in shape after several days when evaporation was prevented, but if placed

Fig. 7.

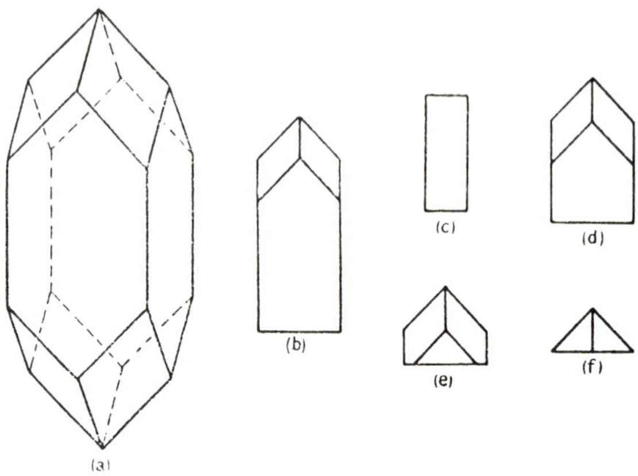

Crystals of pentaerythritol.

Fig. 8.

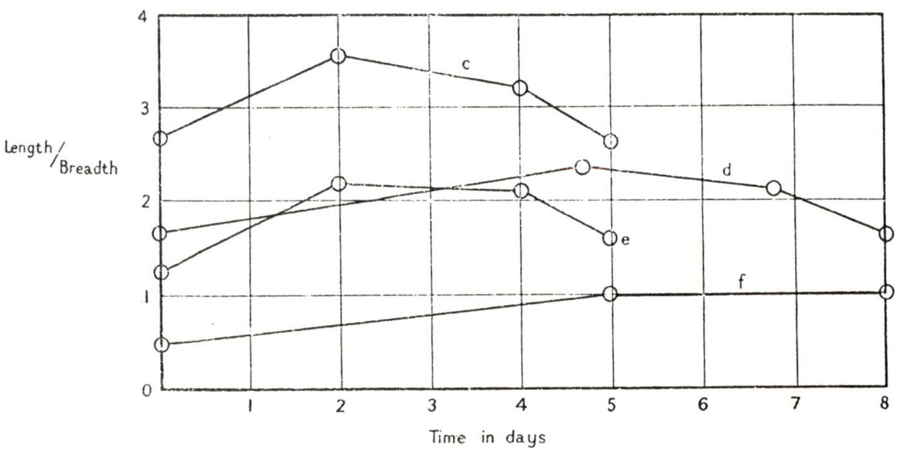

Pentaerythritol growth curves.

in saturated solution in an open beaker, rapid growth took place at one end until, after 2–3 hours, the (001) face was completely replaced by bipyramid faces. This is described in more detail below. The experiment was repeated with a series of crystals (*c*)–(*f*), and measurements of length and breadth were made at intervals. The results are shown in fig. 8. The initial rise in the curves is due to the completion of the ends of the crystals, which then all tend towards a definite limiting shape. This is the bipyramid, though small prism faces were still present on many crystals when the experiment had to be concluded owing to the deterioration of the crystals.

A more detailed study of the growth of a crystal of type (*b*) was made under the microscope. Deposition of new material on (001) did not take place in layers, but instead a large number of minute pyramids developed, *i. e.*, the (001) face was replaced as quickly as possible by (101) faces (fig. 9, *a*). Growth then continued as shown at (*b*). Only after the pyramid faces had been completed on the end of the crystal did any appreciable deposition take place on the (110) faces. The crystal then grew in such a way as to reduce the area of (110) faces relative to (101), the crystal tending towards a bipyramid. Fig. 9 (*b*) is drawn to scale from the actual measurements made under the microscope. A cleavage line (dotted) served as a base-line for measurements along the *c*-axis. These observations show that the (001) face is highly unstable in aqueous solution, and that the relative stabilities of the faces are (101) > (110) > (001). These facts will be discussed in relation to the internal structure of the crystal in a later paper.

Fig. 9.

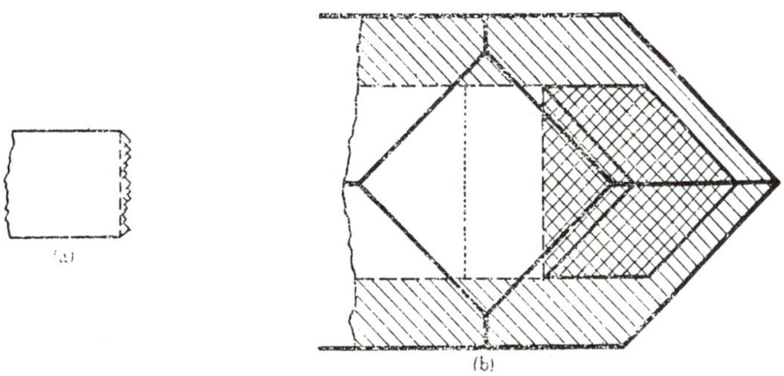

Growth of a crystal of pentaerythritol. In (*b*) the unshaded portion represents the original crystal, and the two kinds of shading indicate two stages in the growth of the crystal.

(iii) *Change in Equilibrium Shape with Change of Solvent.*

Perhaps the most conclusive proof that there is a definite equilibrium shape for a crystal grown from a given solvent is provided by experiments of the following type. A crystal of the substance is grown slowly from solvent I and then transferred to a saturated solution in solvent II from which the substance crystallizes with a markedly different habit. The crystal continues to grow but now develops the habit corresponding to solvent II. Anthranilic acid is a convenient solute for illustrating this point. A crystal grown from aqueous solution, tabular on (100), was suspended in ethyl alcohol solution. Growth took place as shown in fig. 10 (*a*), the final crystal being essentially a bipyramid. Crystals of anthranilic acid from glacial acetic acid are also prismatic like those from water, though with rather different relative dimensions. If, therefore, a crystal grown from alcoholic solution (bipyramid) is regrown in acetic acid solution, it develops into a prismatic crystal as shown in fig. 10 (*b*).

Conclusion.

We have seen that a number of factors can influence the habit of crystals of a (given polymorphic form of a) substance—rate of growth, nature of the solvent, and the presence of impurities. However, all the crystals from a solution in a given solvent, if grown sufficiently slowly, do exhibit a standard shape which is characteristic of the particular solute-solvent combination. The influences tending to restrain the crystals to the equilibrium shape will naturally vary greatly according to the nature of the solute and solvent, and it is to be expected that greater divergences from the equilibrium shape will be observed, and will result from smaller variations in the conditions of crystallization, in some systems than in others. It has been shown that crystals of non-equilibrium shape do

Fig. 10.

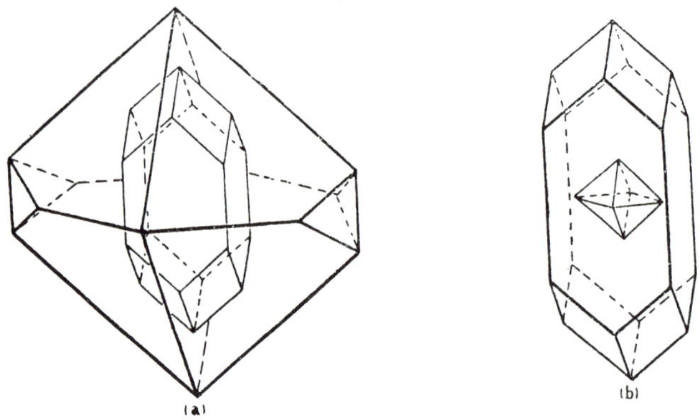

(*a*) Crystal of anthranilic acid from aqueous solution (inset) grown larger in ethyl alcohol solution. (*b*) Crystal from alcohol (inset) grown larger in acetic acid solution.

tend, on further growth, towards the standard shape. Such changes do not, however, take place unless growth is permitted. This fact, together with indirect evidence from the dependence of solubility on particle size, indicates that the criterion of minimum surface free energy, originally postulated by Gibbs as the condition for stability of a crystal, cannot be applied in its original simple form to a macroscopic crystal which has been grown at a finite rate. An essential difference between a drop of fluid and a crystal is that the former can adjust its shape by the rearrangement, within itself, of its molecules. This is not generally possible at ordinary temperatures for a crystal. Thus a crystal of non-equilibrium shape in a saturated solution could rearrange itself to the equilibrium shape only by the passage of molecules *via the solution* from one face on to others or on to other parts of the same face. Such processes do not take place at measurable rates in the cases of macroscopic crystals, except possibly at temperatures near the melting point. The fact that the final shape of large crystals is determined by the relative rates of growth in different directions, a conclusion reached by Gibbs, Valeton and others, would seem indisputable. The problem of accounting for the equilibrium shape therefore consists in discovering what factors determine (1) the choice of planes on which growth takes place, and (2) the relative rates of growth on those planes.

Consider first a crystal built of atoms of one kind. An atom in a face of the crystal has fewer nearest neighbours than one in the interior, and this asymmetry of environment is different for faces of different crystallographic types. We might therefore expect to be able in such cases to relate the nature of the surface to its stability, and hence to the rate of deposition on that surface. If, however, there are atoms of more than one kind in the crystal, faces of different types may differ in another respect, viz., the nature of the atoms in the faces. In such a case stronger interaction between one atomic species and the solvent may well be the decisive factor in determining the relative rates of deposition of solute molecules on the various faces. In Part II. we shall examine the validity of relationships which are supposed to exist between the face developments and internal structures of crystals.

Summary.

Some of the factors influencing the habits of crystals are discussed. In many cases interaction with the solvent is important. The Gibbs condition for the stability of a crystal, viz., that for a given volume the total surface free energy shall be a minimum, is discussed. Evidence of various kinds is adduced to show that there is a definite equilibrium shape for crystals of a given solute crystallizing from a particular solvent, but it is also shown that a macroscopic crystal of non-equilibrium shape shows no tendency to approach this equilibrium shape unless growth is permitted. The equilibrium shape is a function not only of the internal structure of the crystal, but also of the interactions between the atoms (molecules) in the various faces of the crystal with the solvent and other molecules in the solution.

References.

Buckley, Proc. Manchester Lit. Phil. Soc. lxxxiii. p. 31 (1939).
Curie, *Bull. Soc. min. de France,* viii. p. 145 (1885).
France, J. Phys. Chem. xxxix. p. 425 (1938), xlii. p. 1079 (1938), xlv. p. 395 (1941).
Gibbs, 'Collected Works,' i. p. 320.
Herzfeld, Colloid Symposium, vii. p. 51 (1930).
Hulett, *Z. Phys. Chem.* xxxvii. p. 385 (1901).
Laue, *Z. Krist.* cv. p. 124 (1943).
Saylor, Colloid Symposium, v. p. 192 (1927).
Stranski, *Z. Krist.* cv. p. 90 (1943).
Stranski and Kaischew, *Z. phys. Chem.* B, xxvi. pp. 81, 114, 312 (1934), *ibid* B, xxxv. p. 427 (1937) ; *Physik. Z.* xxxvi. p. 393 (1935).
Valeton, *Z. Krist.* lx. p. 31 (1924).
Volmer, 'Kinetik der Phasenbildung,' p. 91 (Dresden, 1939).
Wulff, *Z. Krist.* xxxiv. p. 449 (1901).

Crystal Habit and Internal Structure.—II.

By A. F. Wells, M.A., Ph.D.

Introduction.

In Part I. we mentioned some of the factors which influence the habits of crystals, and the equilibrium shape of crystals was discussed. In the present paper we shall examine some aspects of the relation between the face development and the atomic structure of a crystal. During the past few years a considerable number of papers on this subject has been published, and it is important that this work should be viewed in proper perspective in relation to the more general problem which was outlined in Part I.

The Relation between Bravais Lattice, Space-group and Face Development.

In a crystal the structural units are arranged in conformity with one of the fourteen Bravais lattices. Bravais suggested that the face development of a crystal was closely related to its fundamental lattice, the faces being parallel to the net planes in the lattice and those planes most closely packed with lattice points being the most important forms† on the crystal. Friedel (1904, -05, -07) showed that this is often true. The Bravais "law" implies that the Miller indices of crystal faces will be small integers, and it is well known that forms with high indices are rare. It was not to be expected that the law would be universally valid for it took account only of the internal structure of the crystal, and it was realized that external factors such as the rate of crystallization and the presence of impurities in the solution could affect the habits of crystals.

The Bravais law was elaborated by Donnay and Harker (1937), who showed that not only the type of lattice but also the presence of screw axes and glide planes (other than those inherent in the lattice) should be taken into account in correlating face development with internal structure. The Bravais law states that the importance of a form $\{hkl\}$ is proportional to the reticular density (number of lattice points in unit area) or inversely proportional to the reticular area S_{hkl}. Since $S_{hkl} \cdot d_{hkl} = V$, where d_{hkl} is the spacing of successive planes hkl and V

† A form is a set of crystallographically equivalent faces related to one another by the symmetry elements.

is the volume of the unit cell, the importance of a form should be directly proportional to the interplanar spacing d_{hkl}. Donnay pointed out that since screw axes and glide planes involve translations, their presence alters the effective spacings of planes perpendicular to the former and of certain of the planes perpendicular to the latter. For example, a 3_1 axis changes the effective spacing of the (0001) plane from c to $c/3$, and since this is the only plane affected, the importance of the form {0001} relative to others is altered. It is evident that the planes affected in this way are just those of which certain orders of reflexion of X-rays are absent owing to destructive interference (the systematic absences given in space-group tables).

If the importance of a form $\{hkl\}$ is proportional to the spacing d_{hkl} it is possible to draw up for any crystal a list of planes arranged in order of decreasing d_{hkl}, and the position of a given plane in this " morphological aspect " (M.A.) should express the importance of the form relative to that of others. The Donnay–Harker extension of the Bravais law requires that instead of using in all cases the simplest Miller indices we use the multiple indices of the lowest order of X-ray reflexion compatible with the space-group symmetry. In propounding the new law the authors claimed only that it was an " improved approximation," for it is, of course, subject to the same limitations as regards external factors as was the original law of Bravais. In addition, Donnay and Harker pointed out that if the space-group symmetry were the only influential factor then all cubic crystals, for example, with the same space-group would exhibit the same face development, whereas this is not so.

A morphological aspect is defined as " a possible list of decreasingly important forms for any set of axial elements (axial ratios and inter-axial angles) in any given crystal system." Corresponding to the 230 space-groups there are 97 such morphological aspects (Donnay and Harker, 1940). Only in the cubic system can the planes be listed for a given M.A. without reference to a particular crystal ; in other systems the axial elements affect the order of the planes. In order to determine the space-group from the face development, the M.A. drawn up from observations on the crystals is compared with the above theoretical M.A.'s. In the case of minerals the experimental M.A. is drawn up in a standard way (Tremblay, 1942) by making a statistical survey of the frequency of occurrence and relative sizes of various forms on a large number of crystals from different localities. The presence of a plane as a cleavage was originally taken into account in assessing its position in the M.A., but this practice was discontinued in later work. Donnay and others have shown that in a number of cases there is good agreement between the observed face development and that predicted from the space-group, and that in some cases the space-group could be deduced (or confirmed) from morphological studies (Donnay, 1939). Crystals studied in this way include rhombic sulphur, pyrites, spinel and analcime (Donnay and Harker, 1937), stephanite (Taylor, 1940), garnet (Donnay and Faessler, 1941), scheelite (Donnay, 1942), goethite and lepidocrocite (Peacock, 1942), carborundum (Donnay, 1943), and muscovite (Peacock and Ferguson, 1943).

Now because relationships between space-groups and face developments such as that proposed by Donnay are admittedly of restricted validity, a critical examination of the " exceptions " should indicate some of the other factors determining the stability of crystal faces. We have to enquire (1) how we should expect reticular densities, as defined above, to be related to the stabilities of crystal faces, (2) whether the Donnay–Harker treatment gives a true estimate of relative reticular densities, and (3) what is the significance of the M.A. derived from a survey of observed face developments ?

The Significance of Reticular Density.

This point can only be discussed in detail with reference to particular crystal structures. Reticular density has been defined as the number of lattice points in unit area of a particular plane. We shall show, however, that the criterion of absolute coplanarity is not applicable when estimating the relative stabilities of different faces of a crystal (see the structure of diamond, below), and that there is no simple way of deciding which atoms shall be counted as " surface atoms."

It should perhaps be emphasized that a plane section through a crystal structure will seldom give a true picture of the contour of an actual crystal face. Apart from the mosaic structure and the defects of various kinds which are known to exist in the interiors of practically all crystals, we have to recognize that additional surface irregularities will generally occur. Random imperfections which are quite large on the atomic scale may be present on optically plane surfaces—see, for example, the work of Tolanski (1945) on the topography of crystal faces. In this connection reference may be made to a concept introduced by Stranski (1932). Fig. 1 represents a section through a simple cubic lattice (perpendicular to the *c*-axis) showing the traces of (100) and (110) planes. When counting the nearest neighbours of any atom we shall neglect two which lie respectively above and below each atom in this section. If we draw the (110) plane in the ordinary way we see that the outermost atoms have only two neighbours, while those in the next layer have four. There is also, however, an infinite set of planes $(110)_n$ with steps of different sizes. That with double steps, $(110)_2$, is shown in fig. 1, and mixed types with steps of different sizes can also be visualized. It will be seen that $(110)_2$ is intermediate between a (110) and a (100) plane, and this hybrid character becomes more marked as the size of the steps increases. In what follows we shall assume that a crystal face has the simplest possible type of contour, but it must be remembered that this is an idealized picture of an actual crystal face.

By taking account of glide planes and screw axes, Donnay and Harker imply that the " reticular densities " relevant to a discussion of the relative stabilities of crystal faces refer to the numbers of coplanar atoms per unit area. Their treatment allows only for symmetry operations which move atoms by amounts which are certain sub-multiples of the appropriate cell dimensions, viz. $1/n$ where n is 2, 3, 4, or 6, and which therefore lead to extinction of certain X-ray reflexions. However, any movement of an atom out of a plane affects the density of atoms in the plane if this is taken to mean the number of atoms exactly in the plane, though it does not, except in the cases mentioned above, lead to a space-group extinction. It does, of course, affect the intensity of X-rays from the plane, and if the intermediate atoms are suitably placed may lead to complete extinction of certain orders of reflexion. Since such extinctions are dependent on the value of a variable parameter they are not " space-group extinctions " and are not allowed for in the M.A. of Donnay and Harker. In other words, to obtain a true M.A. we should have to consider which atoms have variable parameters, *i. e.*

Fig. 1.

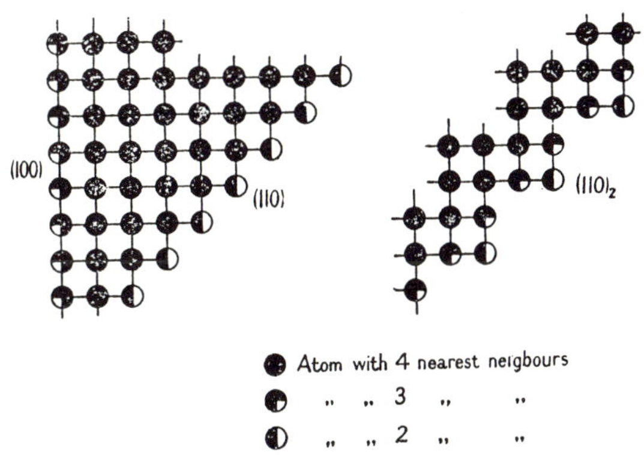

Contours of (100), (110) and $(110)_2$ faces (see text), showing environments of surface atoms.

whether they occupy special or general positions in the space-group. The fundamental difference between lattice halvings and those due to screw axes and glide planes arises from the fact that a lattice centring affects every atom in the unit cell in the same way, whereas this is not so in the case of symmetry elements additional to those inherent in the lattice. In the latter case the number of equivalent atoms related by the operation of the symmetry element is determined by the position of an atom relative to the symmetry element, which also determines the number of variable parameters required to define their positions. The densities of atoms in certain planes may be greatly enhanced if the parameters of the atoms have certain values, and in such a case the M.A. of Donnay and Harker may bear very little relation to the relative importance of the various forms. This point is very evident in the morphology of crystals containing planar molecules, and may be illustrated by reference to the structures of naphthalene, anthracene and benzoquinone.

The Morphology of Naphthalene, Anthracene and Benzoquinone.

These compounds all crystallize in the monoclinic system, space-group $P2_1/a$, and their structures and face developments are very similar. The most prominent forms are in all cases {001}, {110} and {20$\bar{1}$}. In Table I. the first eight planes of anthracene (Robertson, 1933) and benzoquinone (Robertson, 1935) are listed in order of increasing $\sin\theta$ (or decreasing d_{hkl}), the observed forms being marked with asterisks. These lists form the M.A.'s of Donnay and Harker. The agreement with the observed face development is obviously poor, particularly in the case of benzoquinone. The importance of {20$\bar{1}$} and the good (20$\bar{1}$) cleavage are due to the fact that the molecules lie in this plane. Accordingly (20$\bar{1}$) is the strongest X-ray reflection and the higher orders decrease uniformly in intensity. These two criteria together furnish the best test for coplanarity of atoms.

TABLE I.

Anthracene		Benzoquinone		Anthracene	
hkl	$\sin\theta$	*hkl*	$\sin\theta$	*hkl*	$S/\sin\theta$
001*	0·084	001*	0·135	110*	533
011	0·152	110*	0·159	001*	519
		011	0·176		
11$\bar{1}$	0·158			20$\bar{1}$*	410
110*	0·167	11$\bar{1}$	0·197		
002	0·168	111	0·222	210	352
		200	0·223		
20$\bar{1}$* }	0·184	020	0·226		
20$\bar{2}$ }		20$\bar{1}$*	0·238	etc.	
11$\bar{2}$	0·187				

If, therefore, we wished to draw up a true M.A. from the X-ray data it would not be sufficient merely to allow for the space-group absences, as do Donnay and Harker, but also to consider the intensities of the X-ray reflections, thereby taking into account the actual parameters of atoms in general positions. We could, for example, list the *first-order* reflections (others are irrelevant) in order of decreasing $S/\sin\theta$; where S is the geometrical structure factor for the plane (see Table I.). For a first-order reflection S is a measure of the coplanarity of the atoms, but it does not take account of the distribution of the atoms in the plane. For this reason $\sin\theta$ is introduced. Unfortunately, however, S is not a measure of the departures from coplanarity in *absolute* measure but only in terms of the spacing of each particular plane. We shall not pursue this point further as it will soon be evident that we cannot expect to find general relations between space-groups and face developments which will be applicable to all types of crystals.

We mentioned above that the criterion of absolute coplanarity is not of general applicability in estimating the stabilities of crystal faces. This point can be conveniently discussed with reference to the structures of diamond and zinc blende.

The Morphology of Diamond and Zinc Blende.

In the unit cell of diamond atoms occupy the two sets of positions : C_1 (000, $0\frac{1}{2}\frac{1}{2}$, $\frac{1}{2}0\frac{1}{2}$, $\frac{1}{2}\frac{1}{2}0$) and C_2 ($\frac{1}{4}\frac{1}{4}\frac{1}{4}$, $\frac{1}{4}\frac{3}{4}\frac{3}{4}$, $\frac{3}{4}\frac{1}{4}\frac{3}{4}$, $\frac{3}{4}\frac{3}{4}\frac{1}{4}$), *i. e.*, they form two interpenetrating face-centred cubic lattices related by a translation of ($\frac{1}{4}\frac{1}{4}\frac{1}{4}$). The space-group is F*d*3*m*. In addition to those for a face-centred lattice (*h*, *k* and *l* all odd or all even) there are additional halvings due to

Fig. 2.

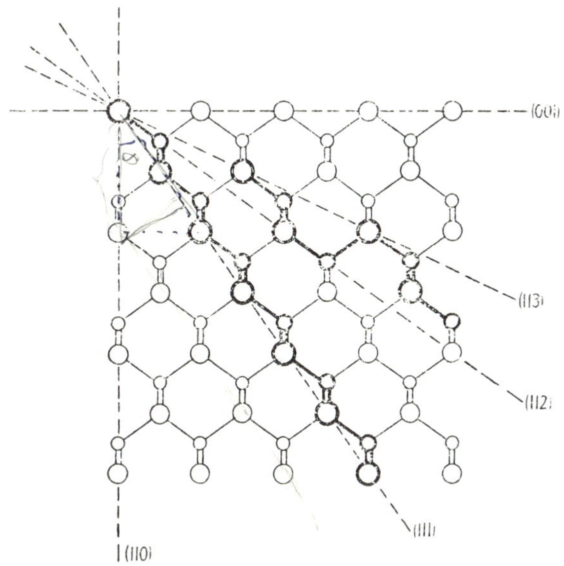

Projection of the crystal structure of ZnS (or diamond) on ($\bar{1}10$), showing the contours of (111), (112) and (113) faces. Zn and S (or C_1 and C_2) atoms are distinguished as large and small circles respectively.

the diagonal glide planes ($h0l$, $h+l=4n$, etc.). Allowing for these various halvings the M.A. is

111	110	113	001	133	112, etc.,
	(220)		(004)		(224)

giving first place to (111), which is the dominant form and also a perfect cleavage. The agreement would appear to be satisfactory, but we shall see that (111) is not the plane most densely packed with *coplanar* atoms, although it comes first in the above series. In the zinc blende structure the Zn and S atoms occupy the above sets of equivalent positions, the space-group is F$\bar{4}$3*m*, and (111) occupies first place in the M.A. as in the case of diamond. However, the common forms on blende are {110},

{111}, and {113}, probably {110} most important, and the cleavage is (110), perfect. These differences between two crystals with the same atomic arrangement call for explanation.

From fig. 2 it will be seen that the S atoms fall exactly *in* some but *between* other planes of Zn atoms. The second column in Table II. gives the numbers of Zn atoms per unit area in the various planes, the third column the type(s) of atom lying *in* the plane, and the fourth column the total numbers of atoms per unit area. It is seen that (110) is the plane most densely populated with atoms (Zn+S), and the next plane in which Zn and S are coplanar is (112). (For these planes the figures in column 4 are twice those in column 2.) The importance of (111) both as a face and a cleavage of diamond shows that the criterion of absolute coplanarity is not the one determining stability (see fig. 2). In spite of the fact that C_1 and C_2 are strictly coplanar for (112) but not for (113), the latter is the more stable face because the contour of a (113) face will obviously be that shown in fig. 2, containing both C_1 and C_2. The difference between the stabilities of (110) and (111) for diamond and ZnS must be associated with the partial ionic character of the bonds in the latter.

TABLE II.

hkl	Zn/a^2	Atoms in Plane	Total Atoms/a^2
111	2·31	2·31 Zn or 2·31 S	2·31
100	2·00	2·00 Zn or 2·00 S	2·00
110	1·41	1·41 Zn + 1·41 S	2·82
113	1·21	1·21 Zn or 1·21 S	1·21
133	0·92	0·92 Zn or 0·92 S	0·92
210	0·89	0·89 Zn or 0·89 S	0·89
112	0·82	0·82 Zn+ 0·82 S	1·64

If we count as surface atoms for (111) all those shown (for this plane) as heavy circles in fig. 2, then one-half are forming four and the other half three bonds (a total of 4·62 atoms/a^2), compared with 2·82/a^2 all forming three bonds in (110), a being the side of the unit cell. Thus for diamond (111) is more stable than (110). In the case of ZnS, however, the outer atoms of (111) are all Zn (or all S) forming only three bonds, so that there is a resultant positive or negative charge (compare (111) of NaCl). On the other hand (110) contains equal numbers of Zn and S atoms all with the same residual charge, so that the surface is electrically neutral. A (111) cleavage of ZnS would mean the separation of oppositely charged surfaces (contrast the effect of a (110) cleavage), whereas any cleavage of diamond leaves two neutral carbon surfaces and a (111) cleavage breaks the smallest number of C–C bonds per unit area.

We shall now consider in turn three structures which all have the same space-group, F*m*3*m*, viz., the face-centred cubic, the rock-salt and the fluorite structures.

The Face-centred Cubic Structure.

In this structure (cubic close-packing) every atom has twelve equidistant nearest neighbours. An atom in the surface of the crystal has a smaller number of neighbours, this number depending on the indices of the face. Not only are the atoms in the outermost layer affected but in general also those in layers below the surface. This may be seen from fig. 3, which shows the plan of the structure on ($\bar{1}10$) and the traces of certain planes which are perpendicular to the plane of the paper. The outermost atoms in a (112) face, for example, have only seven nearest neighbours (two are not shown, there being one above and one below the plane of the paper), those in the next layer have nine, and in the next ten.

Fig. 3.

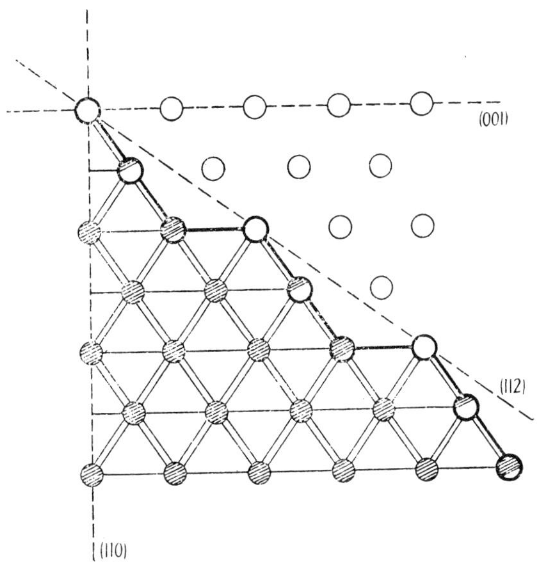

The face-centred cubic structure projected on ($\bar{1}10$), showing the environments of atoms forming a (112) face.

Only when we come to the fourth layer of atoms below the surface is the environment the same as in the interior of the crystal. These differences are shown by the shading of the circles in fig. 3.

In Table III. the planes in this structure are listed in order of decreasing d_{hkl}, and the environments of the surface atoms are given. The indices used in calculating the spacings of the planes are those consistent with the conditions for reflection of X-rays from a face-centred lattice, viz., *h*, *k* and *l* all odd or all even. The planes are arranged in order of decreasing density of packing of atoms in the planes, *i. e.*, they form the M.A. for the space-group F*m*3*m*.

The Rock-salt Structure.

In this structure the co-ordinates of each type of ion are those of cubic close-packing, and in the interior of the crystal each ion is surrounded octahedrally by six of the other kind. Table IV. shows the planes arranged in order of decreasing density of packing of Na *or* Cl ions. It shows, for example, that (111) is the plane most closely packed with ions *of one kind.* This is not, however, the criterion for stability of a face on a NaCl crystal. It is likely that electrical neutrality is the criterion in the case of an essentially ionic crystal, or alternatively, that the surface ions have

TABLE III.

hkl	$d_{hkl}=1/(h^2+k^2+l^2)^{\frac{1}{2}}$	Neighbours of Surface Atom
111	0·577	9
200 (100)	0·500	8
220 (110)	0·353	7 (11)
113	0·301	7 (10)
133	0·229	7 (9) (11)
420 (210)	0·224	6 (9) (11)
224 (112)	0·204	7 (9) (10)

TABLE IV.

hkl	Nature of Plane	Total No. of Ions in the Area a^2	Number of oppositely charged Neighbours of an Ion in the Face
111	Na or Cl	2·31	3
200	Na+Cl	4·00	5
220	Na+Cl	2·82	4
113	Na or Cl	1·21	3 (5) (5)
133	Na or Cl	0·92	3 (4) (4)
420	Na+Cl	1·79	4 (5)
224	Na+Cl	1·63	3 (5)

the maximum number of neighbours of opposite charge. Table IV. shows the environments of the ions in different faces of a crystal of NaCl, the planes being arranged in the same order as in Table III. The third column gives the total numbers of ions in the area a^2, a being the length (in A. units) of the side of the unit cell. The planes fall into two groups, and the figures in which we are interested are those to the right in column 4. We have, therefore, two possibilities for a compound AX with the rock-salt structure. If the condition for stability of a face is that it contains equal numbers of A and X atoms, then (100) is the most stable face. If, for any reason, the most stable face is one consisting of atoms of one kind only, then (111) will occur. This seems to apply to the less ionic CdO. It appears that this crystal is most stable if its outer surface consists of oxygen atoms, for {111} is the dominant form. The same development will be observed if interaction with the solvent or impurity in the solution stabilizes the face consisting of all A or all X ions—see the discussion (below) of the habit of NaF from water and NaCl from solutions containing OH^- ions or $CO(NH_2)_2$.

The Fluorite Structure.

The natures of the faces of a crystal of CaF_2 are shown in Table V. Only three planes are given as they are sufficient to illustrate the point we wish to emphasize. It is assumed that the contour of a (111) face would be that indicated in fig. 4, *i. e.* it is electrically neutral with an outer surface consisting of practically coplanar Ca and F ions. This should be the most stable face if the criterion is, as in the case of NaCl, electrical neutrality. If the condition for stability were the greatest density of F ions then (100) should develop. The mineral fluorspar is almost invariably in the form of cubes, and the same development is

TABLE V.

hkl	Nature of Plane	Numbers of Ions in Area a^2	Environment of Ion in Surface
111	Ca+F	2·3 Ca+2·3 F	Ca (7F) and F (3Ca)
100	Ca *or* F	2 Ca or 4 F	Ca (4F) or F (2Ca)
110	Ca+2F	1·4 Ca+2·8 F	Ca (6F) and F (3Ca)

shown by crystals grown by the slow diffusion of solutions of $CaCl_2$ and NH_4F. The development of (100), instead of the more densely populated and electrically neutral (111) or (110), is in striking contrast to the development of (100) on NaCl, and is referred to again later. The perfect (111) cleavage is consistent with the assumption that the cleavages of simple ionic crystals result in the separation of electrically neutral surfaces by the rupture of the smallest possible number of bonds per unit area, or alternatively the separation of the most densely populated electrically neutral planes. The (111) cleavage of CaF_2 breaks 4·62 bonds/a^2, but a (110) cleavage 5·67 bonds/a^2. Similarly, a (100) cleavage of NaCl breaks 4 bonds/a^2, but a (110) cleavage 5·6 bonds/a^2. (An equation for the cleavability of ionic minerals proposed by Shappell (1936) fails to account for the cleavages of these two simple crystals, though it has some success with more complex structures.)

In general we should expect some relation between the ease of cleavage parallel to a particular plane and the stability of that plane as a crystal face for crystals throughout which all the bonds are of the same kind (NaCl, ZnS, diamond, etc.). Where there are pronounced differences between the types and/or strengths of bonds in different directions in the crystal, good cleavages are often observed, as in the case of many crystals with layer structures. It does not follow, however, that these cleavage planes are stable in contact with the saturated solution of the substance in a particular solvent. We have already noted in Part I. that the (001) cleavage plane of pentaerythritol is highly unstable in a saturated aqueous solution.

Conclusions: Relation between Space-group and Morphology.

These considerations lead to the following conclusions :—

(1) The use of "reticular densities" is not satisfactory in this connection. The Donnay-Harker treatment implies that only atoms

Fig. 4.

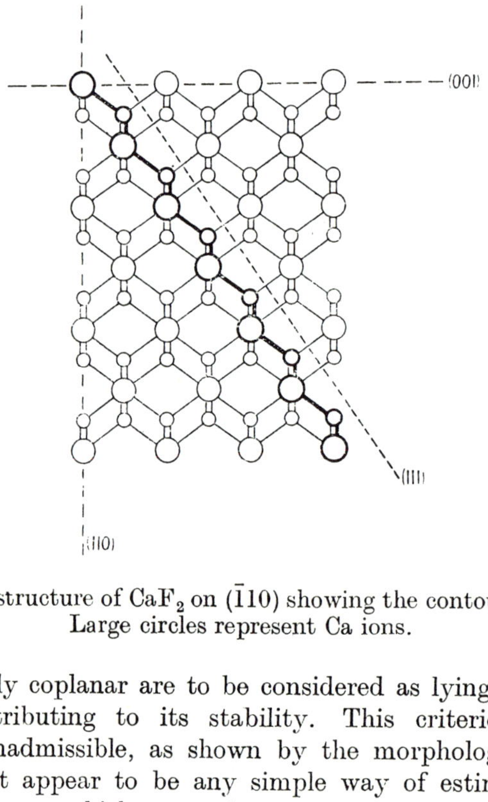

Projection of the structure of CaF_2 on $(\bar{1}10)$ showing the contour of a (111) face. Large circles represent Ca ions.

which are exactly coplanar are to be considered as lying in a particular plane and contributing to its stability. This criterion of absolute coplanarity is inadmissible, as shown by the morphology of diamond. There would not appear to be any simple way of estimating, directly from the space-group, which atoms should be counted as " surface atoms " for a particular face.

(2) The criterion for stability of a face is not necessarily the density of atoms per unit area regardless of their nature, so that if the crystal contains atoms of more than one kind it is necessary to know which type of atom or combination of atoms forms the most stable surface. This will depend on the nature of the crystal. For an ionic crystal the condition appears to be electrical neutrality, provided that there is no

preferential interaction between the solvent and any of the constituent ionic species (see the discussion of NaF, below). If, on the other hand, there is strong interaction between certain atoms or certain parts of the molecules of a solute with molecules of the solvent, then the most stable faces will tend to be those containing the greatest concentrations of these active atoms or groups, and these are not necessarily the most densely populated planes.

(3) Even within the above limitations, a " morphological aspect " which takes account only of space-group halvings (due to lattice-centring, glide planes and screw axes) does not necessarily give a true indication of the relative importances of faces of different types, though it may give a good approximation in some cases.

(*a*) The M.A. does not allow for the difference in stability of pairs of parallel faces in cases of hemihedral symmetry, since their indices have the same numerical values (see CuI, below).

(*b*) If all the atoms in the unit cell are in positions exhibiting the symmetry of the space-group, the M.A. will give a measure of the relative stabilities of the faces *with the proviso* that much higher concentrations of atoms may occur in certain planes owing to the values of the parameters. This is not allowed for if we take account only of the space-group absences, but will show itself in the relative intensities of the X-ray reflexions.

(*c*) If some of the atoms are in special positions of a space-group other effects may arise. For example, the atoms of one kind may be arranged in accordance with the symmetry of another space-group (often of higher symmetry), as may be seen from the lists of Gitterkomplexe in the 'International Tables of Space-groups.' If these atoms are more numerous than the remaining atoms then the M.A. will not place the planes in order of decreasing density of atoms/unit area (*cf.* CaF_2, above). Cuprite, Cu_2O, illustrates this point well. The space-group is P*n*3*m*. In this crystal the Cu atoms are arranged on a face-centred lattice (F) and the O atoms on a body-centred lattice (I). Accordingly, we obtain different M.A.'s depending on whether the criterion is the number of Cu+O, of Cu, or of O atoms per unit area :—

P*n*3*m* - -	011	111	001	112	122, etc.
F(Cu) - -	111	001	011	113	133
I(O) - -	011	001	112	013	111

(*d*) If the structure consists of two or more sets of atoms of different kinds arranged in the same way but the sets being related to one another by a simple translation which does not introduce additional halvings, then the M.A. will correspond to the density of packing of atoms of one kind only. Thus the M.A. for NaCl relates to the density of packing of Na *or* Cl, but not of Na+Cl.

(4) *The Significance of the Experimental M.A.*—We shall not discuss the experimental M.A. in any detail since it would be difficult

to find any theoretical justification for it. The face developments of natural crystals grown from solution may have been influenced by a number of factors which are not taken into account in deriving the theoretical M.A.: (i.) the crystals have grown from solutions or melts of unknown composition, and it is well known that some impurities affect crystal habit, or (ii.) the observed development may represent an intermediate phase in the growth (or dissolution) of a crystal of type (i.), the conditions having changed during growth. (Variations in shape due to rapid crystallization can probably be ignored in the case of minerals.) The fact that it is possible to draw up an "observed M.A." for NaCl, for which the equilibrium shape is certainly the cube, shows that either or both of these factors are involved. The significance of the relative positions of the faces in such a series is therefore not known.

The Crystal Habits of Certain Halides.

We shall now review, in the light of the ideas outlined above, the habits of certain halides, taking these as examples of solutes on the habits of which solute-solvent interaction has in general little effect.

(a) *Halides with the Rock-salt Structure.*

We have seen that (1) the most densely populated plane is the cube face (100), which contains equal numbers of positive and negative ions, and (2) the octahedron face (111) differs from the more closely packed (100) and (110) planes in that it contains ions all of the same type. We should therefore expect that in the absence of any complicating factors crystals with this structure will grow as cubes, but that any strong interaction between either the positive or negative ions and the solvent or added impurity will lead to the development of (111) faces.

All the halides of lithium, sodium, potassium and rubidium, and also CsF and NH_4I, crystallize with this structure. The chlorides, bromides and iodides crystallize as cubes from aqueous solution, melt or vapour. According to the literature, additions of other solutes cause habit modifications of which the following are typical:—NaCl, {111} favoured by NaOH, H_3BO_3, $CaCl_2$ and $FeCl_3$: {110} by $HgCl_2$. Certain organic compounds also favour the development of {111}, notably urea. Experiments on the growth of NaCl crystals (*e.g.* Spangenberg (1924), Schnorr (1928), Neuhaus (1928)) showed that various faces develop on a sphere of NaCl during growth, and that the rates of growth normal to the different faces are: $v_{100}<v_{111}<v_{210}<v_{110}$. The final product is a cube.

Data on the fluorides are less complete and far less satisfactory. Just as the chemistry of fluorine is very different from that of the other halogens owing to the extreme electronegativity of the smallest halogen, so there are some striking differences between fluorides and the other halides. Of particular interest here are the following. Although NaCl, NaBr and NaI crystallize from water as well-defined cubes, NaF separates as ill-defined crystals with poorly-developed faces. Again, when some of the sparingly soluble fluorides such as CaF_2 or MgF_2 are precipitated from aqueous solution, they have a great tendency to form colloidal suspensions or gels. This is also true of some complex fluorides, *e.g.* silicofluorides and titanifluorides. These facts would appear to be related to the strong tendency of fluorine to form hydrogen bonds, and indicate greater interaction between the fluorides and water than between the other halides and this solvent.

An elaborate study has been made by Frondell (1940) of the effects of some 143 "cosolutes" on the habit of NaF crystallizing from aqueous solution. He gives few experimental details, and some of his assertions seem open to question. For example, he states that NaF crystallizes from water as cubes at all temperatures from 22–100° C. It is well known that NaF, unlike the other sodium halides, hydrolyzes ($H_2O+2F^- \rightarrow [HF_2]^- + OH^-$) so that it is difficult—if not impossible—to obtain a neutral solution of this salt. It may easily be verified that octahedral crystals of NaF separate from a boiled saturated solution which is left to stand for some days at room temperature. Such a solution is, however, distinctly alkaline. There is no doubt that the effect of OH^- ions on NaF is much greater than on the other sodium halides, but the statement of Frondell that "the habit-varying effect of the OH^- ion is peculiar to the fluorides NaF, LiF, and KF (RbF and CsF were not studied)" can easily be disproved. Frondell wished to show that the effect of OH^- ions is due to *substitutional* adsorption of OH^- for F^- (these ions having approximately the same radius) in the surface of the growing crystal. The fact that OH^- ions also lead to the development of (111) faces on NaCl and KBr obviously does nothing to strengthen such a theory. Although there are not yet sufficient data to show that the interaction of F^- ions with the solvent (H_2O) has a decisive effect on the habits of fluorides, nevertheless the above facts together with the cubic habit of CaF_2 do suggest that such interaction is much more pronounced than in the case of the other halides.

(b) *Halides with the Cæsium Chloride Structure.*

This structure consists of two interpenetrating simple cubic lattices, each ion being surrounded by eight of opposite charge, situated at the corners of a cube. The compositions of the faces are set out in Table VI. We should expect crystals with this structure to develop (110) faces (dodecahedron). The next most stable face should be (211), for although (100) is more densely populated than (211) it contains ions all of the same sign. The presence of ions or molecules which are preferentially adsorbed on a face consisting of all positive or all negative ions should accordingly encourage the development of cube faces. Groth records the habits of those cæsium and ammonium halides with this structure as follows :—

CsCl.—from aqueous solution {100} or {110}, CsBr and CsI, {110}

Without further data on CsCl it is not possible to say more than that there is general agreement with expectation.

NH_4Cl and NH_4Br.—(Ammonium fluoride has the wurtzite (ZnS) structure and NH_4I the rock-salt structure at ordinary temperatures).

NH_4Cl.—On natural crystals {110} and {211}. The form {211} is commonly the only one on crystals grown from solution, forming distorted icositetrahedra. From aqueous solution containing $FeCl_3$ this is the only form developed, but when crystallized from pure aqueous solution skeletal growths are normally obtained.

NH_4Br.—From aqueous solution {211}. From solutions containing urea, $CrCl_3$ or $PbBr_2$, {100}.

The peculiar habits of NH_4Cl and NH_4Br are presumably connected with the tetrahedral shape of the ammonium ion. At higher temperatures both these salts, like the iodide at ordinary temperatures, have the rock-salt structure, in which the NH_4^+ ion has attained spherical symmetry by free rotation.

TABLE VI.

hkl	Ions in Face	Total Number of Ions in Arbitrary Unit Area	Environments of Surface Ions
110	M+X	3·46	6
100	all M *or* X	2·45	4
211	M+X	2·00	5 (7)
130	M+X	1·55	4 (6)
111	all M *or* X	1·42	4 (7) (7)

The presence of urea in an aqueous solution of NH_4Cl causes a change in habit from skeletal growths to compact cubes, and a similar effect is observed with NH_4Br. In the case of NaCl, which normally crystallizes as cubes, urea causes the development of octahedral faces. The effect is smaller with KCl and, according to Retgers, urea has no effect on the habits of KBr or KI. These facts indicate that although it is true that urea favours the development of those planes which contain ions all with the same charge, any explanation based solely on the electrical nature of the faces is not adequate. Many attempts have been made to interpret changes in habit due to foreign ions or molecules—see, for example, the work of Bunn (1933) and Royer (1934) on the relation between habit modification, mixed crystal formation and oriented overgrowths. It is equally doubtful if purely geometrical theories based on similarities in interatomic distances in certain planes of the crystalline solute and of the *crystalline* foreign material can be of general validity. In some cases (*e. g.* the effect of dyestuffs on alum, K_2SO_4, etc.) it is likely that the effects are due to interactions between particular atoms or groups of the adsorbed molecule and atoms in the crystal face, *i.e.* that intra- as well as inter-molecular distances are important. In cases where comparatively large concentrations of habit-modifying material are present, it may well be necessary also to take into account the interactions between this material and the solute and solvent molecules in the solution.

(c) *The Tetramethylammonium Halides.*

The chloride, bromide and iodide are isomorphous : tetragonal, space-group P4*nmm*/, $c : a = 0{\cdot}7108$, $0{\cdot}7131$ and $0{\cdot}7223$ respectively. The chloride will not be further considered here. The bromide crystallizes as bipyramids {101} (fig. 5, *a*), but the iodide with {100} and {111} (fig. 5, *c*). These habits are related to the internal structures of the crystals in very much the same way as for the ammonium halides. It is seen that $a \gtrsim \sqrt{2}c$, so that the crystals can be referred to pseudo-cubic axes, and the structure may be described as a deformed CsCl structure.

Fig. 5.

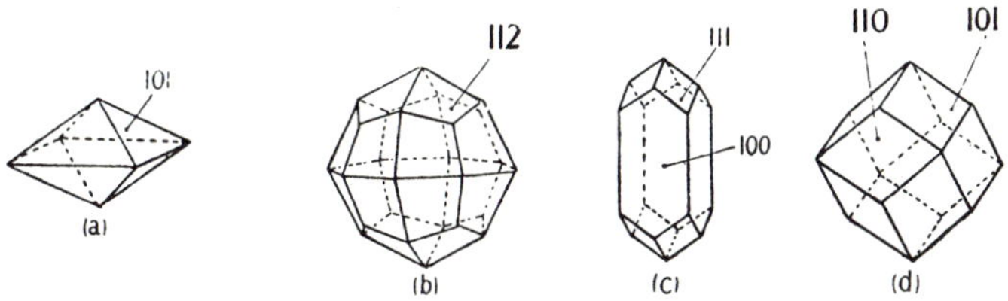

(*a*) Tetramethylammonium bromide, (*b*) equivalent cubic form, (*c*) tetramethylammonium iodide, (*d*) equivalent cubic form. Pseudo-cubic indices are shown in larger figures.

When referred to the pseudo-cubic axes the faces have the indices shown below :—

	Tetragonal.	Pseudo-cubic.
Tetramethylammonium bromide - -	101	112
,, ,, iodide - -	{100	110
	{111	101

i. e., the bromide develops the same (112) faces as NH_4Br but only eight, forming a tetragonal bipyramid, instead of the 24 forming an icositetrahedron, while the iodide develops 8 (111) and 4 (100) which, when equally developed, form a dodecahedron—compare the dodecahedral development of cæsium halides. It is interesting that while the iodide develops these (111) and (100) faces, which are the most densely populated planes containing equal numbers of positive and negative ions, the bromide develops the next most densely populated (101) planes, just as NH_4Cl commonly grows (211) instead of the expected (110).

(d) *Certain Halides with Close-packed Structures.*

It is convenient to consider together two groups of halides in which (*a*) the halogen atoms and (*b*) the halogen and certain of the metal atoms are arranged in cubic close-packing. Although the chlorine ions in NaCl are in the positions of cubic close-packing, this structure cannot be regarded as a close-packed lattice of Cl^- ions with Na^+ ions in the interstices—witness the Cl—Cl distances in

	LiCl	NaCl	KCl	RbCl
Cl—Cl - -	3·64	3·98	4·44	4·65 A.

On the other hand, the approximate constancy of the I—I distances in the following halides shows that we are justified in regarding them as based on a close-packed halogen lattice :—

	CuI	SnI_4	GeI_4	NiI_2
I—I - -	4·28	4·32	4·20	4·18 A.

(GeI_4 is isomorphous with SnI_4; NiI_2 crystallizes with a non-cubic structure which we shall not deal with here.)

Halides of the first group include

(1) those with the zinc-blende structure, *e. g.* CuCl, CuBr and CuI,

(2) molecular halides, *e.g.* SnI_4.

If we refer the cubic close-packed halogen lattice to the usual face-centred cubic unit cell of side a, the relative densities of atoms in the simpler planes are as follows :—

hkl - - -	111	100	110	113	133	210
Atoms/a^2 - -	2·31	2·00	1·41	1·21	0·92	0·89

According to the literature CuCl forms tetrahedra from solution in hot hydrochloric acid and $\{111\}+\{1\bar{1}1\}$ when prepared by the action of sulphur dioxide on cupric chloride, and both CuBr and CuI crystallize as tetrahedra. SnI_4 forms octahedra {111} with small {210} from solution in carbon disulphide.

Two points call for explanation—the development of tetrahedra instead of octahedra in the case of the cuprous halides, and the presence of {210} on SnI_4, this face being very much less closely packed with halogen atoms than the intermediate (100), (110), (113) and (133) planes. So far we have not taken into account the metal atoms. In a close-packed arrangement of atoms the largest holes are between six and four close-packed atoms respectively, and we shall refer to these as " octahedral " and " tetrahedral " holes. In CuI one-half of the tetrahedral holes are occupied. In fig. 6 are shown projections of the structure on $(\bar{1}10)$ with (111) and $(\bar{1}\bar{1}\bar{1})$ planes perpendicular to that of the paper. Since we are viewing the structure along the edge of a tetrahedral CuI_4 or ICu_4 co-ordination group, only three of the four nearest neighbours of each atom are visible. If we assume an outer layer of halogen atoms for both (111) and $(\bar{1}\bar{1}\bar{1})$, then in a (111) plane each of these atoms would be attached to three Cu atoms, instead of four as in the interior, but in the parallel $(\bar{1}\bar{1}\bar{1})$ plane to only one Cu atom (fig. 6, *a*). Alternatively the $(\bar{1}\bar{1}\bar{1})$ face could have an outer surface of

Cu atoms, each attached to three halogen atoms (fig. 6, *b*). In either case the structure and stability of (111) is different from that of ($\bar{1}\bar{1}\bar{1}$) and the tetrahedral development results.

In SnI_4 the arrangement of the metal atoms, in one-eighth of the tetrahedral holes, is such that there is no difference between parallel pairs of (111) faces. In a crystal consisting of discrete molecules we must, however, in estimating the relative stabilities of faces take into

Fig. 6.

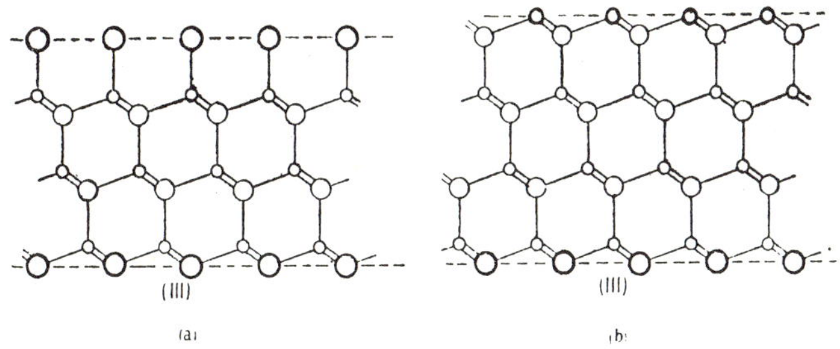

Projections of the structure of CuI on ($\bar{1}$10) showing possible contours of (111) and ($\bar{1}\bar{1}\bar{1}$) faces (see text).

Fig. 7.

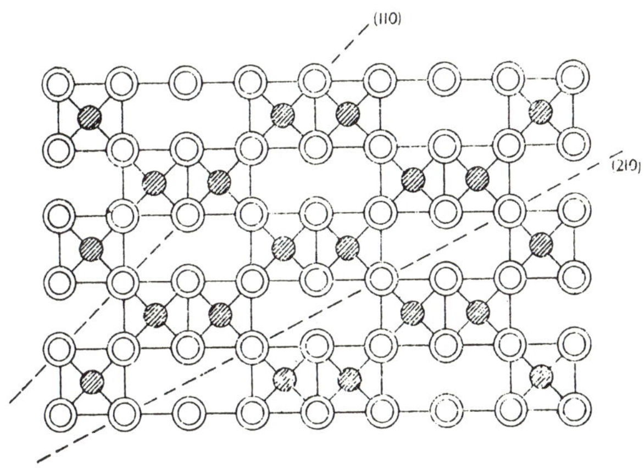

Projection of the structure of SnI_4 on (001). Shaded circles represent Sn atoms.

account the relation of the various faces to the molecules. The projection of the structure of SnI_4 on (001) is shown in fig. 7, where it is seen that the trace of the (110) plane, for example, cuts through many molecules, so that a (110) face would have a very irregular contour. This is not so for the (100) or (210) planes. In other words, the relative densities of *molecules* in the various possible faces are different from those of *halogen atoms*. As regards the numbers of molecules of SnI_4 per unit area, the most densely populated planes in this structure are (111), (100) and (210), with respectively 2·31, 2·00 and 1·79 molecules/a^2, where a is the side of the unit cell. These figures should be compared with those given above. At present it is not possible to predict whether a combination of {111} and {100} or of {111} and {210} would be more stable.

The characteristic feature common to a number of complex halides $A_xB_yX_z$ is that the larger positive ions A together with the halogen ions X form a close-packed array with the smaller positive ions B in the interstices. In such halides the ratio $z:x=3$. Examples are $KMgF_3$, K_2PtCl_6 and $Cs_3Tl_2Cl_9$. Little is known of the habit of $KMgF_3$, and $Cs_3Tl_2Cl_9$ is rhombohedral and therefore cannot be directly compared with the cubic halides under discussion. The compositions of the three most densely populated planes are shown below. Accordingly it is found that K_2PtCl_6, K_2SiF_6, Rb_2SiF_6, etc., all with the same structure, crystallize from water as octahedra.

hkl	Proportions of A and X in Face	Total number of Atoms per Unit Area
111	A : 3X	2·31
100	A : X	2·00
110	A : X	1·41

SUMMARY.

The recent work of Donnay and Harker and others on the relation between the face developments and space-groups of crystals is reviewed. The morphologies of anthracene, diamond, zinc blende, rock-salt and fluorite are considered in some detail. Apart from the fact that morphological relationships which do not take account of the solvent are necessarily only of limited validity, it is shown that there are other complications, of a purely crystallographic nature, which severely restrict the usefulness of the Donnay–Harker generalization. The crystal habits of some metallic halides are discussed in relation to their crystal structures.

References.

Bunn, Proc. Roy. Soc. cxli. A, p. 567 (1933).
Donnay, Amer. Min. xxiii. p. 5, *ibid.* xxiv. p. 184 (1939).
Donnay, Trans. Roy. Soc. Canada, xxxvi. sec. 4, p. 37 (1942).
Donnay, *ibid.* xxxvii. sec. 4, p. 43 (1943).
Donnay and Faessler, Univ. Toronto Studies, Geol. Ser. xlvi. p. 19 (1941).
Donnay and Harker, *Compt. rend. Acad. Sci. Paris,* cciv. p. 274 (1937).
Donnay and Harker, Amer. Min. xxii. p. 446 (1937).
Donnay and Harker, *Naturaliste canadien,* lxvii. p. 33 (1940).
Friedel, *Compt. rend.* cxxxix. pp. 221, 314 (1904).
Friedel, Bull. Soc. franc. Min. xxviii. p. 6 (1905).
Friedel, *ibid.* xxx. p. 326 (1907).
Frondell, Amer. Min. xxv. p. 339 (1940).
Neuhaus, *Z. Krist.* lxviii. p. 15 (1928).
Peacock, Trans. Roy. Soc. Canada, xxxvi. sec. 4, p. 107 (1942).

Peacock and Ferguson, Univ. Toronto Studies, Geol. Ser. xlviii. p. 65 (1943).
Robertson, Proc. Roy. Soc. cxl. A, p. 79 (1933).
Robertson, *ibid.* cl. A, p. 106 (1935).
Royer, *Compt. rend.* cxcviii. pp. 185, 585 (1934).
Schnorr, *Z. Krist.* lxviii. p. 1 (1928).
Shappell, Amer. Min. xxi. p. 75 (1936).
Spangenberg, *Z. Krist.* lix. p. 375 (1924).
Stranski, *Z. phys. Chem.* xvii. B, p. 127 (1932).
Taylor, Amer. Min. xxv. p. 327 (1940).
Tolansky, Proc. Roy. Soc. clxxxiv. A, pp. 41, 51 (1945).
Tremblay, J. Wash. Acad. Sci. xxxii. p. 327 (1942).

19

Reprinted from *Am. Mineralogist,* **32**(11–12), 593–606 (1947)

THE RELATIVE IMPORTANCE OF THE SEVERAL FACES OF A CRYSTAL

M. J. BUERGER, *Massachusetts Institute of Technology, Cambridge, Massachusetts.*

ABSTRACT

A number of rules have been proposed to correlate the relative importance of the several faces of a crystal with the geometrical features of the crystal lattice. Chief of these are the rule of simplest indices, Bravais' rule, and Donnay-Harker's rule. The latter has had considerable success in formulating the relative importance of the faces of a crystal, but it is completely empirical, as are all the other rules. This paper suggests a rational basis for the rule. It is shown that if the crystal grows by the accretion of molecular chunks which are equivalent by an operation with a translation component, then Bravais' rule can be derived from crude surface energy considerations. These considerations also explain the stability of crystal faces and predict that crystals should form without faces when they grow from media of equal surface tension. The same theory accounts for the known deviations from Bravais' rule by ionic crystals.

INTRODUCTION

The faces displayed by a crystal represent a boundary between the crystal and the medium which deposited the crystal. Thus it should be obvious that the character of the collection of faces displayed by a crystal is determined by both the crystal and the depositing medium, or, more generally, by the crystal and its environment.

There is plenty of experimental evidence that the relative sizes of the several faces of a crystal can be influenced by the environment depositing the crystal. Buckley[1] and Bunn[2] have been active in this field and have provided rival structural explanations for the mechanism of the influence of impurities in the depositing solution on the variation of crystal habit. It is also known that temperature, pressure, and rapidity of growth (i.e., other aspects of the general environment of growth) produce variations in crystal habit. While the nature of the external influences other

[1] Buckley, Harold E., The influence of RO_4'' and related ions on the crystalline form of sodium chlorate: *Zeits. Krist.* (A) **75,** 15–31 (1930).

[2] Bunn, C. W., Adsorption, oriented overgrowth and mixed crystal formation: *Proc. Roy. Soc.,* **141A,** 567–593 (1933).

than impurities have not been extensively discussed, it is not difficult to advance reasonable explanations for them.

It has also been recognized that the crystal itself has an influence on its habit, and, in a crude empirical way, the general characteristics of the influence are known. Yet those crystallographers interested in this aspect of morphology have been content with the empirical relation and have not attempted to give any rational explanation of it. It is the chief purpose of this paper to suggest such an explanation.

Attempts to Formulate the Influence of the Crystal

Attempts to formulate the characteristics of the crystal's influence on its own form development arose early in the history of crystallography. All such attempts were empirical formulations which sought to relate the form development to some aspect of the lattice geometry (as distinct from structure or bonding).

Perhaps the crudest approximation is the rule that faces with simplest indices are the best developed and that the importance of the faces decreases with increasing complexity of the indices. This approximation dates from Haüy's time[3] and has even been called the "Law of Haüy." There are other formulations which are essentially equivalent to this rule. Thus, Barker's *Principle of Simplest Indices*[4] provides that the important faces of a crystal can be assigned simple indices, and this implies, in turn, that importance in development of a face is related to the simplicity of the indices. Again, Goldschmidt's Law of Complication,[5] which recently achieved some popularity in America,[6] implies this relation.[7] The simplest-indices rule has obvious shortcomings. For example, it requires all isometric crystals to have identical form development, and it requires *all* crystals to have form developments which are related to one another by homogeneous deformations, which is obviously contrary to observation.

A much better approximation was discovered by Bravais.[8] Bravais' rule states that the relative importance (implying both frequency of occurrence and area) of the development of the faces of a crystal is in the

[3] Haüy, René Just, *Cristallographie*, 2nd ed., vol. **2**, p. 155.

[4] Barker, T. V., *Systematic Crystallography* (Thomas Murby & Co., London 1930) p. 2 *et seq.*

[5] Goldschmidt, V., Über Entwickelung der Krystallformen: *Zeits. Krist.*, **28**, 1–35 and 414–451 (1897).

[6] Peacock, M. A., Calaverite and the law of complication: *Am. Mineral.*, **17**, 317–337 (1932).

[7] Buerger, M. J., The law of complication: *Am. Mineral.*, **21**, 702–714 (1936).

[8] Bravais, M. A., Études cristallographiques: *Jour. de l'École Polytechnique*, 167 (**1851**).

same order as their reticular density (or, what is the equivalent, in the same order as their interplanar spacing).

An alternative form of Bravais' rule is attributed to Fedorov. This is that the habit of a crystal is approximately that of the shape of a cell of its polar (reciprocal) lattice. This can be easily shown to follow directly from Bravais' rule, but it is less comprehensive in that it merely gives the order of development of the three important pinacoidal enclosing forms.

As Bravais' rule came into vogue in America, Goldschmidt's "Law of Complication" waned. The protagonists of Bravais' rule enthusiastically proclaimed it "The Law of Bravais" and gave the impression that it was a law of nature on the same plane as one of the laws of thermodynamics. This generality of the "law" was supported by analyses of the form developments of certain individual crystals favorable to the rule.

That the "Law" of Bravais was not quite as perfect as its proponents claimed was suspected as the numerous exceptional crystals turned up. A much better approximation was then discovered which explained many of the exceptions. This was announced as a new "law" of nature by Donnay and Harker,[9] who pointed out that many more crystals came into conformity with the requirements of Bravais' rule if "interplanar spacings" are interpreted to mean not only the spacings of the lattice but also to include modifications of spacings as required by those space group operations which contain translation components. This required a recognition of the same extinction conditions that occur in x-ray diffraction by crystals.

It might be pointed out here that an interesting relation exists between form development, as required by the Bravais-Donnay-Harker rule, and the powder photograph of the crystal. The order of importance of the crystal faces is exactly that of the order of appearance of the corresponding lines on the powder photograph (proceeding from $\theta=0$ to $\theta=90°$), provided that no orders of a line other than the first are considered.This proviso merely takes account of the fact that one form is represented by one line, since orders of a face have no meaning.

It is important to recognize that the Bravais-Donnay-Harker rule is strictly empirical and represents an attempt to find something about the crystal which is related to the comparative development of crystal faces. It is undesirable to accept the rule as a "law," particularly in view of the embarrassing history of the designation "law" in connection with crystal form development. As a generalization it rests on the sum total of the individual cases investigated, and among these there are some serious

[9] Donnay, J. D. H., and Harker, David. A new law of crystal morphology extending the Law of Bravais: *Am. Mineral.*, **22**, 446–447 (1937).

exceptions.[10,11] Furthermore, in the cases which support the rule, there is not necessarily complete agreement between the relative reticular density sequence and the order-of-importance sequence. Nevertheless, it must be recognized that there is a strong tendency for many crystals to conform in at least a general way to the requirements of the rule. What is the explanation of the agreement?

Pinacoid Development for a Simple Crystal

To get a preliminary picture of the possible significance of the Bravais rule, consider the growth of a simple crystal. The crystal is assumed to be simple in that,

(1) the crystal is triclinic,
(2) the crystal is built up of discrete molecules, and
(3) there is one molecule per unit cell.

The growth process consists of the locating of one molecule at each point of the triclinic lattice. For further simplication, suppose that the molecule has the shape of the parallelepiped unit cell. Growth then consists of the addition of blocks to the solid structure.

No assumptions are made about the specific character of the bonding between molecules in the crystal other than that ionic bonding is temporarily excluded (for reasons which will develop subsequently) and that the latent bond strength density is, on the average, uniformly distributed about the surface of the molecule. When a molecule lands on the crystal surface so as to continue the crystal structure, then the energy of the bond between molecule and crystal is (to first approximation) proportional to the area of the surface joining the molecule to the crystal. If several sites on the crystal are available to molecules, the preferred site (neglecting differential thermal effects) is the one with maximum bond strength. With the simplifying assumptions of this section, this is the site which offers the greatest area of attachment between crystal and molecule.

Consider, first, the growth process of three pinacoids, which represent, in simplified form, a series of three planes of different reticular density. Except when a new layer starts, the general conditions of the three pinacoids is shown in Fig. 1. New molecules add themselves to each layer by falling into a corner of the step on the growing surface. The corner on each of these pinacoids exposes an identical area to a molecule about to take its place in a corner. This area is the sum of the areas of the three

[10] Donnay, J. D. H., Tunell, G., and Barth, T. F. W., Various modes of attack in crystallographic investigations: *Am. Mineral.*, **19**, 456–458 (1934).

[11] Donnay, J. D. H., and Harker, David, A new law of crystal morphology extending the law of Bravais: *Am. Mineral.*, **22**, 463–465 (1937).

pinacoidal bases of the unit cell and is a measure of the first coordination of the molecule arriving at the surface. It is easily demonstrated that a molecule landing on any growing rational plane whatever (provided, as above, that the molecule is not starting a new layer) lands in a corner of the same character and achieves a first coordination measured by the area of the same three pinacoidal bases of the cell. This can be generalized for molecules of more general shape than that of a unit cell parallelepiped,

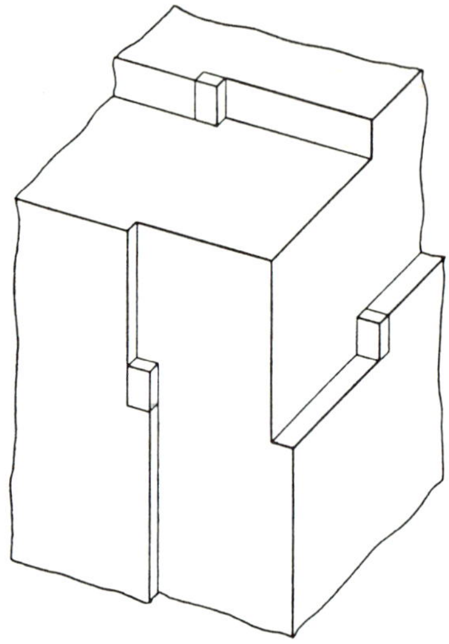

FIG. 1

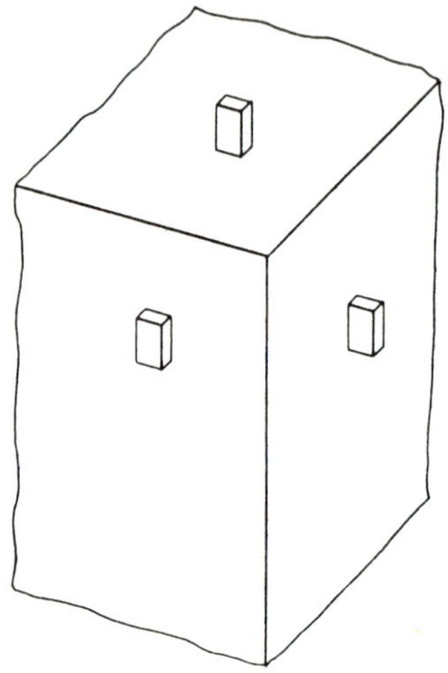

FIG. 2

with the result that it can be said that any molecule landing on a rational plane in such a way as to continue the general growth of that growing plane, does so in such a way as to achieve a first coordination measured by half the external surface of the molecule. Thus, in the growth of any rational plane, there is no differential in the immediate coordination of the system crystal-molecule and therefore, from the point of view of the energy realized by the coordination of the molecule and the plane, no plane of the crystal has an advantage over any other. Therefore if molecules rain evenly on all surfaces of a crystal, equal volumes might be expected to be added to each pinacoid (or other plane) in a given time, and consequently the planes might be expected to advance at an even rate.

But consider the situation as a layer starts: Fig. 2 shows a molecule in place on each of the three pinacoids of the simple crystal. It is evident that the surfaces of attachment are the pinacoidal bases of the cell taken one at a time. Since the volume of the cell is any base times its corresponding interplanar spacing, the area of any base is inversely proportional to its spacing. Thus the surfaces of attachment of the first molecules are in the inverse order of the interplanar spacings of the pinacoids to which they attach themselves. The most likely attachment of the three possible ones is that which has the greatest coordination of attachment, namely the molecule which rests on the pinacoid of *least* interplanar spacing. It might thus be expected that the probability of starting a new layer is highest for the pinacoid of least spacing. Since the bottleneck of starting a new layer continually recurs, the pinacoid of least spacing must be the pinacoid of most rapid growth. It is well known that the most rapidly growing faces tend to eliminate themselves. Consequently the pinacoid of least spacing tends to be the least well developed.

The reasoning given above evidently accounts for Fedorov's rule, which, freely interpreted and expressed in the language of the Bravais' rule, states that a crystal tends to have a habit development such that the decreasing importance of the pinacoids is in the order of decreasing spacing.

Coordination and Surface Energy

The simplified analysis just given does not lend itself easily to generalization, chiefly because coordination other than that due to the immediate coordination is difficult to evaluate. Fortunately it is easy to reformulate coordination in terms of surface energy. It should be evident that surface energy is caused by the lack of saturation of all bonds in the crystal due to the free surface. The chief contribution to this bond energy comes from bonds at the surface, and thus from surface coordination, but secondary coordinations are also involved. Thus, surface energy and total realized coordination energy across the free surface are complementary.

Returning, now, to the simple case discussed in the last section, the situation can be summarized as follows: If several sites are available to a molecule, the preferred site is that which permits the molecule to attain the greatest immediate coordination. Evidently this statement ought to be corrected so that its last part reads, "which permits the molecule to attain the greatest total coordination." In view of the complementary nature of realized coordination and surface energy, this can also be stated in the following terms: If several sites are available to a molecule, the preferred site is the one for which the molecule achieves the greatest decrease in its surface energy. Furthermore, growth is least rapid for the plane which requires, for the starting of each new layer, the attachment of a molecule in the position which leads the molecule to retain the most surface energy (which is the complementary way of saying that the molecule achieves the least total coordination or least bonding energy).

Surface Energy and Bravais' Rule

Bravais' rule is usually stated in such a way as to involve the "importance" of the development of a face. An equivalent, but somewhat more convenient form of the rule for the present requirements is that the faces of a crystal have growth velocities in the inverse order of their interplanar spacings. If this is true, and if the analysis of the last section is also true, this implies that the residual surface energies of the attached molecules which start new layers are in the inverse order of the interplanar spacings of the faces to which they are attached. That this is at least roughly correct can be appreciated from Fig. 3, where a crystal is

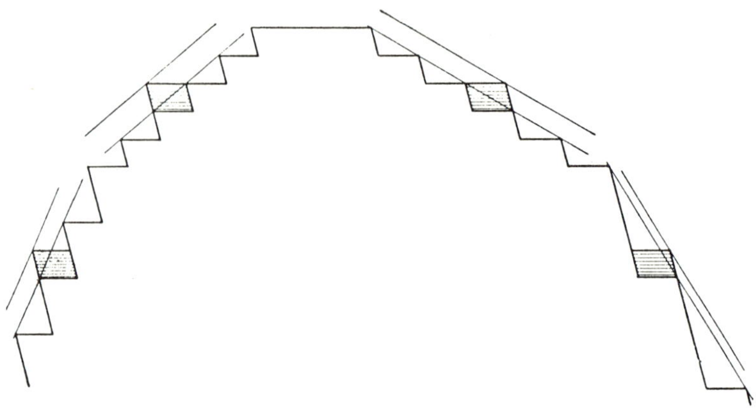

Fig. 3

represented as being made up of parallelepiped cells. (If ellipsoids are substituted for these parallelepipeds, the following feature still holds.) It will be observed that the molecules shown shaded, which start the new layers, project above the general levels of the planes by an amount equal to the interplanar spacing, d_{hkl}. When a molecule starts a new layer, it thus constitutes an excrescence on the layer, the height of which is roughly a measure of the residual surface energy of the molecule which starts the new layer.

This analysis is quite crude, but it appears to be essentially correct for the simple case taken for study. In this case, the Bravais' rule appears to operate because the residual surface energy of the added molecule is a function of the way the molecule fits into the surface, and this, in turn, is a function of d_{hkl}.

That this crude analysis is essentially correct can be appreciated by considering the contours of potential in the neighborhood of a location on a simple crystal surface. In Fig. 4, the full line *AB* may be taken to represent a contour in the energy field of a plane with indices (11). At *B*, the main contribution comes from molecule *a*. For a plane of index (12), a molecule *b* is also part of the surface, and it also contributes a potential at *B* of a magnitude comparable with that contributed by molecule *a*. For this second plane, therefore, point *B* has a greater energy than it had for the first plane. Plane (12), therefore, offers a molecule which starts a

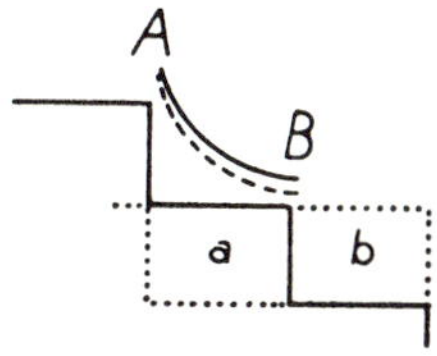

FIG. 4

new layer by adding itself above *a* a greater drop in energy than does plane (11). The presence of molecule *b* thus increases the realized coordination (or decreases the residual surface energy) of the molecule which starts a new layer.

THE STABILITY OF FACES

The crude analysis just given for the differing rates of growth of the faces of a crystal can also be applied in answering the question as to why high-index faces do not accept molecules in such a way as to become low-index faces. For example, Fig. 5 shows a surface with intercepts 1, 3.

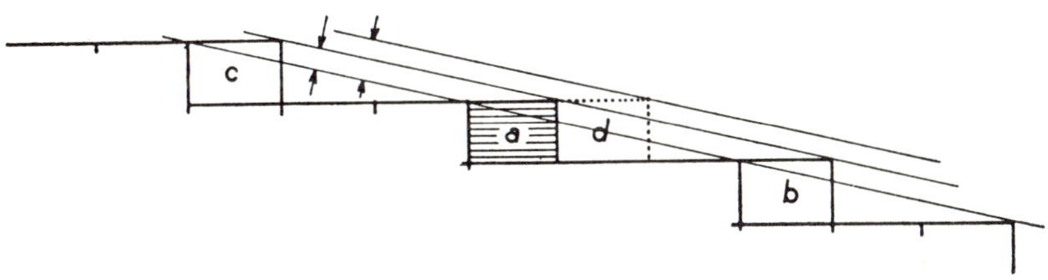

FIG. 5

When the plane grows, molecule *a* is first added to the surface, temporarily changing the slope of that region of the surface to intercepts 1, 2. Why does not another molecule, *d*, then add itself to the surface next to *a*, thus changing the intercepts of the surface to 1, 1, and so on, until the crystal consists of pinacoids only? It will be observed, however, that if both molecules *a* and *d* are added to the surface, the surface excrescence due to *a* plus *d* is twice as high as that due to *a* alone, and hence molecules added to the surface after *a* arrives preferably attach themselves at neighboring positions similar to *a*, such as *b* and *c*. Thus, a plane rejects molecules which attempt to add themselves in positions which would change the index of the plane to one of less complexity.

A very similar analysis shows that a plane rejects molecules which at-attempt to add themselves in positions which would change the index of the plane to one of greater complexity. Thus a plane tends to avoid variations in its slope because of surface energy barriers. In this way a crystal plane, once started, is stable, and maintains its slope, and therefore its identity.

Crystals Without Faces

Crystals without faces would be anomalous to some crystallographers, yet they are not only possible but normal under certain circumstances. From the last section it is evident that the stability of a face is a function of surface energy. Furthermore, the very existence of crystal faces depends on surface energy. Therefore, if a crystal is grown in a medium whose surface tension is equal to its own, no faces can develop.

This condition is exactly realized when a crystal grows in an aggregate of identical crystals, as in the case of recrystallization. Under such circumstances, no crystal form development is to be expected, and Harker and Parker[12] have shown that the shapes actually assumed by crystals grown by recrystallization tend to be polyhedra bounded by dihedral angles of 120° whose orientations are independent of any vectorial properties of the crystal.

The condition is nearly realized when a close-packed crystal forms from a melt of the pure metal. In this case, the melt is nearly close-packed, and the forces on an atom at a metal-melt interface are nearly alike on both sides, thus causing the interfacial tension to nearly vanish. For this reason, copper crystals and zinc crystals formed from pure melts have typically rounded forms. On the other hand, when a metal crystal and its melt differ considerably in density, as in the case of bismuth, then the interfacial tension between crystal and melt is considerable. Under these circumstances crystals with well-developed faces form from a pure melt.

The Donnay-Harker Rule

In the development given above, it was assumed that the crystal was characterized by molecules which were translationally equivalent. The

[12] Harker, David, and Parker, Earl R., Grain shape and grain growth: *Trans. Amer. Soc. Met.*, **34**, 156–195 (1945).

residual surface energy of the molecule was due to the projection of this molecule above the general level of the nearly smooth surface. It is evident that the same general state of affairs will hold if the molecules are equivalent by other operations having a translation component, namely screw axes and glide planes. Under such circumstances, some planes related by the translation-bearing operation (for example, in symmetry *Pc*, planes (*h0l*) with *l* odd) have submultiple spacings, and therefore have lower excrescences when molecules add themselves to the surfaces. Consequently the surface energy conditions of molecules added to these planes are equivalent to those of planes with corresponding submultiple spacings. These planes tend to eliminate themselves by rapid growth as compared with other planes. Thus the Donnay-Harker rule also follows from surface energy considerations.

Deviations from the Bravais-Donnay-Harker Rule

Notable deviations from the generalized Bravais rule have been found. These include certain individual nonconforming planes[13] in certain crystals and also crystals which do not conform at all. Such exceptions should be a warning that the effect is not referred to the appropriate variable. Specifically, it suggests that the variable in the crystal's effect on its own habit is not as simple as interplanar spacings based on either planes rational with respect to lattice translations or rational with respect to translation-bearing symmetry operations in general. A general explanation of certain deviations from Bravais' rule is therefore that the planes of growth need not always be rational. When this is the case, there is no need for the habit of a crystal to conform with Bravais' rule.

In the following sections, some of the causes of deviation from Bravais' rule are discussed.

Deviations Due to Growth from Ionic Solution.—Among the crystals which do not conform to the Bravais' rule at all are the class of crystals which are notably ionic. Since the ionic crystals which have been studied for conformity with the rule have been grown from solutions which ionize the solute, it is evident that molecules cannot possibly be deposited on the surface of the crystal from such solutions. Rather the solutions must deposit particles which are of two kinds, namely the two oppositely charged ions, and these are not equivalent by any symmetry operation of the crystal. For this reason the excrescences on the surface of the crystal caused by adding the particle to the surface, as the plane grows, is smaller

[13] For example, see: Taylor, E. D., The morphology of columbite crystals: *Am. Mineral.* **25**, esp. p. 134 (1940); also Nuffield, E. W., and Peacock, M. A., Studies of mineral sulphosalts: VIII, *Univ. of Toronto Studies,* Geol. Ser., No. **49,** 33 (1945).

than it would be if the entire motif were added. It is thus no longer possible to measure the size of the projection by d_{hkl}, and consequently the crystal does not conform to Bravais' rule.

As outlined above, the Bravais rule works because the crystal builds up by the addition of translationally equivalent units, and this requires that the new layers added are always rational lattice planes. When the entire motif is not deposited as a unit, then the crystal cannot be built up by the addition of layers which are rational planes, for some layers must be interleaved between rational layers. Such interleaved layers can be fictitiously described as rational provided a fictitious translation-bearing operation is assumed between them. This fictitious operation also brings some positive and negative ions into coincidence with one another. Therefore, if a subcell is chosen from the lattice which is composed of such fictitious translations (three which bring positive and negative ions into coincidence) the crystal comes into fictitious conformity with Bravais' rule.

Deviations Due to Growth by Accretion of Molecules Which Are Not Crystallographically Equivalent.—In discussing the rational basis for Bravais' rule with Dr. J. D. H. Donnay about four years ago, the writer pointed out that the form development of superstructures ought to follow that required by Bravais' rule for the basic structure rather than that required for the superstructure itself. Dr. Donnay tested this prediction by studying the form development of orthorhombic low-chalcocite, which was known to the writer to be related as a superstructure to hexagonal high-chalcocite.[14] In a private communication to the writer, Dr. Donnay reported that the morphology of high-chalcocite did indeed follow that of the simpler structure upon which it is based.

This general behavior is to be expected on the basis of the theory of form development presented in this paper. Suppose, for example, that a crystal is composed of two kinds of chemically equivalent molecules which are located in the crystal in non-equivalent sites. Unless a polymerization has taken place in the solution from which the crystal is grown, the growth of the faces of the crystal can be expected to take place by addition of the individual molecules to the faces. This means that motif fractions rather than translation-equivalent motifs land on the crystal faces during growth. Now Bravais' rule can be rationalized along the lines discussed in this paper only if growth is by motif units which are equivalent by an operation with a translation component, a feature which does not obtain in the case under consideration. Consequently Bravais'

[14] Buerger, M. J., and Buerger, N. W., Low-chalcocite and high-chalcocite: *Am. Mineral.*, 29, 55–65 (1944).

rule cannot be expected to hold in this case. Of course the reason for this is that the surface energy due to the excrescence can be measured by d_{hkl} only if the added molecular units are translation-equivalent. When the unit added is a motif fraction, the excrescence due to the addition of the molecule is less than it would be if the entire motif were added and its height follows a different rule. Indeed it follows, at least roughly, the rule that its height is approximately that which it would be if all molecules were translation-equivalent, and this is so only if the structure is referred to the cell and symmetry of the basic structure.

In a rough way, the discussion just given applies to cases where two chemically non-identical molecules, which do not associate in solution, add themselves to the surface of a growing crystal. The crystals in these two cases correspond with *distortion structures* and *substitution structures* respectively in the theory of *derivative crystal structures*.[15] Provided association does not take place in the solution, crystals with derivative structures can be expected to display a form development which deviates from the Bravais-Donnay-Harker rule in the direction of the habit to be expected from the basic structure.

Deviations Due to Molecular Asymmetry.—In the rationalization of Bravais' rule it was assumed that the molecule was at least as symmetrical as the cell of the lattice. When this is not so, then it does not follow that the surface energy of the attached molecule is measurable simply by d_{hkl}. Thus, in the accretion of polar molecules to the surfaces of polar crystals, one should not expect the same rate of growth for all faces of identical spacing. For example, if the crystal lacks a center, one would not expect the same rate of growth for *hkl* and $\overline{hkl}$, in spite of the fact that d_{hkl} and $d_{\overline{hkl}}$ are identical. The reason, of course, is difference in surface energy.

Deviations Due to Non-uniform Bond Density.—One of the simplifying assumptions which were made was that the added molecule has a uniform distribution of bond density. When this is the case, the coordination energy realized when the molecule adds itself to the crystal is a purely geometrical function of the manner of attachment of the molecule to the crystal. If, however, some aspect of the molecule's surface is characterized by a concentration of strong bonds, the preferred mode of attachment of the molecule is no longer a simple function of geometry, but is influenced by the location and strength of the strong bonds. The coordination energy realized when the molecule attaches itself is therefore no longer a simple function of d_{hkl}, and it cannot be expected that Bravais' rule will be strictly obeyed. The deviation from Bravais' rule to be expected in such

[15] Buerger, M. J., Derivative crystal structures: *J. Chem. Phys.*, **15**, 1–16 (1947).

cases is in the direction of the reduction of importance of faces parallel to the surface of strong bond concentration in the molecule starting a new layer, and an increase in importance of faces parallel to the surface of low bond concentration in the molecule starting the new layer.

Variation of Habit Due to Impurities

It is not the function of this paper to discuss in any detail the influence of the environment on the form development of the crystal. Nevertheless it is desirable to draw attention at this point to the fact that the influence of impurities on form development can be considered from the viewpoint just adopted for considering the influence of the "normal" molecules on the habit of their crystal. If impurities are present, then both the impurity particles (which may be molecules or ions) and normal molecules may compete for positions of high coordination on the crystal surfaces. Since impurity and "normal" molecules are different, a site of high coordination for one particle is not necessarily identical with a site of high coordination for the other. Nevertheless, realized coordination energy, or its complement, surface energy, is the feature which governs the preference of sites in each case. This should be the guiding principle in seeking to predict which faces have their growths most retarded by the adsorption of impurity particles.

The influence of the impurity is more complex than that of the normal particle, however, for the impurity particle may be included as a particle in a layer being formed. Its effect on slowing down the growth of the particular face may not be felt until the next layer is about to cover the impurity in the first layer. When this occurs, it may be impossible for the second layer to continue its growth until the foreign particle has been removed by energy competition and replaced by a normal particle. A new second layer may fortuitously start elsewhere on the first layer before this can occur.

Conclusion

Those who have hitherto interested themselves in the influence of the crystal on its own form development have followed a purely empirical course and have arrived at a correlation of form development with lattice geometry. In this paper there has been an attempt to rationalize these empirical findings along lines of simple energy considerations.

Such considerations lend rational support to the Bravais' rule and its generalization. On the other hand, they throw emphasis on another aspect of crystal growth. Whereas Bravais' rule emphasizes purely the lattice geometry of the growing crystal, energy considerations make it plain that emphasis should be focussed on the characteristics of the particles which arrive at the crystal surfaces during the growth process. When there is a simple relation between the characteristics of the arriving particles and the lattice geometry the whole growth effect can be formulated in terms of lattice geometry alone. When the arriving particles have characteristics other than mere geometrical shape to be taken into consideration, then the habit effect cannot be made functional upon lattice geometry only, and deviations can be expected from Bravais' rule.

Thus one of the most fundamental factors in determining crystal habit is the characteristics of the particles which arrive at the crystal surfaces, including both their geometry and physics of bonding. When the particle which would normally be added to the crystal to continue its structure must compete with "impurity" particles, then the corresponding characteristics of the impurity particles constitute another fundamental factor in determining habit. Both lattice geometry and bonding characteristics of the crystal constitute a third factor. Only in special cases can all three factors be formulated as a function of lattice geometry alone.

Part VII

MORPHOLOGY IN PHASE SPACE

Editor's Comments on Papers 20 Through 22

Max Born wrote of Brillouin's *Wave Propagation in Periodic Structures* (part of which we reproduce as Paper 20) that it dealt with a general method and not with any particular branch of physical science. The Brillouin zones were introduced as the fully general volumetric counterparts of the Bravais space lattices. In this work, Brillouin treated problems of both pure physics and applied engineering by developing a new branch of mathematics, precisely the mathematical relationships between form and structure. It is at once obvious that the first Brillouin zone for the face-centered cubic lattice corresponds to the typical habit of galena; for the body-centered cubic lattice, garnet; for the primitive cubic, rock salt; etc. But in place of a serial ranking in order of prominence of the faces (the morphological aspect of the morphologists), the zones provide for the simultaneous appearance of distinct forms, as in the (100) and (111) of the fcc zone, the base and prism for the hexagonal close-packed zone, etc. Mott and Jones pointed out the extension of the principle to the diamond and other structures involving translation symmetry. "It is often sufficient to pick out the planes of small indices with large values of S_n [the geometric structure factor] by noting the strong lines in the X-ray powder photograph . . .. All planes for which this factor vanishes are excluded, and from the remainder, the forms of the various zones are obtained by simple geometry. . . ."[1]

Paper 21, by William Hume-Rothery, is taken from *Electrons, Atoms, Metals and Alloys*, which was written in the form of a didactic dialogue between a Young Scientist and an Older Metallurgist. In these pages, Young Scientist employs the concepts of quantum physics to explain Jones's theory for the electron or Hume-Rothery compounds.[2] The Brillouin zones are shown to be the stable forms bounding the energy states of the crystal. The paper is included as an example of the relationship between external form and internal structure in momentum space. In this space the shape that will accommodate the requisite proportion of free electrons at minimum total energy determines the structure. That the same shape in direct space for the same structure is that calculated for the minimum surface energy (cf. Paper 28, Figure 3, for example) is a consequence of the equivalence of the mathematics underlying the relationships employed. "These planes play the same role for all waves," Brillouin wrote, "whatever their particular nature may be."

One of the architects of the modern quantum theory, Alfred Landé (b. 1886), in Paper 22 joins Friedel in questioning the literally realistic approach to the theory of crystals. Paper 22 is included because, in appealing to formalism, i.e., to the equivalence of wave and particle, as warranted by their mathematical equivalence, Landé opens the door to a comparison of Brillouin's surfaces of discontinuity in *k*-space with the morphologist's surfaces of discontinuity in direct space.

REFERENCES

1. N. F. Mott and H. Jones, *The Theory of the Properties of Metals and Alloys* (1936), Dover, New York, 1958, p. 156.
2. H. Jones, "The Phase Boundaries in Binary Alloys, II, The Theory of the α-β Phase Boundaries," *Proc. Phys. Soc.*, **49**, 250–257 (1937).

20

Reprinted through the permission of the publisher from pp. 136–139, 148–155 of L. Brillouin, *Wave Propagation in Periodic Structures*, 2nd ed., Dover Publications, Inc., New York, 1953

THREE-DIMENSIONAL LATTICES

L. Brillouin

[*Editor's Note*: In the original, material precedes this excerpt.]

33. Zones in Three Dimensions and Bragg Reflection; Ewald's Construction

The discussion given in Chap. VI can be used without any change for three-dimensional problems. We used vector notation, which works just as well for three as for two components. The areas S_d and S_b are to be replaced by the volumes V_d and V_b as in Eq. (32.5), and atomic planes in the direct lattice replace the atomic rows occurring in two dimensions, as explained in Eq. (32.14).

Triply periodic functions are expanded in triple Fourier series in exactly the same way as in Sec. 26 for two dimensions. The coefficients $C_{h_1h_2h_3}$ with three integral indices h_1, h_2, h_3 will simply be written C_h for convenience. The proof of the periodicity of ν as a function of **a** is carried out exactly as in Sec. 27 and may be briefly repeated here on account of its importance.

A point in the direct lattice is given by

$$\mathbf{R} = l_1\mathbf{d}_1 + l_2\mathbf{d}_2 + l_3\mathbf{d}_3 \tag{33.1}$$

where l_1, l_2, and l_3 are integers and $\mathbf{d}_1$, $\mathbf{d}_2$, and $\mathbf{d}_3$ are the lattice vectors. An arbitrary vector in the direct lattice is given by

$$\mathbf{r} = \xi_1\mathbf{d}_1 + \xi_2\mathbf{d}_2 + \xi_3\mathbf{d}_3 \tag{33.2}$$

where ξ_1, ξ_2, and ξ_3 are the components of $\mathbf{r}$ in the directions $\mathbf{d}_1$, $\mathbf{d}_2$, and $\mathbf{d}_3$, respectively. If a wave is propagating through the lattice and can be observed only at the lattice points **R**, we may express the disturbance ψ by

$$\psi = Ae^{2\pi i[\nu t - (\mathbf{a}\cdot\mathbf{R})]} \tag{33.3}$$

where **a** is the wave vector and is regarded as a vector of the reciprocal lattice. The function ψ is evidently periodic in **a** and **R**. The vector

$$\mathbf{a}'_h = \mathbf{a} + h_1\mathbf{b}_1 + h_2\mathbf{b}_2 + h_3\mathbf{b}_3 \tag{33.4}$$

where h_1, h_2, and h_3 are integers, will describe the motion at the lattice points just as well as **a**; for

$$\begin{aligned}(\mathbf{a}'_h \cdot \mathbf{R}) &= (\mathbf{a}\cdot\mathbf{R}) + h_1(\mathbf{b}_1\cdot\mathbf{R}) + h_2(\mathbf{b}_2\cdot\mathbf{R}) + h_3(\mathbf{b}_3\cdot\mathbf{R}) \\ &= (\mathbf{a}\cdot\mathbf{R}) + h_1l_1 + h_2l_2 + h_3l_3 = (\mathbf{a}\cdot\mathbf{R}) + \text{integer}\end{aligned}$$

Since the same motion can be represented by **a** or any arbitrary $\mathbf{a}'_h$, the frequency must be a periodic function of **a** also, with the periodicity $\mathbf{b}_1$, $\mathbf{b}_2$, and $\mathbf{b}_3$ of the reciprocal lattice.

We wish to set up conventions for eliminating all but one of the values for **a**, *i.e.*, define zones to which **a** is to be restricted, as was done in Sec. 27 for the two-dimensional case. The method is exactly the same, and we take the first zone to be a volume centered upon the origin O of the reciprocal lattice and limited by plane perpendicular bisectors of vectors in the reciprocal lattice. The first zone has a volume V_b equal to the volume of the elementary cell in the reciprocal lattice, and any point in space can be brought back into the first zone by **H** translations in the reciprocal lattice.

$$\mathbf{H} = h_1\mathbf{b}_1 + h_2\mathbf{b}_2 + h_3\mathbf{b}_3$$

Higher zones surrounding the first one are built in a similar way, and some actual examples will be given below.

The choice of these particular planes, bisecting and perpendicular to the vectors of the reciprocal lattice, is based on the analysis previously given in Secs. 28 and 30 for two dimensions. The theory of waves propagating through a continuous medium with a small periodic perturbation is identical with the two-dimensional theory, and one must distinguish between

$$\left.\begin{array}{ll} \textit{Case 1, where all} & |\mathbf{a}'_h| \neq |\mathbf{a}| \\ \textit{Case 2, where one certain } \mathbf{a}'_h\textit{, say } \mathbf{a}'_n\textit{, is} & |\mathbf{a}'_n| \approx |\mathbf{a}| \end{array}\right\} \qquad (33.5)$$

where $\mathbf{a}'_h$ is a vector (33.4). Case 2 arises when two vectors $\mathbf{a}'_n$ and $\mathbf{a}$ having almost the same length can be found, such that their difference is a vector of the reciprocal lattice.

$$\mathbf{a}'_n - \mathbf{a} = \mathbf{H} = h_1\mathbf{b}_1 + h_2\mathbf{b}_2 + h_3\mathbf{b}_3 \qquad (33.6)$$

A geometrical interpretation of this condition was presented in Sec. 29 where it was proved that the circumstances leading to case 2 were identical with those yielding Bragg's reflections in the crystal. This is always true if atomic rows are replaced by atomic planes in the direct lattice.

Another geometrical interpretation was given by P. P. Ewald. From point O of the reciprocal lattice he draws a vector $\mathbf{a}$ to point P. A sphere is drawn with radius $|\mathbf{a}|$ about the point P at the end of the vector $\mathbf{a}$. If it happens that this sphere passes through or near to a second point $\mathbf{B}$ of the reciprocal lattice, we obtain a discontinuity in the frequency as a function of $\mathbf{a}$.

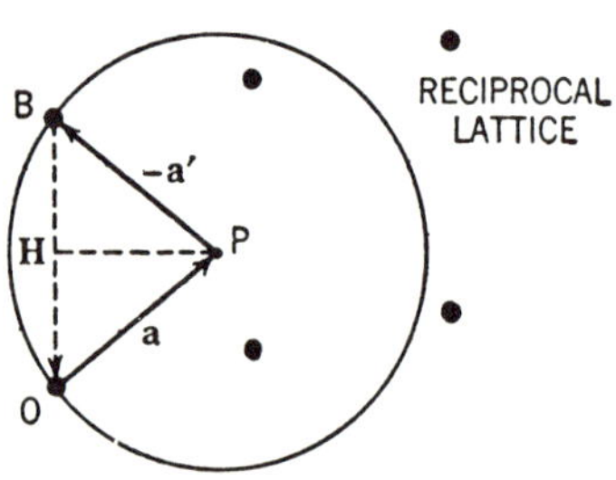

FIG. 33.1.

The reason for this is that there is some $\mathbf{a}'_n$ that just fulfills Eq. (33.6). There is no way of telling from the motion of the lattice points which of vectors $\mathbf{a}$ and $\mathbf{a}'_n$ should be preferred. Since the two vectors have different directions, we must assume that the motion is given by a superposition of the two waves. One of these is an incident wave and the other a reflected wave (Bragg wave). There will be two values for the frequency for this single value of $|\mathbf{a}|$ and hence a discontinuity in the ν vs. $\mathbf{a}$ curve.

We note that when Bragg reflection occurs, the vector terminates on a plane that is the perpendicular bisector of the line $\mathbf{H}$ joining two lattice points of the reciprocal lattice. This justifies the rule for constructing zones. We construct planes that are perpendicular bisectors of lines joining lattice points with the origin. The smallest polyhedron bounded by such planes is the first zone. The area bounded by the first zone and the second smallest polyhedron is the second zone and similarly for higher order zones.

[*Editor's Note*: Material omitted at this point.]

37. Some Examples of Zones in Three Dimensions

The face-centered and body-centered cubic lattices were discussed in Sec. 32 (Fig. 32.2) where it was proved that they are mutually reciprocal. The *face-centered cubic lattice* is of special

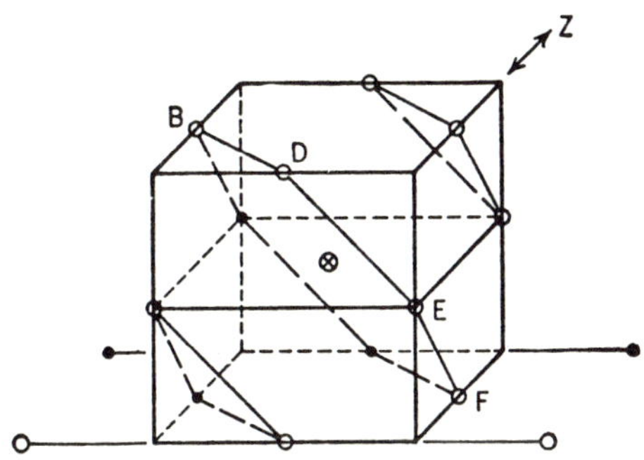

FIG. 37.1.—Face-centered cubic lattice.

importance since it represents one of the possible structures for close-packed spheres. The lattice can be described as in Fig. 32.2*a*, and also as in Fig. 37.1, which obviously represents the same structure since it is derived from Fig. 32.2*a* by translation of ½*d* parallel to the axis. Looking at this cubic structure from the direction of the arrow and taking the diagonal *OZ* as the vertical axis, we obtain the drawing of Fig. 37.2, which represents the same lattice as the hexagonal structure shown in Fig. 37.3. This is one of the possible structures obtained by piling up equal spheres provided that the distance from the central sphere to its *twelve neighbors* is the same, and this is the case for

$$\frac{c}{d} = 2\sqrt{\frac{2}{3}} = 1.633 \qquad (37.1)$$

where the distance c measured vertically from A to B is slightly less than the distance $BE = \sqrt{3}\,d = 1.732d$. The structure repeats itself after a vertical distance $(3/2)c$.

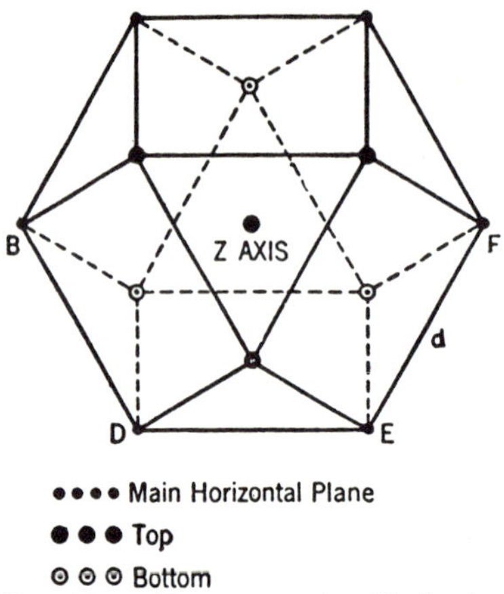

FIG. 37.2.—Face-centered cubic lattice.

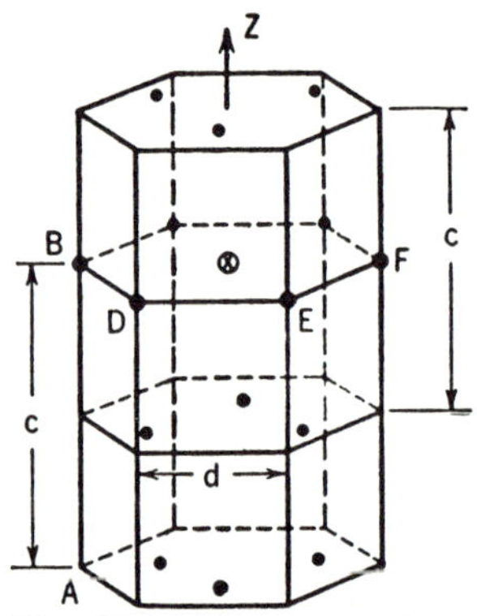

FIG. 37.3.—Face-centered cubic lattice.

There is *another possible structure* for *close-packed spheres* shown in Figs. 37.4 and 37.5. It obviously has exactly the same density as the face-centered cubic lattice. Comparing Fig. 37.3 with Fig. 37.4, we notice that in Fig. 37.3 the lower atoms are diagonally opposite to the upper ones. In Fig. 37.4, on the other hand, the lower atoms lie just below the upper ones, and the structure repeats itself after a vertical distance c. This remark shows immediately that the *hexagonal lattice* of Figs. 37.4 and 37.5 is *not a simple Bravais lattice* built on three vectors $\mathbf{d}_1$, $\mathbf{d}_2$, and $\mathbf{d}_3$, but that it represents a *lattice with basis* according to the definition given in Sec. 31 (Fig. 31.8).

In the case of lattices containing only one type of particle, the following criterion can be used to distinguish between a simple

Bravais lattice and lattices with a basis: Taking a vector OA from one particle O to another A, we construct its negative OA' (Fig. 37.6). If the lattice is a simple Bravais lattice, there must be an atom at the point A' (as in Figs. 37.2 and 37.3). If there is no atom at A' (as in Figs. 37.4 and 37.5), the lattice has a basis. Hence, the face-centered cubic structure, which

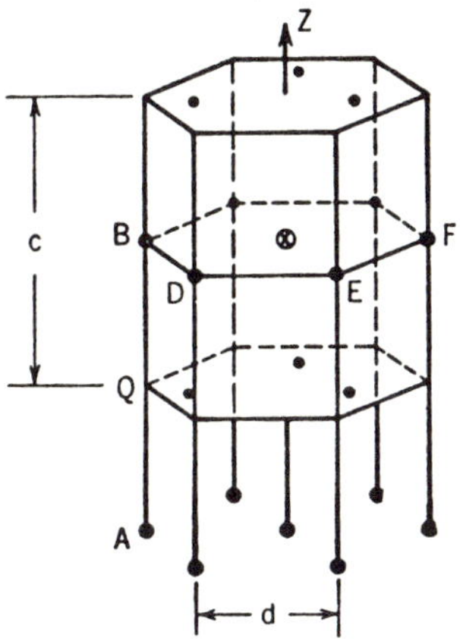

FIG. 37.4.—Hexagonal close-packed lattice.

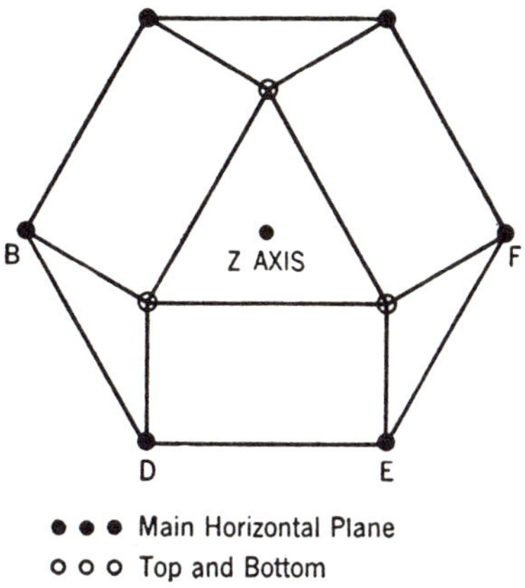

FIG. 37.5.—Hexagonal close-packed lattice.

FIG. 37.6.

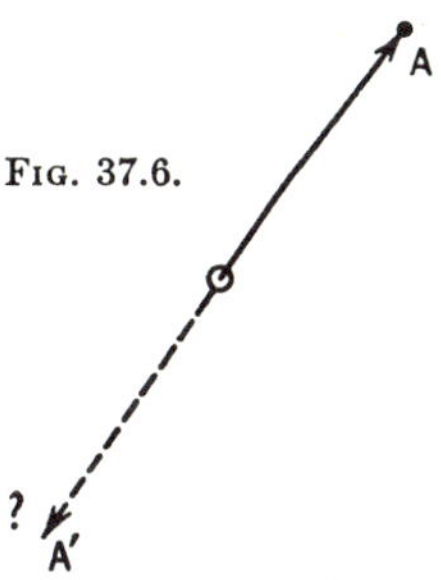

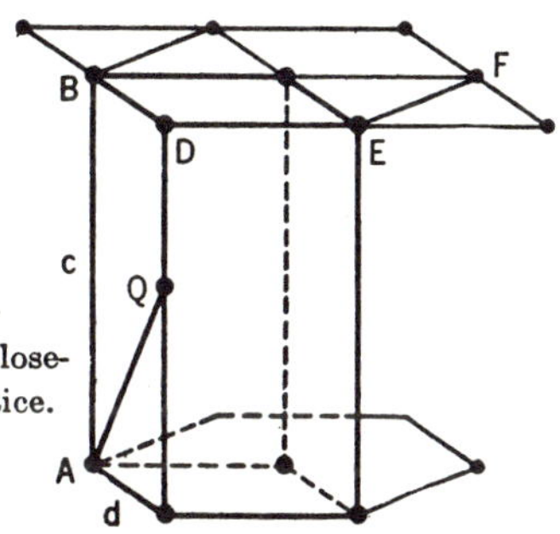

FIG. 37.7.—Hexagonal close-packed lattice.

Note that point Q is *not* on the vertical line going to D but well inside the parallelepiped, as can be seen on Fig. 37.4.

looked like a structure with a basis, actually represents a simple Bravais lattice. The hexagonal close-packed lattice, however, cannot be represented that way and actually possesses a basis. The fundamental cell contains two atoms as shown in Fig. 37.7 where AQ represents the basis vector. In this hexagonal lattice as in the face-centered cubic lattice of Fig. 37.3

$$AB = c = 1.633d, \qquad BE = 1.732d$$

for spherical particles. Both structures can be projected on a vertical plane passing through *DF* as shown in Fig. 37.8. In the reciprocal-lattice discussion of a lattice with a basis, all additional atoms inside the basis must be ignored, and the discussion refers only to the three fundamental vectors $\mathbf{d}_1$, $\mathbf{d}_2$, and $\mathbf{d}_3$. In the hexagonal structure of Fig. 37.7, for instance, we ignore the *Q* points on the level $\frac{1}{2}c$ and look only for lattice planes running through the points on levels *O* or *c*. In a horizontal projection,

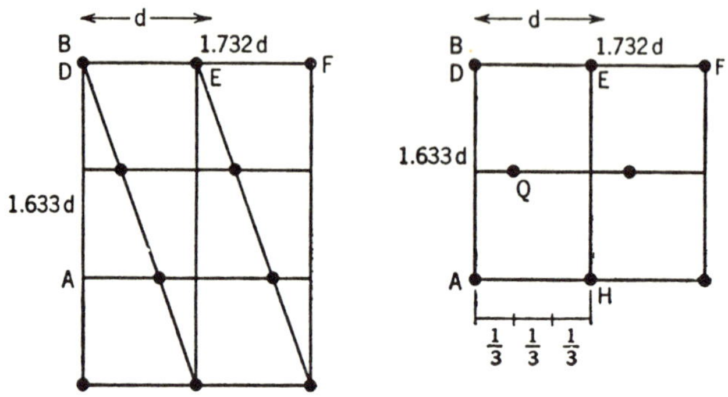

(*a*) FACE CENTERED CUBE (*b*) HEXAGONAL

Fig. 37.8.

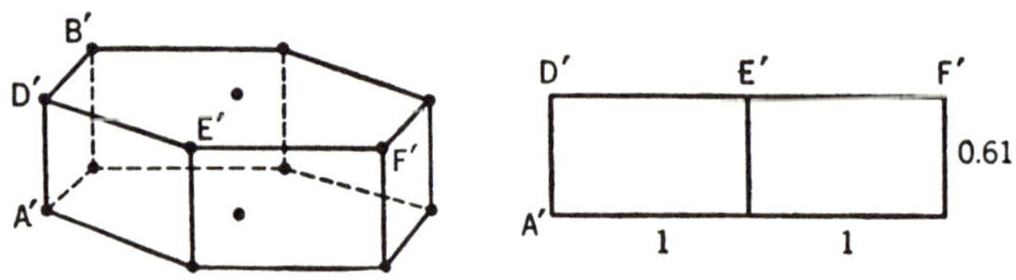

Fig. 37.9.—Hexagonal close-packed lattice.

the hexagonal structure is the same as in Fig. 31.2 for two dimensions and the reciprocal is another hexagonal structure with the side

$$b = \frac{2}{\sqrt{3}\,d} \tag{37.2}$$

as drawn in Fig. 31.3. Looking at the vertical projection in Fig. 37.8*b*, we obtain the fundamental cell represented by *ABEH* with dimensions

$$BE = \frac{\sqrt{3}}{2}d \qquad \text{and} \qquad AB = c = 2\sqrt{\frac{2}{3}}\,d \tag{37.3}$$

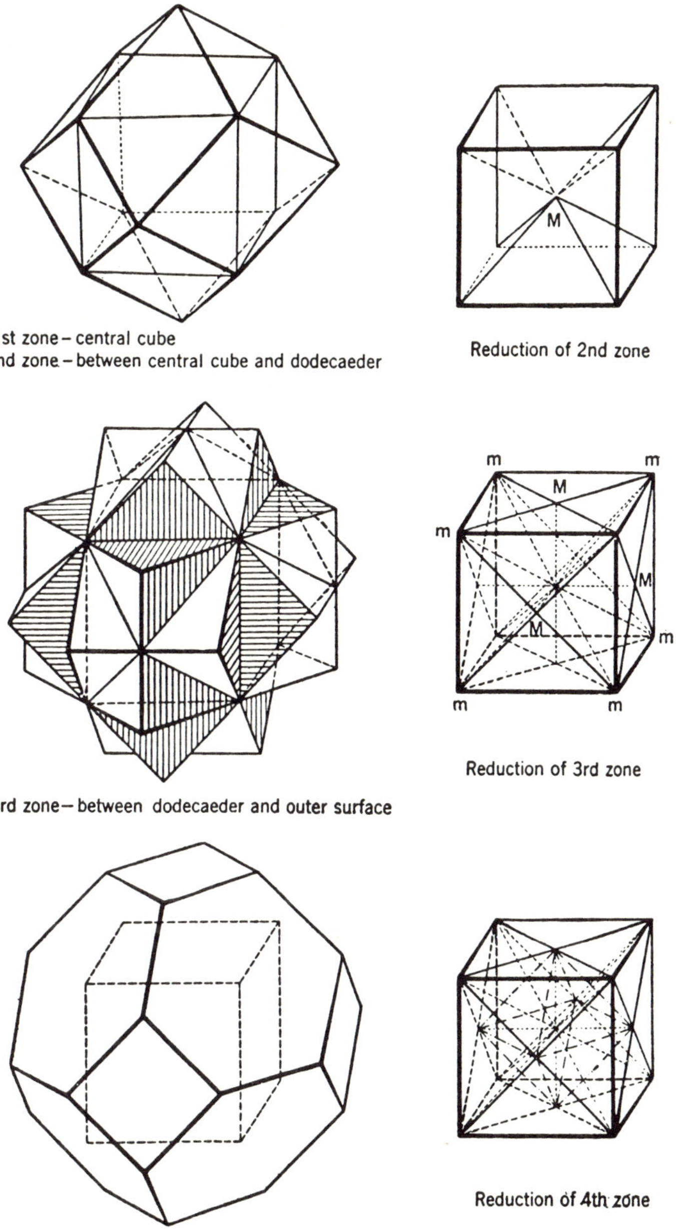

FIG. 37.10.—Cubic lattice, zones 1, 2, 3 and 4.

This gives us the reciprocal lattice shown in Fig. 37.9, a flat hexagonal structure whose vertical projection is also given.

Let us now discuss the *zone structures*. Starting with a *cubic lattice*, we obtain the zones represented in Fig. 37.10, where the first four surfaces have been drawn and the reduction to the first cubic zone is indicated on the right with the three-dimensional mosaics obtained by translations parallel to the side of the cube.

A *face-centered* cubic lattice has a body-centered cubic lattice as its reciprocal. The first zone is shown in Fig. 37.11 and has the same shape as the fourth zone of the cubic lattice. The fundamental cell of the direct lattice has a volume $\frac{1}{4}d^3$ (four

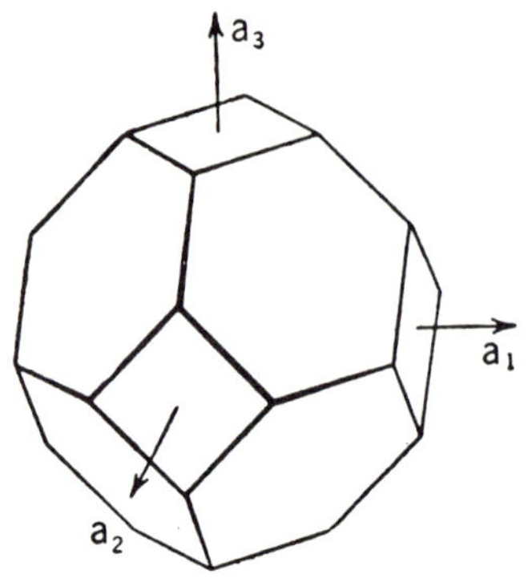

FIG. 37.11.—Face-centered cubic lattice, first zone.

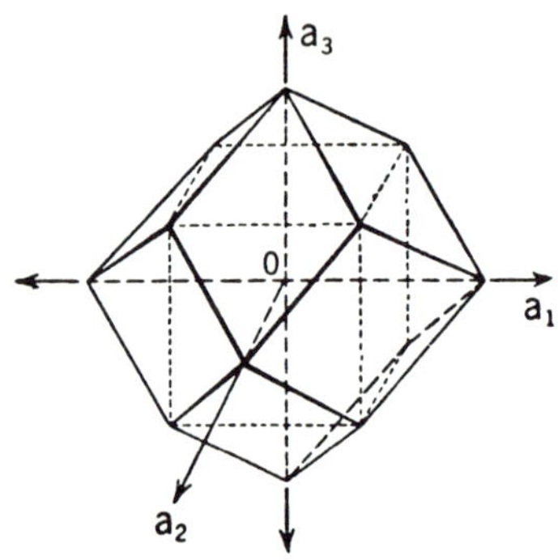

FIG. 37.12.—Body-centered cubic lattice, first zone.

points per cube); hence the first zone and each fundamental cell of the reciprocal lattice have a volume of $4/d^3$.

A *body-centered* cubic lattice has a face-centered cubic lattice as its reciprocal. The first zone is shown in Fig. 37.12 and is similar to the second zone of the cubic lattice. The volumes are $\frac{1}{2}d^3$ for the direct lattice and $2/d^3$ for the reciprocal lattice and the first zone. The second zone is similar to Fig. 37.11.

A hexagonal lattice with arbitrary d and c has a hexagonal reciprocal with $2/\sqrt{3}\,d$ and $1/c$ as shown in Fig. 37.9. The first zone is a hexagonal cell looking exactly like Fig. 37.9. First and second zones are shown in Fig. 37.13.

Let us now consider the transition from a continuous medium to a discontinuous lattice. Starting from a homogeneous isotropic continuum and a certain type of wave (longitudinal elastic waves, for instance), we first introduce a slight periodic perturbation with the face-centered cubic distribution. This introduces discontinuities in the $\nu(\mathbf{a})$ relation on all plane surfaces corresponding to zone boundaries. Instead of letting **a** run from 0 to ∞, we may reduce all vectors **a** inside the first zone, since each higher zone possesses sections that give a continuous $\nu(\mathbf{a})$ branch inside the first zone, and finally we obtain an infinite set of successive branches inside the first zone. This is the three-dimensional equivalent of the process described by Fig. 17.2 for one dimension. Increasing the periodic perturbation, we finally reduce the structure to the discrete atoms located at the points of the face-centered lattice. In this process all the upper branches of the $\nu(a)$ curves rise to infinity and disappear. The only remaining branch is the lower one, and

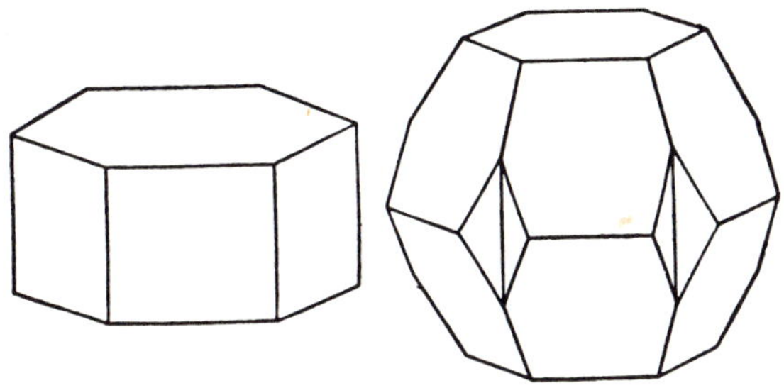

FIG. 37.13.—Hexagonal lattice, first and second zone.

Fig. 17.2 reduces to Fig. 2.4*b*. A very similar process of transformation was discussed at the beginning of Sec. 16.

Let us now follow the same procedure with a hexagonal close packing. The periodicity is hexagonal, but there are two atoms in each cell. The transformation will leave us with *two branches* in the first zone instead of one. The final $\nu(\mathbf{a})$ will be similar to the double curve or the one-dimensional NaCl lattice (Fig. 16.1), instead of the single curve of Fig. 2.4*b*. In Sec. 16 we discussed the transition from the case of two different masses $M_1 \neq M_2$ to two equal masses $M_1 = M_2$, and Fig. 16.1 explained the transformation. A similar discussion could be applied to the transformation indicated in Fig. 37.14 and obtained by straightening the vertical atomic line, thus going from the close-packed hexagonal lattice of height c to the simple Bravais hexagonal lattice of height $\frac{1}{2}c$. The fundamental cell of the latter has a volume one-half that of the original. Its first zone has a double volume, being twice as high as the original one of Fig. 37.13. In this first zone we should find just *one* $\nu(\mathbf{a})$ branch, but as soon as the perturbation corresponding to the zigzag of the vertical lines is introduced, this zone is cut in two, and the single branch splits into two branches in the new first zone. The procedure

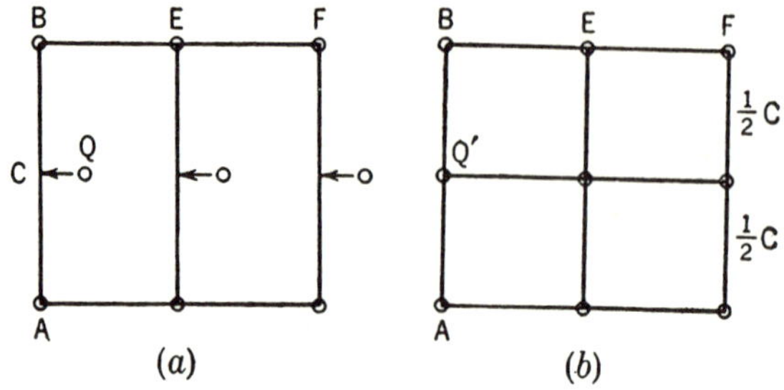

Fig. 37.14.

along the z axis is exactly the same as in Born's discussion of the NaCl reduction.

[*Editor's Note*: Material omitted at this point.]

21

Reprinted through the permission of the publisher from pp. 169–175 of W. Hume-Rothery, *Electrons, Atoms, Metals and Alloys*, 3rd ed., Dover Publications, Inc., New York, 1963

BRILLOUIN ZONES AND ELECTRON DISTRIBUTION CURVES

W. Hume-Rothery

[*Editor's Note*: In the original, material precedes this excerpt.]

OLDER METALLURGIST: From what you have said it would seem that for electrons in a periodic field the $\mathcal{N}(E)$ curve will no longer be a simple parabola as it was in the free electron theory.

YOUNG SCIENTIST: That's quite right, and we must now see what types of $\mathcal{N}(E)$ curve are actually formed. For simplicity we will again deal with a hypothetical two-dimensional crystal, and we will suppose that the crystal structure gives rise to a first Brillouin zone which is a square, as in Fig. 59(*a*). We may now imagine that we introduce increasing numbers of electrons into the crystal, and so gradually fill up the lowest energy states—as I explained before we imagine the positive charge on the ions to increase gradually so that the crystal as a whole is electrically neutral.

OM: I suppose all this will be at the absolute zero, where an assembly of $\mathcal{N}$ electrons occupies the $\frac{\mathcal{N}}{2}$ lowest energy levels.

YS: That's right. All this theory is developed first for the absolute zero. We then find that for the first few electrons which are introduced, the behaviour is very like that of the free electron theory, and the occupied states lie inside a circle, as shown in Fig. 59(*b*). The corresponding portion of the $\mathcal{N}(E)$ curve is then a parabola of the form $\mathcal{N}(E) \propto E^{1/2}$, and resembles that of the free electron theory. The shaded region in Fig. 60(*a*) extends over this part of the $\mathcal{N}(E)$ curve.

OM: Then as more and more electrons are introduced, are the boundaries of the occupied states still circular?

YS: No. As the limit of the occupied states approaches the boundaries of the Brillouin zone, the line bounding the occupied states—one may call it an *energy contour* because it joins states of equal energy—becomes distorted from a circle. The corresponding portion of the $\mathcal{N}(E)$ curve then rises above that of the free electron theory, as shown in Fig. 60(*b*). In a proper three-dimensional crystal the general process is the same. The two-dimensional circle is replaced by a sphere, and this becomes gradually distorted as the occupied states approach the zone boundary, and again the $\mathcal{N}(E)$ curve rises above that of the free electron theory.

OM: Now, what happens when the surface of occupied states touches the zone boundary?

YS: See if you can't think that out for yourself.

OM: Well, it would seem that if there is an abrupt increase in energy on crossing the zone boundary, then when the occupied states reach the surface of the zone, the next electrons added will go on

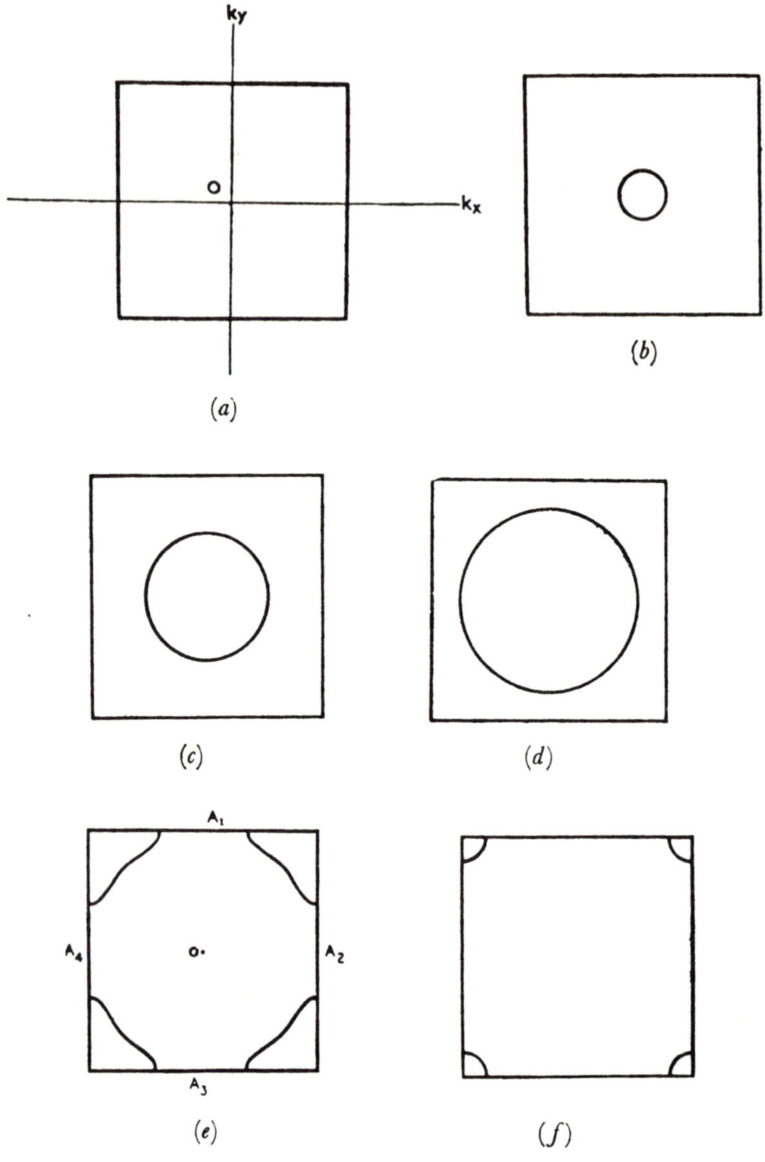

Fig. 59—To illustrate the filling up of electron states for a hypothetical two-dimensional crystal of which the first Brillouin zone is a square

For the first few electrons the Fermi surface is a circle whose radius increases with the number of electrons. When the occupied electron states approach the boundaries of the zone the circles are distorted and curve round so that they are eventually perpendicular to the zone boundary.

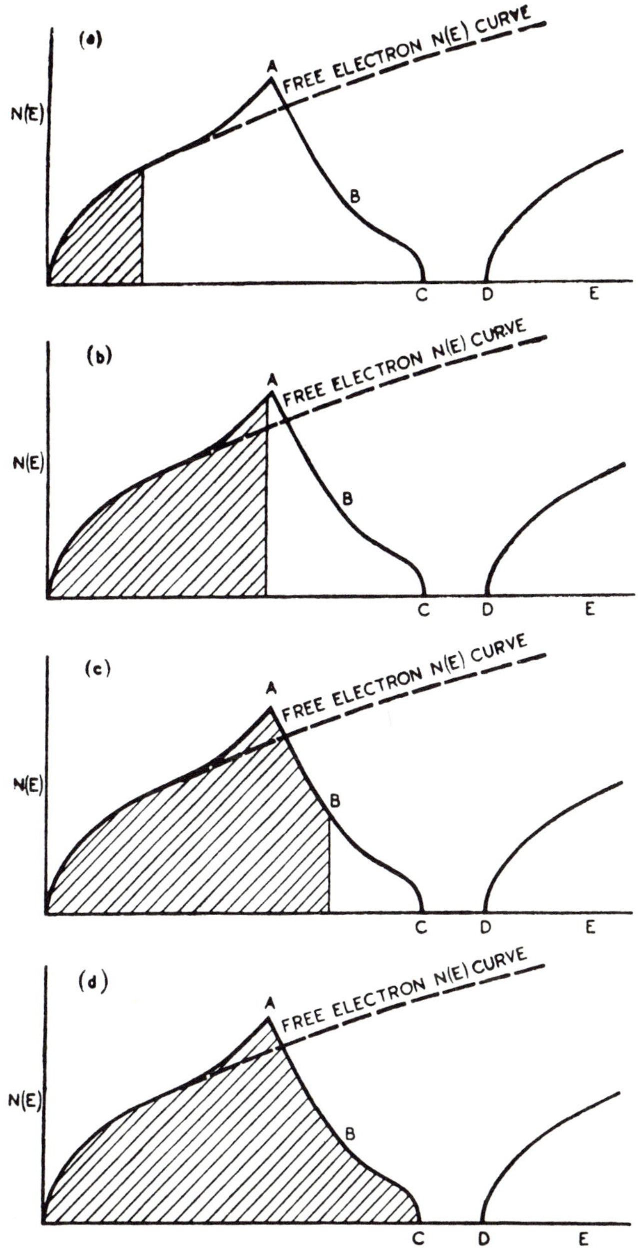

Fig. 60—To illustrate the N(E) curve for the Brillouin zone of Fig. 59. The case shown is one where the first and second zones do not overlap

filling up the states in the first zone, but they will not begin to enter the states in the second zone until the energy is sufficiently great to exceed the energy gap at the zone boundary.

YS: That's right.

OM: In that case the additional electrons will not at first enter states in the directions OA_1, OA_2, OA_3, and OA_4 (Fig. 59(*e*)), and so the $N(E)$ curve will fall because there will no longer be new energy states in all directions.

YS: That's right. The occupied states for this part of the $N(E)$ curve are shown in Fig. 60(*c*). The abrupt fall in the $N(E)$ curve at A is the result of there being no states in this range of energy for certain directions in k space. Fig. 59(*e*) shows the way in which the energy contour bounding the occupied states has continued to spread out inside the zone. Now see if you can say what will happen if we introduce still more electrons.

OM: It seems to me that there will be two possibilities. If the energy steps at the boundaries of the first zone are sufficiently great, the electrons will gradually fill up all the states of the first zone, and then there will be a sudden increase in the energy before any electrons enter the states of the second zone—in this case the $N(E)$ curve must sink to zero, because there will be some energies for which no electron states are possible.

YS: Yes, and the resulting condition of affairs is shown in Fig. 60(*d*). Here the portion ABC of the $N(E)$ curve corresponds with the filling up of the first zone states as shown in Fig. 59(*e*) and (*f*). The break CD on which the $N(E)$ curve is zero has a length equal to the difference between the energy of the highest state of the first zone and the lowest state of the second zone.

OM: That is quite clear and simple, but if the energy gap at the surface of the zone is relatively small, it would seem that the electrons might begin to enter the lowest states of the second zone before the states of the first zone were all filled.

YS: That's very true. It is in fact the characteristic of nearly if not all metallic structures that the lowest energy states of the second zone have a lower energy than the highest energy states of the first zone. In this case as the number of electrons is increased, they begin to enter the second zone as shown in Fig. 61, and the total $N(E)$ curve is the sum of that for the first zone and that for the second zone. In such a case it is said that the first and second zones overlap, and the opening up of the states in the second zone then produces a sudden rise in the $N(E)$ curve as shown in Fig. 62. Then on introducing more and more electrons, the first zone becomes completely filled, and the total $N(E)$ curve is that of the second zone alone. When the occupied states reach the boundary of the second zone, there will be a sudden fall in the $N(E)$ curve for the same reason as that for the fall at A in Fig. 60. If the second and third zones overlap there will be a sudden

Fig. 61—To illustrate one of the Fermi surfaces in a structure for which the first and second Brillouin zones overlap. The electrons have begun to fill up the states of the second zone before those of the first zone are completely full

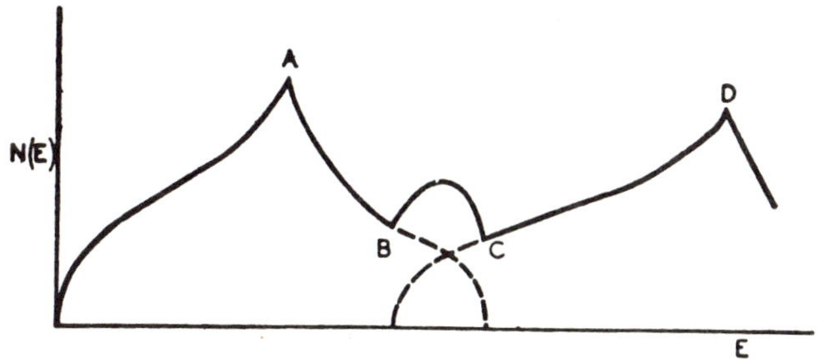

Fig. 62—To illustrate the N(E) curve for overlapping zones. The point B represents the stage at which the electrons begin to enter the second zone. At this stage the first zone is not completely full, and the total N(E) is obtained by the summation of the curves for the separate zones. The point C represents the stage at which the first zone is completely filled, and additional electrons then fill up the states of the second zone which may overlap the third zone and so on

rise in the $N(E)$ curve when the electrons begin to enter the states of the third zone. In this way the theory predicts $N(E)$ curves with a whole series of peaks and valleys, and each type of crystal structure gives rise to a characteristic $N(E)$ curve for a given energy gap at the surface of the zone. Fig. 63, for example, shows some early $N(E)$ curves calculated by Mott and Jones for the face-centred cubic and body-centred cubic structures, on the assumption of an energy gap deduced from a study of the optical properties of copper. Later work has shown that the finer details of these curves are quite incorrect, but the peaks and valleys on the curves are a clear confirmation of the results of X-ray spectroscopy. When the number of electrons is small . . .

OM: Wait a minute. In your description you have continually spoken of introducing increasing numbers of electrons into the crystal, but you have not given me any indication of the magnitudes involved.

YS: You understand how in Fig. 63 the kinks in the $N(E)$ curve occur at the stages at which the electron states in a new zone are being opened up (for example, point B in Fig. 62), or at which a zone is completely filled (for example, point C in Fig. 62). You will notice how the rounded peak resulting from an overlap is distinguished from the sharp peak at the point where the Fermi surface touches the face of a zone.

OM: Yes, that's all right. But how many electrons go to a zone?

YS: In simple translational lattices, each complete zone contains two electrons per atom. In Fig. 63 I have marked the numbers of

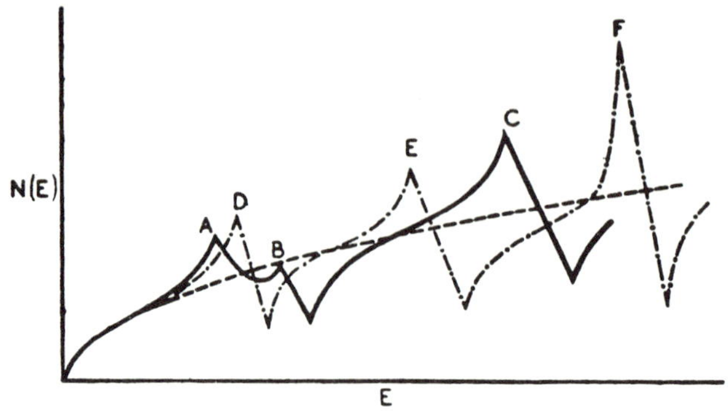

Fig. 63—N(E) curves for the face-centred cubic structure (full line) and the body-centred cubic structure (dotted line) after Mott and Jones. The points A, B, and C correspond with 1.36, 2.10, and 5.9 electrons per atom respectively, and the points D, E, and F correspond with 1.48, 4.21 and 7.7 electrons per atom respectively. These curves are of historical interest as representing one of the earlier theories of copper and silver alloys. They depend on simplifying assumptions, and later work has shown that some of these are incorrect, with the result that the details of the curves are of little value. The curves do, however, show clearly that different crystal structures give rise to N(E) curves whose peaks and valleys are at quite different electron concentrations

electrons per atom corresponding to different points on the curves, assuming the electron states to be fully occupied up to the points concerned. The number of valency electrons per atom is often called the *electron concentration*. In an alkali metal as in copper for example, the electron concentration is 1.0, and in zinc it is 2.0. In an equiatomic alloy of copper and zinc the electron concentration is 1.5.

OM: The various peaks and valleys in Fig. 63 seem to occur at quite different electron concentrations.

YS: Yes. That's the whole point. With a given energy gap at the surface of the Brillouin zone, each crystal structure gives rise to a characteristic $N(E)$ curve, and in this way we can begin to see how for a given number of electrons per atom, the electronic energy may be less for one structure than for another, and we can begin to see which structures should be the most stable so far as the electrons are concerned.

OM: Do you mean that the same principle (page 137) of high $\mathcal{N}(E)$ curve, lower energy still applies?

YS: Yes. If you compare the curves for the body-centred and face-centred cubic structures in Fig. 63, you will see that for the face-centred cube, the first peak occurs at 1.36 electrons per atom, whilst for the body-centred cube the first peak is at 1.48 electrons per atom. So as the electron concentration is increased from 1.36 to 1.48 the $\mathcal{N}(E)$ curve for the face-centred cube is falling rapidly, whilst that for the body-centred cube is continuing to rise. In this way we can predict that so far as the electron concentration is concerned, the body-centred cubic phase will become more stable as the electron concentration is increased from 1.36 to 1.48.

OM: That's interesting, but it doesn't seem to be of much use because the electron concentration of a metal must be a whole number, 1, 2, 3, 4 . . . according to the valency.

YS: What you say is quite right so far as the pure elements are concerned. But in alloys, the electron concentration need not be a whole number. In a copper-zinc alloy containing 50 atomic per cent zinc, the composition will be CuZn, and there will be three valency electrons to two atoms, and an electron concentration of 1.5. The above $\mathcal{N}(E)$ curves then indicate that so far as electron concentration is concerned, the body-centred cubic structure will be more stable than the face-centred cubic structure for the equiatomic copper-zinc alloy.

OM: Oh, I see! I suppose that might be the reason why β brass has a body-centred cubic structure.

YS: I had meant to keep the present discussion to the pure elements. But we may anticipate a little of Part III, and say that the above $\mathcal{N}(E)$ curves were the first beginnings of a theory of β brass. As we shall see later (p. 302a), many difficulties have been encountered, and the theory is still uncertain, but it is reasonable to claim that the above $\mathcal{N}(E)$ curves did seem to explain why β brass has a body-centred and not a face-centred cubic structure.

You will see therefore that the simple theory of the periodic field has taken us a long way. It has given us an $\mathcal{N}(E)$ curve which for small electron concentrations has the same general form as that of the free electron theory, and this is in agreement with the soft X-ray spectroscopic work on the alkali metals. It has shown that for higher electron concentrations the $\mathcal{N}(E)$ curves will have a number of peaks and valleys in agreement with Skinner's work on elements of higher valency. Finally, the particular curves of Fig. 63 did seem to lead to an understanding of the theory of β brass. But I must emphasise that later work shows the details of Fig. 63 to be incorrect, and the diagram should now be regarded only as showing the kind of curve indicated by the theory.

22

Reprinted by permission from *Am. Scientist*, Journal of Sigma Xi, the Scientific Research Society of America, 37, 414–416 (1949)

THE PHYSICAL SIGNIFICANCE OF THE RECIPROCAL LATTICE OF CRYSTALS

By ALFRED LANDÉ
Ohio State University, Columbus, Ohio

SINCE the discovery of Laue in 1912, the diffraction of X rays through crystals has been considered as an interference effect of waves, controlled by the Laue and Bragg rules for the direction of the diffraction maxima and supplemented by rules connecting the relative intensity of the maxima with the Fourier wave analysis of the matter distribution within the basic lattice cell. In spite of the great success of the classical wave theory it must not be forgotten, however, that the quantum theory considers waves only as a one-sided working hypothesis opposed by the hypothesis of mechanical particles. The uncertainty principle delimits the applicability of the classical theories in microscopic dimensions. There are many large-scale phenomena, however, among them the diffraction of a broad ray of light (or a cathode ray) through a large-sized crystal, which may be derived from either classical hypothesis. A mechanical theory of X-ray diffraction, resting on a re-appraisal of the Bragg reflection rule (section 1, below), was first established in 1923 by P. Duane. A similar re-interpretation of the Laue interference rules (section 2) lends a new physical significance to the concept of the *reciprocal lattice* of a crystal. The latter is usually considered as an abstract geometrical construction in spite of its wide usage in structural analysis. If the present article should contribute to lifting the reciprocal lattice from its abstract mathematical pedestal to a more realistic position in the physics of crystals, it will have attained its goal.

1. *Bragg diagrams.* Quantum theory contends that whenever the classical wave theory can explain an experiment, the same experiment must also be explainable in terms of particles whose energies and momenta are related to the wave length and frequency by the relations $E = h\nu$ and $p = h/\lambda$. No macroscopic experiment can ever decide which of the two theories is correct. (Microphenomena show that neither is correct.) The classical wave theory explains the X-ray diffraction maxima as resulting from the Bragg interference condition

$$2d \sin \theta = n\,\lambda \quad \text{(Bragg's relation)}. \tag{1}$$

in which d is the distance between successive lattice planes, and θ is the angle between the latter and the incident as well as the reflected ray. When (1) is divided by $d \cdot \lambda$ and multiplied by Planck's constant h it reads $2\frac{h}{\lambda}\sin\theta = n\frac{h}{d}$. With introduction of $h/\lambda = p$ as the momentum of the incident photons (or electrons) and with $h/d = P_1$ as an elementary momentum parallel to d characteristic of the crystal, the Bragg relation finally translates into

$$2p \sin \theta = nP_1 \quad \text{(Duane's relation)}. \qquad (2)$$

The left-hand side now is the change of the normal momentum component of an incident particle (changing from $+p \cdot \sin \theta$ to $-p \sin \theta$). According to the conservation law of momentum, the right-hand side must be the change of momentum of the whole crystal normal to the reflecting plane. This momentum change is $P = nP_1$. We thus learn that in reaction to an incident particle *the crystal changes its momentum component normal to a lattice plane only by amounts P which are integral multiples of a basic momentum quantum,* $P = nP_1$, where $P_1 = h/d$. The order *n* of the Bragg interference appears now as a *quantum number.* Because of its large mass, the kinetic energy of the crystal as a whole during the reflection changes only by a negligible amount so that we have a quantized *elastic* reflection of a small particle from a heavy body. Remember that the wave theory also considers the whole crystal as a physical unit responsible for the interference effect.

At the time of its publication Duane's mechanical theory of diffraction (2) seemed artificial as compared with the more natural wave interference theory (1). Modern quantum theory, however, considers both particles and waves as legitimate working hypotheses, and Duane's conception must be accepted as a serious physical theory, "reciprocal," rather than in contradiction, to the usual wave theory. It is not quite correct when a famous book on Röntgenstrahl Interferenzen remarks:

(I) "The space lattice structure of crystals . . . . was proved by the X-ray experiments of Friedrich and Knipping, which at the same time brought the *final decision* in favor of the wave nature of X rays" (italics ours).

Indeed one may counter this statement by the reciprocal statement:

(II) The experiments of Friedrich and Knipping proved that crystals are mechanical systems able to gain and lose momentum only in certain quantized amounts, $P = nP_1$, in certain selected directions, and the same experiments brought the *final decision* that X rays are particles subjected to the conservation laws of mechanics. From the standpoint of quantum theory both (I) and (II) are one-sided assertions; quantum physics is built on the (limited) equivalence of the two classical theories in explaining the observed facts. There is no point in arguing whether (I) or (II) describes the physical situation adequately.

2. *Laue diagrams.* We now discuss the two statements (I) and (II) from the standpoint of the Laue interference rules,

$$a(\cos \alpha - \cos \alpha_0) = h_1\lambda, \text{ etc.}, \qquad (3)$$

in which *a, b, c* are three vectors producing a tricline space lattice cell, whereas α, β, γ and α_0, β_0, γ_0 are the angles between *a, b, c* and the incident and the diffracted X ray respectively whose wave length is λ. The integers h_1, h_2, h_3 describe the "order" of interference. The same rules are geometrically represented by the *Ewald sphere construction*

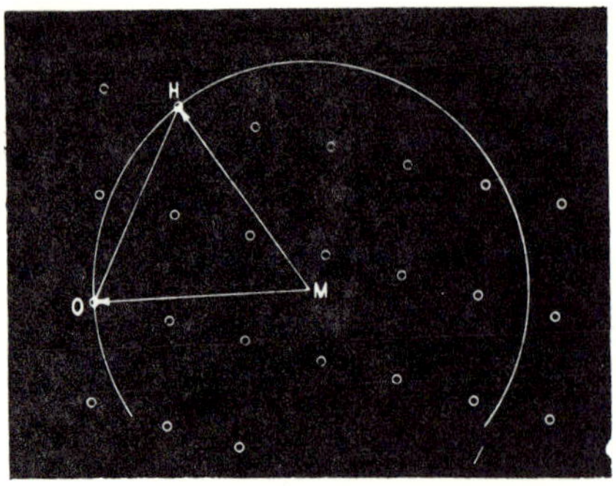

Fig. 1. Ewald sphere construction.

in the reciprocal lattice with basic vectors A, B, C of dimension cm^{-1} (Fig. 1). According to Ewald, draw the vector MO of length λ^{-1} in the direction of the incident X ray so as to end in any reciprocal lattice point O; describe a sphere with M as "middle point" and MO as radius; when this sphere runs through, or very close to, another lattice point H, then MH is the direction of a possible diffracted ray. We have the vector relation MO + OH = MH, in which the reciprocal lattice vector OH is a composition of the basic vectors A, B, C, namely, $h_1A + h_2B + h_3C$. This geometrical construction obtains a physical significance when it is carried out in a diagram h times as large, i.e., in momentum space, so that the numbers h_1, h_2, h_3 indicate integral numbers of quanta h. The vectors MO and MH represent the momentum vectors h/λ of the photon before and after the reaction with the crystal. Hence OH is the momentum given out by the crystal during the reaction. Experience as represented by the Ewald sphere construction thus tells us that:

(II′) a crystal is a mechanical system which can change its momentum only by such quantized amounts and directions as are indicated by vectors from one to any other point of the reciprocal lattice, supposing that the latter is enlarged by a factor h so as to represent a lattice in momentum space.

Those who consider the wave interpretation (namely, that a crystal is a periodic lattice in space) as physical, and the mechanical interpretation (II′) as artificial or far-fetched, may imagine for a moment the situation which would have prevailed if the diffraction of cathode rays had been discovered prior to that of X rays, at a time when cathode rays were "finally decided" to consist of particles. The deflection of various electron velocities then would soon have revealed that crystals always gain or lose certain quantized momenta which could conveniently be represented by a three-dimensional lattice in momentum space. Much later a theorist may have found that a rather unphysical interpretation of the same deflections was possible in terms of matter waves diffracted by a hypothetical space lattice reciprocal to the physical momentum lattice. We know now, of course, that both interpretations have equal rights before the laws of the quantum theory.

Part VIII

SPIRAL GROWTH

Editor's Comments on Papers 23 Through 25

23 BURTON and CABRERA
Crystal Growth and Surface Structure, Part I

24 CABRERA and BURTON
Crystal Growth and Surface Structure, Part II

25 FRANK
The Influence of Dislocations on Crystal Growth

The 1951 paper of Burton, Cabrera, and Frank in the *Philosophical Transactions of the Royal Society* (vol. 243, A866, p. 299) established the requirement of dislocations for ordinary crystal growth and the spiral as the major shape of the growth mechanism. The theory was the culmination of years of effort associated with the names of Frenkel, Kossel, and Stranski, among many others, and is the acknowledged basis of the present theory of crystal growth. From its inception it has directed attention away from the topic of this volume, the relationship between external form and internal structure. The most original development in the theory of crystal growth since Wulff's paper of 1901, the spiral growth theory divorced growth from considerations of surface energy and tied the development of any face to the random distribution of dislocations. Rather than the formal treatment of the Royal Society paper, the theory is here presented as it appeared originally in the form of three separate papers at the Conference on Dislocation Theory in the *Discussions of the Faraday Society*.

Burton and Cabrera (Papers 23 and 24) distinguish *"close-packed"* from *"stepped"* surfaces, defining the *"close-packed"* as those in which all molecules are equidistant from a parallel plane; all others are stepped. The steps or terraces have a finite number of *kinks* at ordinary temperatures, and these kinks are nuclei for growth. In the general case, growth of a crystal with a random distribution of dislocations is shown to be in agreement with experiment.

In Paper 25, by Frank, the dislocation theory originally developed to account for the anomalously low strength of real crystals is extended to the geometry of crystal growth. The screw dislocation of Burgers exposes a terrace along which growth can proceed, winding the line of the terrace around itself in a continuing spiral. Frank shows that the mutual interference of dislocations results in the growth considered by Burton and Cabrera.

The theoretical treatment of real surface planes and their separation into "close-packed" or *flat* and *stepped* should be distinguished from the application to crystallographic lattice planes. Lattice planes are distinguished solely by their net reticular density (see Paper 4, p. 24). Burton and Cabrera distinguish (110) for the primitive cube as a close-packed plane, and (120) as a stepped plane with (100) steps; yet (110) may also be considered as a stepped plane with (100) steps, and differs from (120) solely in the density of points along the plane—a difference in degree and not in kind. The Burton, Cabrera, and Frank theory, to the extent that it is related to structure and symmetry, is an approximation of Bravais's rule, since Bravais related morphological importance to net reticular density, a quantitative measure of the degree of closeness of packing of the lattice points of a given crystallographic plane. It is as a theory of the dependence of the mechanism of growth upon dislocations that this work received general acceptance.

23

Reprinted from *Discussions Faraday Soc.*, No. 5, 33–39 (1949)

CRYSTAL GROWTH AND SURFACE STRUCTURE

Part I

BY W. K. BURTON * AND N. CABRERA

Received 15*th February*, 1949

Introduction. As a preliminary to the study of the rate of growth of crystals, we consider in Part I of this paper the influence of surface structure on the rate of advance of the growing surface. If, for the time being, we confine our attention to crystals with perfect lattices, it is found that crystal surfaces can be divided into two classes, (*a*) close-packed † surfaces and (*b*) non-close-packed or " stepped " surfaces, which possess essentially different properties. A surface is close-packed if, when it is as flat as possible, all the surface molecules are at the same distance from a plane parallel to it ; in all other cases the surface will present a stepped appearance, as in Fig. 1, the height of each step being of molecular dimensions. By way of illustration, in the simple cubic system (100) surfaces, (111) surfaces and (110) surfaces are close-packed, all other surfaces are stepped. In a stepped surface

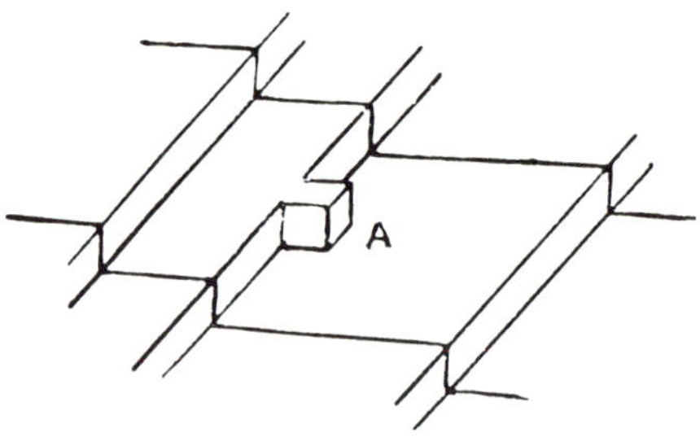

FIG. 1.

the terraces in Fig. 1 are pieces of close-packed surface. In the case of a (1*n*0) surface of a simple cubic crystal, the terraces will be (100) surfaces. This point of view is fruitful because it can be shown that for a stepped surface the portions between the steps play almost no part in growth phenomena at low supersaturations. In fact, deposition on a close-packed surface can only take place by surface nucleation : small " islands " of molecules collect on the surface and grow so as to produce a new layer, a process which is very slow at low supersaturations. On the other hand, deposition on the edge of a step A (Fig. 1) can take place without there being a linear nucleation process. Hence the growth problem for a stepped surface is essentially solved, once the corresponding problem for steps has been solved. If a crystal grows at all, some kind of steps must exist at some time in the surface. These steps may be of the kind already mentioned, or they may be boundaries of two-dimensional nuclei. Growth essentially depends on the existence of " kinks " in these steps. Easy growth is guaranteed if these

* Seconded from I.C.I., Ltd., Butterwick Research Laboratories, The Frythe, Welwyn, Herts.

† Note that our definition of " close-packed " differs somewhat from current usage.

kinks are always present, and this criterion can be reduced to the question of the existence of kinks, when the external concentration is the equilibrium value. For if kinks are present at equilibrium, then when the external concentration is raised there are already suitable deposition points available. If there are no kinks at equilibrium, then these must be created, and a large hindrance to growth appears. It can be shown that the concentration of kinks in a step in equilibrium is high, and that the concentration of kinks in a close-packed surface at equilibrium is negligible. This again speaks in favour of our classification of surfaces into stepped and close-packed surfaces.

In this paper we are concerned with the equilibrium structure and rate of growth of an infinite surface. It is, of course, clear that an infinite surface is in equilibrium with the same external concentration (e.g., vapour concentration) whatever the surface. A finite surface will not be in equilibrium in the same sense as in the case of an infinite surface, and consequently, in general, some change will tend to take place. But changes of orientation can take place only by means of processes which occur at the boundary of the surface, and hence for a surface of observable size the change will occur at an unobservably slow rate, the associated relaxation time tending to infinity with the size of the crystal. Therefore, if we confine ourselves to a region on a finite surface which is almost flat, then its structure will be the same as that in an infinite surface having the same orientation.

The two basic equilibrium problems are now (*a*) the equilibrium structure of an infinite step, and (*b*) the equilibrium structure of an infinite close-packed surface.

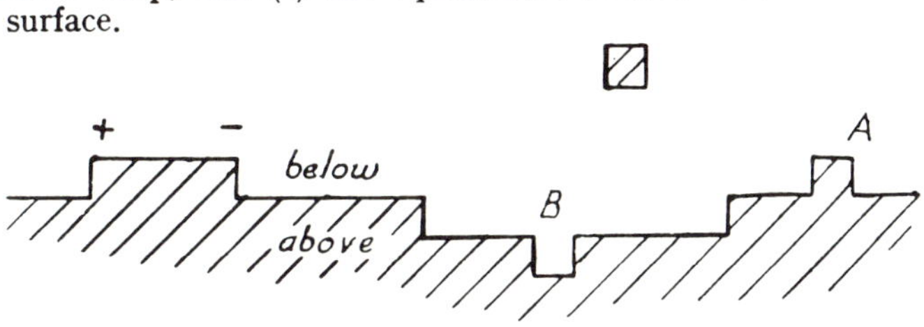

FIG. 2.

The Infinite Step. Consider then, a semi-infinite layer of molecules on an infinite close-packed plane crystal surface bounded by a connected line. This we call a step: it can have any mean direction. At $T = 0°$ K, the step will be perfectly straight, but as the temperature is increased it will consist of a number of "kinks," separated by certain distances as in Fig. 2, a certain number of adsorbed molecules (A) and a certain number of holes (B). We need to know only the concentration of kinks to form a picture of the mean structure of the step. This idea, introduced by Frenkel [1] simplifies the treatment of the problem very much: we call the kinks "Frenkel kinks" (F.k.). It is clear that the concentration of kinks in a step will depend on its orientation, and that there will be orientations for which the number of kinks is a minimum. For instance, for the (001) face of a simple cubic crystal, the (10) steps will have the smallest number of kinks. This minimum number will tend to zero with T. Accordingly, if we can show that a (10) step contains a large number of kinks at $T > 0$ under equilibrium conditions we know that steps of all orientations also contain a large number of kinks.

If we use a simple cubic model with nearest neighbour interactions (Kossel crystal) it is easy to find the equilibrium concentration of adsorbed molecules,

[1] Frenkel, *J. Physics, U.S.S.R.*, 1945, **9**, 392.

holes and kinks in a (10) step. Let the energy of interaction between neighbouring molecules be φ. Then the energy necessary to form an adsorbed atom in the step (Fig. 3) will be φ. The energy to form a hole is also φ, since an energy 2φ is required to form a hole and an adsorbed molecule. The energy to form a kink is, however, only $\frac{1}{2}\varphi$, since from Fig. 4 and Fig. 5 only an energy 2φ is required to form four kinks. There is no change in energy in going from Fig. 4 to Fig. 5. The numbers of positive and negative kinks (Fig. 2) are, of course, equal. We conclude that the probability for having a hole or an adsorbed molecule at a given place on the step are both given by

$$n = \exp(-\varphi/kT) \quad . \quad . \quad . \quad . \quad (1)$$

and that the probability for having a kink at a given place on the step is given by

$$n_+ = n_- = \exp(-\varphi/2kT) \quad . \quad . \quad . \quad . \quad (2)$$

If $T \sim 600°$ K and we take a typical value of φ as 0·2 eV, we find that there is a kink for every ten molecules in the step, and an adsorbed molecule or hole for every hundred molecules.

We have, of course, simplified the problem very much: there is a considerable probability for the existence of kinks of multiple height, particularly for steps which deviate a great deal from the [10] direction (in the simple cubic case). The complete theory has been developed elsewhere.[2]

FIG. 3.

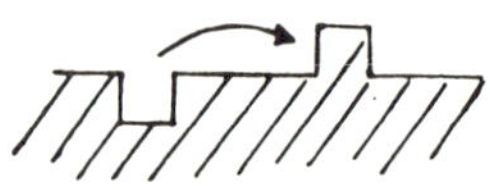

FIG. 4.

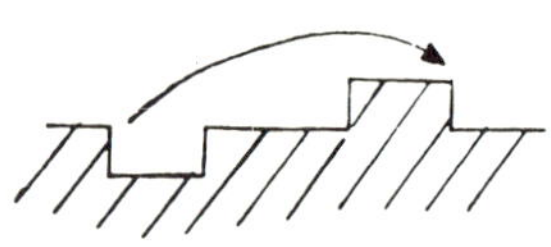

FIG. 5.

In view of the fact that steps of different orientations have different concentrations of kinks in them, the edge-free energy per unit length of a step varies with the orientation and is a minimum for the (10) step. It might be thought that steps other than (10) steps are not in real equilibrium and that there must be a tendency for these steps to change into (10) steps. If we are considering infinite steps, this conclusion would be erroneous, for it can be shown that steps of all orientations are in equilibrium with the same external concentration of molecules, despite the fact that the concentration of Frenkel kinks varies with the orientation.

Frenkel[1] has treated the kinetical problem of the transformation of any step into a (10) step, assuming on energy grounds that the others are not in equilibrium. He obtains in this way a time of relaxation independent of the length of the step. This result is incorrect because the "torque" which applies to the steps not in a [10] direction is evaluated by Frenkel by taking the derivative of the potential energy with respect to the orientation angle. But his formula contains only points corresponding to the equilibrium positions of the steps, and not the intermediate non-equilibrium positions through which the step would have to pass in order to change at all. In fact for an infinite step, each equilibrium position is surrounded by infinitely high potential barriers which cannot be surmounted.

For a finite step, the situation is different. Such a step can only be in equilibrium with a supersaturated or undersaturated external phase, and then the equilibrium is unstable and subject to stringent restrictions as regards shape. The sharper the corners of a finite step, the greater the

[2] Burton and Cabrera (to be published elsewhere).

rate of evaporation, and an arbitrarily oriented step tends to become a (10) step during the evaporation. The time required for this process to take place increases with the length of the step, because the processes which permit the transformation occur only at the corners.

It must not be thought, however, that all the considerations which apply to kinks in a step apply to steps in a surface. It is still true that surfaces of all orientations are in equilibrium with the same vapour concentration in the same sense as for steps, but the fact that for some surfaces the concentration of steps is large does not imply that double steps, treble steps, etc., will be frequent.* The difference between the energy of two single steps and one double step is proportional to the step length, and is very large if the interactions are not of the nearest neighbour type. Similarly, there is no question of steps being formed by thermal fluctuations,* as kinks are formed in a step, since the energy of formation of a step is proportional to its length and is enormous for long steps. Thus a stepped surface tends to become as flat as possible, and at equilibrium, only single steps will appear. It follows that the macroscopic steps which have been observed, e.g., on metals by Graf,[3] on growing crystals have nothing to do with equilibrium problems, but are essentially kinetic in origin. If a surface is produced with macroscopic steps in it, it is obvious that the rate of approach to macroscopic equilibrium is negligibly small and the structure is essentially frozen in.

Close-packed Surfaces. The circumstance which makes stepped surfaces so easy to treat is that the steps themselves present a one-dimensional problem. In each position on the step we have a variety of possible states: occupation by a kink of positive, negative or zero height. Each of these possible states can occur independently at each point, and hence the probability for the occurrence of a compound state affecting more than one position is the product of the probabilities for the individual states at each of the individual positions. We have assumed so far that those parts of the crystal surface between steps can be ignored, and this assumption is shown to be reasonable in the following discussion. However, if there are no steps in the surface, which is the normal case in a close-packed perfect crystal surface, then we are presented with an essentially different two-dimensional problem. We assume that in the close-packed surface of a crystal there can be differences of level, i.e., that " jumps " can occur in the surface. The presence of jumps provide suitable places for evaporation and condensation, provided that the jumps are not due merely to the presence of adsorbed molecules and holes. The problem is to estimate the number of jumps at equilibrium as a function of temperature.

In this case the jumps themselves cannot be assigned independently, since it is possible to have twice as many jumps in a surface as there are molecules. Fig. 6 shows a picture of part of a surface ; the small squares represent molecules seen from above. The heights of these molecules above some arbitrary plane can, of course, be assigned independently, but the distribution of jumps across the full lines in the figure cannot. For suppose we start at the molecule A and follow any closed path such as ABCDEF, then although we can have any jump we choose between neighbouring molecules on this path, providing we do not close it, the necessity for finishing at A at the same level at which we started implies that the magnitude of any jump on a closed path must be fixed by the magnitude of the others.

[3] Graf, *Z. Elektrochem.*, 1942, **48**, 181.

* At least, if the interactions are all attractive.

So there are innumerable sets of relations, corresponding to all the closed paths on the crystal surface, between the jump probabilities. In fact, to specify the probability for the existence of a jump at a given point involves the knowledge of the state of the surface at every other point. So we are faced with a so-called co-operative phenomenon. This makes the Frenkel kink picture employed previously almost unworkable. We must therefore look for some other method.

We have made preliminary calculations on the basis of a model which is somewhat oversimplified: we suppose the levels in the crystal surface to be capable of two values only. The method employed is that due to Montroll,[4] Kramers and Wannier,[5] Onsager,[6] Onsager and Kaufman[7] and Wannier,[8] originally devised for the treatment of ferromagnetism, using the two-dimensional Ising model. Just as there is a transition or critical temperature associated with an infinite specific heat in the case of the two-dimensional ferromagnet, so there is in the case of this crystal surface model.

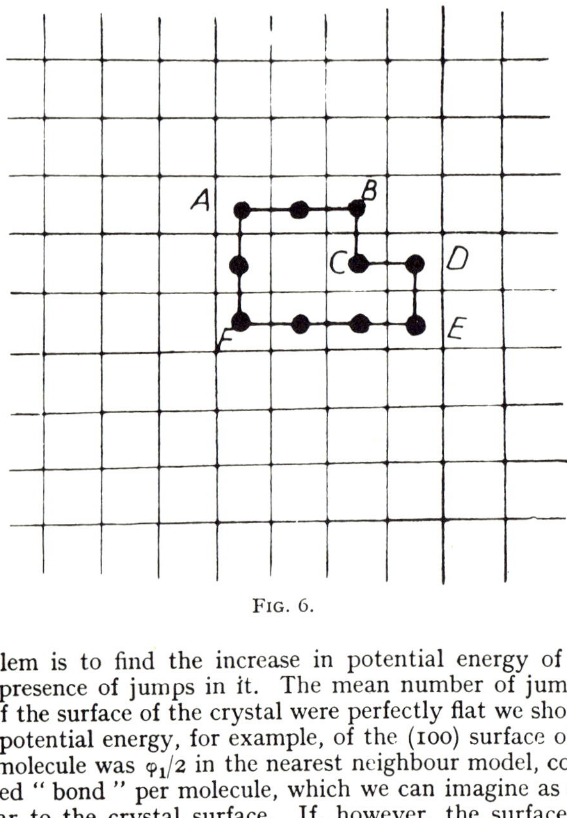

FIG. 6.

The problem is to find the increase in potential energy of the surface due to the presence of jumps in it. The mean number of jumps can then be found. If the surface of the crystal were perfectly flat we should say that the surface potential energy, for example, of the (100) surface of the Kossel crystal per molecule was $\varphi_1/2$ in the nearest neighbour model, corresponding to one unused " bond " per molecule, which we can imagine as sticking out perpendicular to the crystal surface. If, however, the surface is not flat, then there will be additional unused bonds sticking out parallel to the surface, and each of these bonds will contribute $\varphi_1/2$ to the potential energy

[4] Montroll, *J. Chem. Physics*, 1941, **9**, 706.
[5] Kramers and Wannier, *Physic. Rev.*, 1941, **60**, 252, 263.
[6] Onsager, *Physic. Rev.*, 1944, **65**, 117.
[7] Onsager and Kaufman, *Report Int. Conf. on Fund. Particles and Low Temperatures* (Cambridge, July, 1946), Vol. II.: Low Temperatures, Physical Society (1947).
[8] Wannier, *Rev. Mod. Physics*, 1945, **17**, 50.

of the surface. If we take our zero of energy to correspond to a flat surface, then if we evaluate the potential energy per molecule of the surface at equilibrium and divide it by $\varphi_1/2$ we get a figure for the number of unused bonds in the surface which are parallel to it. This figure, s, represents the "roughness" of the surface. This is the quantity we aim to evaluate as a function of temperature. We expect it to go from 0 to 1 as T goes from 0 to ∞.

The crystal surface model we are considering is equivalent to a square lattice of units which we call atoms capable of two states which we designate by $+1$ and -1. If two neighbouring atoms have the same state their interaction energy is zero, otherwise it is $\varphi_1/2$. The first possibility describes the two molecules in the crystal surface when they are at the same level, the second when their levels are different. For the sake of generality we assume, following Onsager, that the interactions can be different in the two directions [10] and [01]: in the case of the (100) surface we shall equate them.

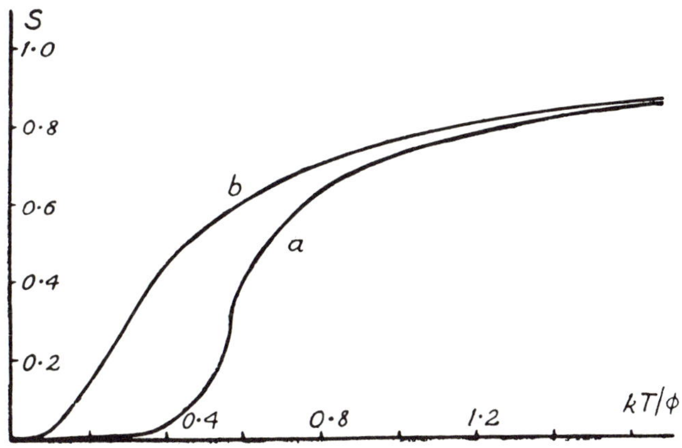

FIG. 7.

The problem is solved by studying the effect on the partition function for the surface by adding an extra row of molecules. The final result is, for the (100) surface

$$s = 1 - \tfrac{1}{2}\,(1 + 2k_2 K_1/\pi) \coth H\,, \qquad (3)$$

where $H = \varphi_1/2\boldsymbol{k}T\,;\ k_2 = 2\tanh^2 H - 1\,;$

$$K_1 = K(k_1) = \int_0^{\frac{\pi}{2}} \frac{d\omega}{(1 - k_1^2\sin^2\omega)^{1/2}}\,;\ k_1^2 + k_2^2 = 1 \qquad (4)$$

K_1 is the complete elliptic integral of the first kind. A graph of s against T is shown in Fig. 7 (a). The curve possesses a vertical tangent at $T = T_c$ given by $k_1 = 1$ or $k_2 = 0$, i.e., by $\sinh H_c = 1$,

or

$$H_c = \varphi_1/2\boldsymbol{k}T_c = \ln\cot\frac{\pi}{8} \sim 0{\cdot}9 \qquad (5)$$

If we had assumed the jumps to be independent, the result would have been

$$s = \frac{2\exp(-\varphi_1/2\boldsymbol{k}T)}{1 + \exp(-\varphi_1/2\boldsymbol{k}T)}\,, \qquad (6)$$

which gives rise to curve b in Fig. 7.

In the absence of square symmetry, Onsager [6, 7] has shown that that preceding condition becomes

$$\sinh H_c \sinh H_c' = 1 \ , \ (H_c' = \varphi_2/2kT_c) \ , \quad . \quad . \quad (7)$$

so that we can apply our result to the (110) surface of our model. In the case of this surface we have a rectangular lattice with energy of interaction $\varphi_1/2$ in one direction and $\varphi_2/2$ in the other, where φ_2 is the strength of the second nearest neighbour interaction. Consequently, if the second nearest neighbour interactions are small enough, the critical temperature can be as low as we like.

The value of T_c has been worked out for all surfaces in the symmetrical case.[8] The result is

$$\text{gd } H_c = \pi/Z \ , \quad . \quad . \quad . \quad . \quad (8)$$

where gd x is the Gudermannian function, and Z is the number of nearest neighbours in the surface, of a given molecule in the surface, when it is as flat as possible. There are three cases: (*a*) surface lattice triangular, $Z = 6$; (*b*) surface lattice square, $Z = 4$; (*c*) surface lattice hexagonal, $Z = 3$. For these cases we have respectively from (8),

$$\exp (2H_c) = 3 \ , \quad . \quad . \quad . \quad . \quad (9)$$

$$\sinh H_c = 1 \ , \quad . \quad . \quad . \quad . \quad (10)$$

$$\cosh H_c = 2 \ . \quad . \quad . \quad . \quad . \quad (11)$$

Eqn. (9) enables us to give the critical temperature for the (111) face of a simple cubic crystal. The surface in this case has a triangular lattice and the nearest neighbour interactions in the surface are due to the second nearest neighbours in the crystal lattice. From (9) we get

$$2H_c' = \varphi_2/kT_c \sim 1{\cdot}1 \ .$$

In Table I we give T_c under the assumption that $\varphi_1 = 0{\cdot}2$ eV, $\varphi_1/\varphi_2 = 8$. If T is much below T_c, the concentration of jumps is almost entirely due to the presence of adsorbed molecules.

TABLE I

Lattice	Surface	T_c
Simple cubic . .	(100)	1000° C
	(110)	400° C
	(111)	−30° C
Face-centred cubic . .	(111)	1700° C

We see that the critical temperature gives the boundary separating the régime where the jump concentration is negligible and surface nucleation is required for growth, $T < T_c$, from that régime $T > T_c$, where the jump concentration is high, and no nucleation is required for growth. If $T > T_c$ the surface will grow rapidly and disappear leaving only the slow growing faces for which $T < T_c$. At ordinary temperatures then, the transition phenomena will only occur for non-habit faces.

Broadly speaking, we have shown that close-packed surfaces will not grow at low supersaturations (see Part II) because of the need for surface nucleation. However, we have assumed that the crystal lattice is perfect. It is clear that the structure of a close-packed surface at equilibrium is extremely sensitive to lattice imperfections, and as Frank [9] has pointed out the presence of a few dislocations terminating in the surface ensures the presence of suitable sites for condensation or evaporation. In fact, dislocations again produce a stepped surface. The perfect close-packed surface will rarely, if ever, occur in reality.

H. H. Wills Physical Laboratory,
University of Bristol.

[9] Burton, Cabrera and Frank, *Nature*, 1949, **163**, 398. Frank, This Discussion.

24

Reprinted from *Discussions Faraday Soc.*, No. 5, 40–48 (1949)

CRYSTAL GROWTH AND SURFACE STRUCTURE

Part II

BY N. CABRERA AND W. K. BURTON *

1. Introduction. In Part I of this paper we studied the microscopic equilibrium structure of crystal surfaces and distinguished two kinds, stepped surfaces and close-packed or saturated surfaces. This second type is essentially the same as the first one, when the temperature is above a certain critical value. Below this temperature the close-packed surfaces behave in quite a different way.

In Part II we treat the kinetic problem of the growth of these surfaces from the vapour ; in all cases it is necessary to consider the diffusion of adsorbed atoms on the surface of the crystal. First, we treat the growth of stepped surfaces ; quantitative formulæ are given as a function of the inclination. The growth of close-packed surfaces below their critical temperature by the mechanism of two-dimensional nucleation is also considered. Treatments of two-dimensional nucleation have been given by several authors, especially by Becker and Döring,[1] neglecting the effect of the diffusion of adsorbed atoms. We discuss in this paper the influence of this diffusion and we show that in spite of the fact that the predicted rate of growth is perhaps different from that given by Becker and Döring, it will certainly not account for the experimental fact that in the small number of cases where a critical supersaturation for growth has been observed, it is of the order 0·01 at most. Finally, a treatment of the initiation of growth of an imperfect crystal containing a random distribution of dislocations is given, and is shown to be in agreement with experiment. The results of this work will be given here in a preliminary way ; the complete treatment will be published elsewhere.

2. Growth of Stepped Surfaces. In Part I we showed that each step in the surface contains in equilibrium a very high concentration of Frenkel kinks, much higher than the concentration of adsorbed atoms in the edge of the step. The equilibrium concentration of adsorbed atoms on the surface is

$$\frac{1}{a^2}\, e^{-W_s/kT}, \quad . \quad . \quad . \quad . \quad . \quad (1)$$

where a is the interatomic distance and W_s the evaporation energy from the step into the surface.

When the vapour pressure is increased by a factor α (saturation ratio $= \alpha$; supersaturation $= \alpha - 1$), the concentration of adsorbed atoms remains practically equal to that in equilibrium near the steps, because of the high concentration of Frenkel kinks in them. The saturation ratio α_s on the surface increases from the value 1 near the steps to a maximum value between the steps. The current of atoms condensing into the steps will be controlled essentially by the diffusion of adsorbed atoms on the surface.

[1] Becker and Döring, *Ann. Physik*, 1935, **24**, 719.

* Seconded from I.C.I., Ltd., Butterwick Research Laboratories, The Frythe, Welwyn, Herts.

The actual variation of α_s can be obtained solving the corresponding diffusion problem, the result being

$$\alpha - \alpha_s(x) = (\alpha - 1)\frac{\cosh \beta x}{\cosh \beta x_0}, \quad . \quad . \quad . \quad (2)$$

where xa is the distance counted from half-way between the steps, $2x_0a$ is the mean distance between them, and

$$\beta^2 = \exp\{-(W_1 - U)/kT\}, \quad . \quad . \quad . \quad (3)$$

W_1 being the energy necessary to take an adsorbed atom from the surface into the vapour ($W = W_s + W_1$ is the total evaporation energy) and U the activation energy for diffusion on the surface. Therefore β^2 is the ratio between the probabilities for an adsorbed atom to evaporate into the vapour and to diffuse on the surface. Usually $W_1 > U$; if U is very small, β will also be very small, and the concentration of adsorbed atoms on the surface will be practically uniform and equal to that in equilibrium, unless x_0 is very big. As U increases the non-uniformity on the surface becomes more important. If $W_1 < U$, the diffusion on the surface does not play any role, the condensation into the steps takes place directly from the vapour. We do not think that this is likely to occur.

In order to illustrate the values that β can take, let us consider a face-centred cubic crystal with a stepped surface consisting of terraces (1, 1, 1) and steps in any direction, and let us call φ the interaction energy between nearest neighbours. Then $W_1 \sim 3\varphi$ and $U \sim \varphi$, therefore

$$\beta = \exp\{-\varphi/kT\} \sim 0{\cdot}05,$$

if, e.g., $\varphi/kT \sim 3$.

The current j of atoms condensing per sec. per cm. into each step will be equal to the current of atoms condensing from the vapour on the surface between two steps; therefore, from (2),

$$= 2\frac{\nu}{a}e^{-W/kT}\frac{1}{\beta}\tanh \beta x_0, \quad . \quad . \quad . \quad (4)$$

where ν is the frequency of vibration of the adsorbed atoms. The velocity of advance of the steps,

$$v = a^2 j,$$

will be a function of the distance between steps. For $x_0\beta \ll 1$ the velocity of the steps is proportional to the distance between them; as x_0 increases above β^{-1}, v tends to a maximum value v_∞ given by

$$v_\infty = 2(\alpha - 1)\nu a \exp\{-(W + W_s + U)/2kT\}, \quad . \quad . \quad (5)$$

which represents the velocity with which a *single* step would move. For the example considered above

$$\exp\{-(W + W_s + U)/kT\} = \exp\{-5\varphi/kT\};$$

putting $\nu \sim 10^{12}$ sec.$^{-1}$, $a \sim 10^{-8}$ cm., $\varphi/kT \sim 3$ and $\alpha - 1 \sim 10^{-2}$, we get $v_\infty \sim 10^{-5}$ cm./sec.

The total current J of atoms adsorbed per cm.^{2} per sec. is

$$J = Nj,$$

where $N = 1/2x_0a$ is the number of steps per cm., therefore

$$Na \geqslant \tfrac{1}{2}\beta; \quad J_0 = (\alpha - 1)\frac{\nu}{a^2}e^{-W/kT} \quad . \quad . \quad . \quad (6)$$

$$Na \leqslant \tfrac{1}{2}\beta; \quad J = 2NaJ_0/\beta \quad . \quad . \quad . \quad (7)$$

For values of $Na \sim \beta$, the current J will depend on the distribution of steps in the surface. The curve, Fig. 1, has been calculated assuming a random distribution. For $Na \sim 0$ and $Na \sim 1$ the stepped surface transforms into close-packed surfaces of quite different character (see § 3). We see that the rate of growth of stepped surfaces is practically given by formula (6), down to small values of *Na*.

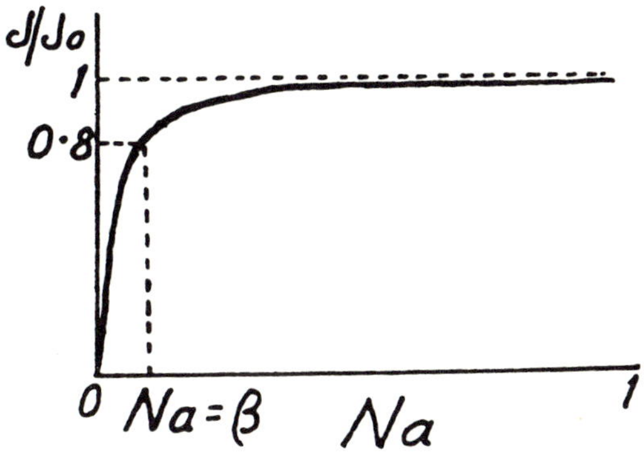

FIG. 1.—Rate of growth of stepped surfaces as a function of the number *N* of steps per cm.

Recent experimental work by Graf,[2] and Mahl and Stranski,[3] suggests that the stepped surfaces present a striated structure during growth from the vapour, the distance between the striations being of the order of 10^{-4} cm. We showed in Part I that this striated structure cannot be in equilibrium with the vapour, the surfaces always having a tendency to be as flat as possible. We believe that the formation of these striations has an essentially kinetical character and is probably related to the fact that the velocity of displacement of the steps is a function of the distance between its nearest neighbours. A correct theory has not yet been obtained.

3. Growth of Close-packed Surfaces. Let us consider now the surfaces without steps. The surfaces with small indices will correspond to this type. We considered in Part I the equilibrium structure of these surfaces, and we have shown that they will remain practically flat if the temperature is below a certain critical temperature, but they will contain a considerable number of Frenkel kinks if the temperature is above it. It is easy to show that for close-packed surfaces above their critical temperature the rate of growth is again given by (6). This type of surface can occur, for instance, as the limiting case of a stepped surface when $Na \sim 1$.

SURFACE NUCLEATION PROCESS.—Let us now consider the growth of habit surfaces, where Frenkel kinks do not occur. A long time ago Gibbs, and later on Volmer, suggested that the growth of these surfaces requires a two-dimensional nucleation process.

The theory of nucleation, especially in the case of formation of three-dimensional liquid or solid nuclei from the vapour, has been developed by Volmer,[4] Stranski [5] and Becker and Döring.[1] In that case the diffusion in the vapour plays a small role and the supersaturation is practically the same all over the volume. The tendency for small nuclei to evaporate is very big and the supersaturation required for them to grow has to be high

[2] Graf, *Z. Elektrochem.*, 1942, **48**, 181.
[3] Mahl and Stranski, *Z. Metallkunde*, 1943, **35**, 147.
[4] Volmer, *Z. physik. Chem.*, 1926, **119**, 277.
[5] Kaishew and Stranski, *Z. physik. Chem. B*, 1934, **26**, 317.

($\alpha - 1$ of the order of 4). The number of nuclei of critical size formed per sec. per cm.^{3} turns out to be of the form,

$$I = B\ e^{-A/kT}, \qquad . \quad . \quad . \quad . \quad (8)$$

where $B \sim 10^{20}$ cm.$^{-3}$ sec.$^{-1}$ and A is the increase in free energy necessary for the formation of a nucleus of critical size; A is tremendously high for all values of α smaller than 4.

Frenkel [6] and Becker [7] applied the same ideas to cases in which there is a diffusion of the atoms condensing into the nuclei, such as occurs in precipitation in a supersaturated alloy. They assumed that the only change in formula (8) to be made in this case is to multiply by a factor $\exp(-U/\boldsymbol{k}T)$, where U is the activation energy for diffusion. Actually this assumption is not entirely correct; it implies that the supersaturation is the same all over the volume, which is not true if diffusion exists.

The surface nucleation required for the growth of habit surfaces can be treated in a similar way, and the diffusion of atoms on the surface has also to be taken into account.

Before studying two-dimensional nucleation, let us consider the growth of a single nucleus on the top of a habit face. We assume a circular shape of radius ρa; if μ is the number of atoms contained in it,

$$\rho = \sqrt{\mu/\pi}.$$

If the total energy of a nucleus of $\mu = \pi\rho^2$ atoms is

$$\pi\rho^2 W_s - 2\pi\rho\gamma,$$

where γ is the edge energy per atom, the mean evaporation energy from the nucleus is

$$W(\rho) = W_s - (\gamma/\rho) \qquad . \quad . \quad . \quad . \quad (9)$$

Let α be the supersaturation ratio in the vapour; assuming the nucleus to be big enough, it will not change appreciably before a steady distribution of adsorbed atoms around it has been formed. Under these conditions the diffusion problem can be solved, and the supersaturation ratio $\alpha_s(r)$ on the surface around the nucleus is given by

$$\begin{aligned} r < \rho,\ \alpha - \alpha_s(r) &= [\alpha - \alpha_s(\rho)]\frac{I_0\,(\beta r)}{I_0\,(\beta\rho)}, \\ r > \rho,\ \alpha - \alpha_s(r) &= [\alpha - \alpha_s(\rho)]\frac{K_0\,(\beta r)}{K_0\,(\beta\rho)}. \end{aligned} \qquad . \quad . \quad (10)$$

I_0 and K_0 are the Bessel functions of first and second kind with imaginary argument. $\alpha_s(\rho)$ is the supersaturation ratio near the edge, which by definition is

$$\alpha_s(\rho) = \alpha_s^e(\rho) + \frac{j(\rho)}{4\pi\rho\nu}\,e^{(W_s+U)/\boldsymbol{k}T},$$

where

$$\alpha_s^e(\rho) = \exp\{\gamma/\rho\boldsymbol{k}T\}$$

is the supersaturation ratio which would be in equilibrium with the nucleus, and $j(\rho)$ is the current of atoms condensing per sec. into the nucleus. The current is then calculated as the number of atoms condensing from the vapour per sec. all over the surface. The result is

$$j(\rho) = 4\pi\rho\nu[\alpha - \alpha_s^e(\rho)]\frac{e^{-W/\boldsymbol{k}T}}{\beta^2[2\rho \cdot I_0(\beta\rho) \cdot K_0(\beta\rho) + 1]} \qquad . \quad (11)$$

[6] Frenkel, *Sowjet Physik*, 1932, **1**, 498.
[7] Becker, *Ann. Physik*, 1938, **32**, 128; *Proc. Physic. Soc.*, 1940, **52**, 71.

The radial velocity is

$$v(\rho) = aj(\rho)/2\pi\rho.$$

Current and velocity change sign when $\alpha = \alpha_s^e(\rho)$, which defines the critical nucleus to have a radius ρ_c given by

$$\rho_c = \gamma/\boldsymbol{k}T \ln \alpha \quad . \quad . \quad . \quad . \quad . \quad (12)$$

For $\rho < \rho_c$ the nucleus evaporates, for $\rho > \rho_c$ the nucleus grows. The maximum velocity of growth of the nucleus, for $\rho \gg \rho_c$ and $\beta\rho \gg 1$, reduces of course to formula (5). If ρ_c is big, such that $\beta\rho_c > 1$, the velocity is given by

$$v(\rho) = v_\infty\left(1 - \frac{\rho_c}{\rho}\right), \quad . \quad . \quad . \quad . \quad . \quad (13)$$

valid for $\rho \gg \rho_c$. For $\beta\rho_c < 1$, the velocity curve becomes steeper near the critical size. This is illustrated in Fig. 2. The concentration $\alpha_s(\rho)$ near the edge of the nucleus is seen to be practically the same as $\alpha_s^e(\rho)$; it is a little bigger when $\rho < \rho_c$ and smaller when $\rho > \rho_c$. The correction increases as ρ decreases but remains small.

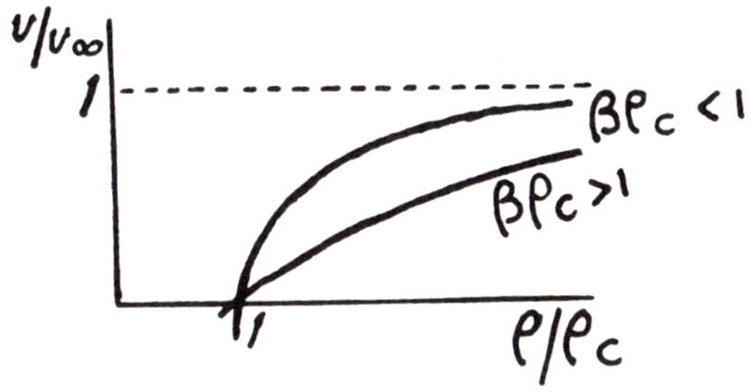

FIG. 2.—Radial velocity of a nucleus bigger than the critical size.

Now let us consider the current J of atoms condensing per cm.^{2} per sec. on a close-packed surface. In a steady state, when a constant supersaturation $\alpha - 1$ is maintained in the vapour, there will be on the surface a stationary distribution n_μ of nuclei of different sizes. They grow until they collide with other nuclei in the same atomic layer and disappear ; we can take this into account by assuming that there is a certain maximum value M for μ. Given the actual distribution n_μ, there will be a surface supersaturation ratio $\alpha_s(x, y)$, a function of position. The total current J will be equal to

$$J = (\alpha - \bar{\alpha}_s)\frac{\nu}{a^2}e^{-W/\boldsymbol{k}T} \quad . \quad . \quad . \quad . \quad . \quad (14)$$

where $\bar{\alpha}_s$ is the mean value of α_s all over the surface. When α is very near 1, the number of nuclei on the surface is small, and therefore most of the surface has a supersaturation ratio $\alpha_s \sim \alpha$; consequently $\bar{\alpha}_s \sim \alpha$ and the current J will be very small. As α increases the number of nuclei increases and the mean value α_s decreases, becoming $\bar{\alpha}_s \sim 1$ when the proportion of big nuclei is high and the distances between them small. We expect therefore that J as a function of α will be represented by a curve such as is illustrated in Fig. 3, where the straight line corresponds to the formula (6).

The calculation of $\bar{\alpha}_s$ as a function of α is a very difficult problem ; it requires, of course, the knowledge of the distribution n_μ of nuclei of different size as a function of α. Nevertheless for small values of $\alpha - 1$ for which the current J remains small (region OA in Fig. 3), it can be estimated using a method proposed by Becker and Döring [1] which we cannot develop here.

We assume that the growth of any nucleus is due fundamentally to the condensation of single adsorbed atoms; this assumption is correct only for a small density of nuclei, and therefore when $\alpha - 1$ is small. Under these conditions, the number I of nuclei of any size μ formed per sec. per cm.^{2} from nuclei $\mu - 1$ can be calculated if the ratios $q(\mu)/a(\mu)$ of the mean probability of evaporation $q(\mu)$ to that of growth $a(\mu)$ for every nucleus are known. The current J of atoms condensing per sec. per cm.^{2} is then calculated from $J = IM$, where M is the maximum value of μ. In general,

$$q(\mu)/a(\mu) = \alpha_s^e(\mu)/\alpha_s(\mu),$$

the ratio of the supersaturation ratio which would be in equilibrium with the nucleus to that actually existing near it.

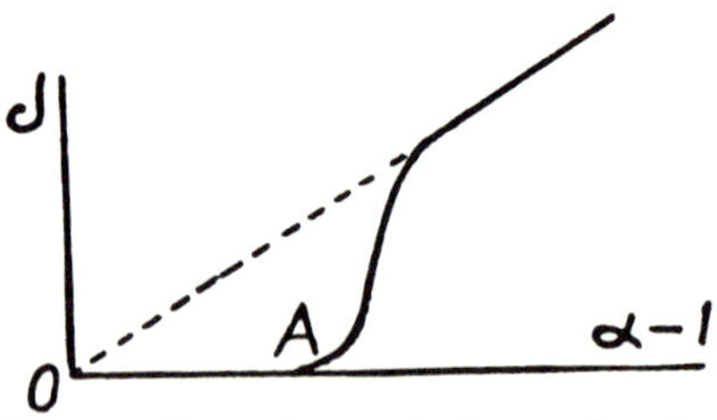

FIG. 3.—Rate of growth J as a function of $\alpha - 1$.

Let us first assume that the influence of the diffusion on the surface is unimportant, and therefore $\alpha_s(\mu) = \alpha$ for all nuclei. Under these conditions one gets for the current J the expression,

$$J = \frac{\nu}{a^2} e^{-(W+U)/kT} e^{-A_c/3kT}, \qquad (15)$$

where

$$A_c = -\pi\rho_c^2 kT \ln \alpha + 2\pi\rho_c\gamma = \gamma^2\pi/kT \ln \alpha,$$

is the increase in free energy necessary for the formation of a critical nucleus. The factor $\frac{1}{3}$ multiplying A_c comes from the calculation of M, which happens to contain a factor $\exp\{\frac{2}{3}A_c/kT\}$; this of course is assuming that the surface itself is much bigger than M. It is easy to see that expression (15) will give a negligible rate of growth, unless α is of the order of 2. Actually, according to Volmer and Schultze [8] a linear rate of growth is observed above $\alpha - 1 \sim 10^{-2}$. For this supersaturation, taking

$$\gamma \sim \varphi, \text{ and } \varphi/kT \sim 3,$$

the exponent in (15) becomes $A/3kT \sim 10^3$, and therefore no growth should occur at all.

Let us now consider the influence of diffusion. This is a very difficult problem, for which only a qualitative answer has been found.

There are two conflicting effects. First of all, the nuclei bigger than the critical size, which are therefore on the average growing, decrease the concentration of adsorbed atoms in the neighbourhood of their edge, with the result that the current of condensation in these nuclei is now much lower than before. On the other hand, the nuclei smaller than the critical size, which are in the average evaporating, may tend to increase the concentration of adsorbed atoms around their edge, and consequently the probability for them to grow further is higher than it was before. From the study of the evaporation of a *single* nucleus we showed that there was an increase of concentration near its edge. Nevertheless, we do not think that the same considerations apply to the assembly of nuclei in the nucleation process. In that case owing to the fact that the distribution of nuclei n_μ is a decreasing function of μ more nuclei are coming to the size μ per sec. by growth from $\mu - 1$ than nuclei coming by evaporation from $\mu + 1$, and this difference increases when μ decreases; therefore we think that the supersaturation ratio $\alpha_s(\mu)$ near the nuclei μ must tend to the value α for values of μ not very small compared with the critical size μ_c.

[8] Volmer and Schultze, *Z. physik. Chem. A*, 1931, **156**, 1.

As an illustration, let us suppose that

$$\alpha_s(\mu) = \alpha$$

for nuclei μ smaller than a certain size μ_0, and

$$\alpha_s(\mu) = \alpha_s^e(\mu)$$

for bigger nuclei. Then the total current J can be shown to be

$$J = \frac{\nu}{a^2} e^{-(W+U)/kT} e^{-A_0/kT} \quad . \quad . \quad . \quad . \quad (16)$$

where

$$A_0 = -\pi\rho_0^2 kT \ln \alpha + 2\pi\rho_0\gamma,$$

is the increase in free energy necessary for the formation of a nucleus of size $\mu_0 = \pi\rho_0^2$. This formula will give a bigger current than (15) if $A_0 < A_c/3$, therefore if $\mu_0 < 0{\cdot}04 \times \mu_c$. This is reasonable if we consider that for supersaturations of the order $\alpha - 1 \sim 10^{-2}$, assuming always

$$\gamma/kT \sim \varphi/kT \sim 3,$$

μ_c is of the order 10^5. On the other hand, it can be shown that formula (16) will account for the observed rate of growth at supersaturations of the order 10^{-2}, only if $\mu_0 < 10$ which is certainly too small. We conclude from these considerations that, in spite of the fact that the diffusion perhaps changes the current given by the simple nucleation theory, it does *not* agree with the current experimentally observed.

Role of Dislocations in Crystal Growth.—According to Frank,[9] the surface of any real crystal must contain a certain number of dislocations, with a screw component, terminating in the surface and producing steps which do *not* disappear during growth. Under these conditions the two-dimensional nucleation is no longer necessary. The current of condensation in the special case of a random distribution of these dislocations can be estimated in the following way.

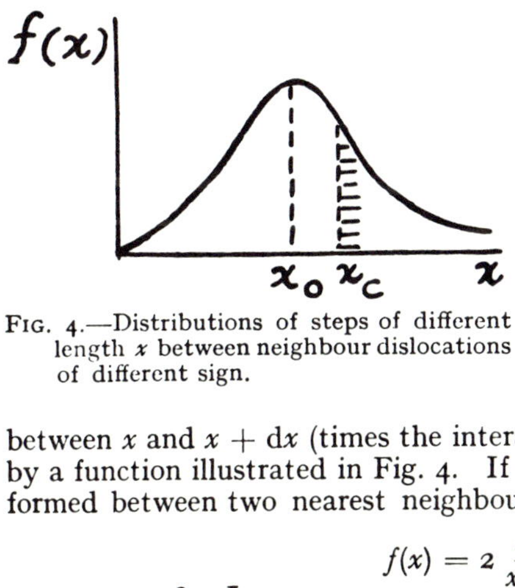

Fig. 4.—Distributions of steps of different length x between neighbour dislocations of different sign.

Let us suppose that we have N dislocations per cm.^{2} distributed at random. We assume also that there is practically the same number of dislocations of both signs. Steps will then exist between pairs of dislocations of different sign. During growth the coupling between dislocations will change, but the steps will be formed nearly always between neighbouring dislocations of different sign. The proportion $f(x)\mathrm{d}x$ of steps of length between x and $x + \mathrm{d}x$ (times the interatomic distance a) will be represented by a function illustrated in Fig. 4. If we assume that the steps are always formed between two nearest neighbour dislocations of different sign, then

$$f(x) = 2\frac{x}{x_0^2} e^{-\left(\frac{x}{x_0}\right)^2}, \quad . \quad . \quad . \quad . \quad . \quad (17)$$

where $x_0^2 = \dfrac{2}{\pi}\dfrac{1}{Na^2}$.

Let $\alpha - 1$ be the supersaturation in the vapour. Then all the steps of length x bigger than the diameter x_c of the critical nucleus, given by

$$x_c = \frac{2\gamma}{kT \ln \alpha} \sim \frac{2\gamma}{kT(\alpha - 1)}, \quad . \quad . \quad . \quad . \quad (18)$$

[9] Frank, This Discussion.

will grow freely, until they collide with other steps. The length of the steps remains always of the order of x. The steps of length $x < x_c$ will have a very small probability for growth; we shall neglect their contribution to the total current.

The current of atoms condensing per sec. into the steps of length x, $j(x)$ will also be a function of the distance to the nearest neighbour steps; nevertheless, provided the condition $\beta x_0 \gg 1$ is satisfied (see §1), we can use formula (5), that is to say,

$$j(x) = \frac{2\nu}{\beta} e^{-W/kT} (\alpha - 1)x.$$

The total current J of atoms condensing per sec. per cm.² in surface will now be given by

$$J = \frac{N}{2} \int_{x_c}^{\infty} j(x) f(x) dx. \quad . \quad . \quad . \quad . \quad (19)$$

Fig. 5 illustrates the current obtained from (19) as a function of $(\alpha - 1)$ and for a given value of N. For $x_c > x_0$ ($\alpha - 1$ very small), J is given by the expression

$$J = \frac{Na^2\nu}{\beta a^2} e^{-W/kT} \exp\left\{-\left(\frac{x_c}{x_0}\right)^2\right\},$$

where x_0 and x_c are given by (17) and (18). For $x_c < x_0$ the current tends to a linear law of the form

$$J = \frac{N^{1/2} a}{\beta} \frac{\nu}{a^2} e^{-W/kT} (\alpha - 1). \quad . \quad . \quad . \quad (20)$$

The critical supersaturation (point C in Fig. 5) for which the current becomes practically linear is given by $x_0/x_c \sim 2$,

$$\alpha - 1 \sim 2\sqrt{2\pi} \frac{\gamma}{kT} Na^{1/2}. \quad . \quad . \quad . \quad (21)$$

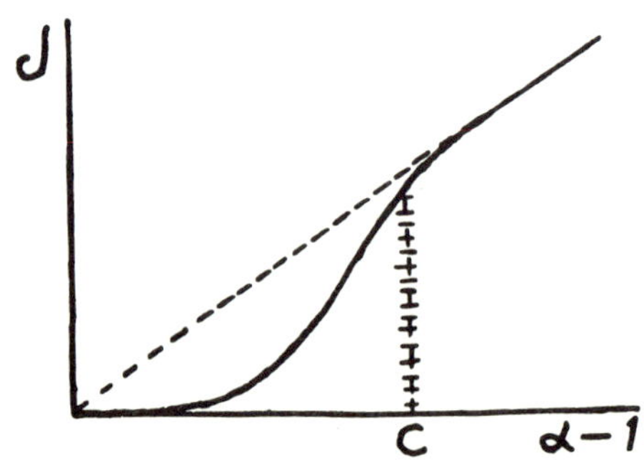

FIG. 5.—Rate of growth of a close-packed surface containing N dislocations per cm.² as a function of the supersaturation $\alpha - 1$.

The general shape of the curve represented in Fig. 5 agrees with the experimental results of Volmer and Schultze,[8] for the growth of iodine crystals at 0° C. The critical supersaturation is in their case $\alpha - 1 \sim 10^{-2}$, which agrees with the value given by (21), taking $\gamma/kT \sim 3$ and assuming $N = 10^8$ cm.$^{-2}$ which agrees with the value generally assumed to explain the mechanical properties of crystals.

Volmer and Schultze (loc. cit.) observe also a linear law of growth, as a function of $\alpha - 1$, for $\alpha - 1 > 10^{-2}$. The experimental value of the rate of growth for $\alpha - 1 = 10^{-2}$ is of the order of 10^2 atomic layers per sec.

Formula (20) is strictly speaking only applicable to simple monoatomic substances; in order to apply it to complicated structures as iodine, we have just to calculate W and ν in such a way to account for the saturation vapour pressure of iodine. Using the experimental values of Gillespie and Fraser,[10] one obtains $W = 0{\cdot}7$ e$\sqrt{}$, $\nu = 0{\cdot}4 \times 10^{17}$ sec.$^{-1}$.

Putting these values, and $N \sim 10^8$ cm.$^{-2}$, $\beta \sim 10^{-2}$, into formula (20) one obtains a rate of growth of the order of the experimental value.

Department of Physics,
University of Bristol.

[10] Gillespie and Fraser, *J. Amer. Chem. Soc.*, 1936, **58**, 2260.

25

Reprinted from *Discussions Faraday Soc.*, No. 5, 48–54 (1949)

THE INFLUENCE OF DISLOCATIONS ON CRYSTAL GROWTH

BY F. C. FRANK

Received 16th February, 1949

The kinetic theory of the nucleation of new phases, developed especially by Volmer,[1] by Farkas,[2] by Kaischew and Stranski [3] and by Becker and Döring,[4] indicates that under typical conditions * the self-nucleation from vapour of new crystals, new liquid drops and fresh two-dimensional monolayers of molecules on a " saturated " crystal face require respectively supersaturations of the vapour by factors of typically 10, 5 and 1·5 respectively in order to proceed at appreciable rates. Experimentally, the first two of these figures are apparently correct : but the third is much larger than the actual supersaturation required to cause further growth of a crystal already formed. In fact, the existence of a critical finite supersaturation for further growth has only been established for a few materials, and then for individual faces of individual crystals, being different from case to case ; at the most it is about 1 %. Volmer and Schultze,[5] who found a critical supersaturation of 0·8 % for the growth of an iodine crystal from the vapour, interpreted this as the critical supersaturation for two-dimensional nucleation : but the quantitative discrepancy is far too great (for details of the growth rate formulæ, see the contributions of Burton and Cabrera to this Discussion).

However, this discrepancy is not in the least surprising. One ought not to expect that any visible crystal will exhibit a completed perfect face needing fresh two-dimensional nucleation in order to grow. Investigation of the mechanical properties of solids shows that no macroscopic specimen ever exhibits the theoretical strength of the perfect crystal ; and this enormous discrepancy (a factor of 100, say, and *more* for " good "

[1] Volmer and Weber, *Z. physik. Chem.*, 1926, **119**, 277. Volmer, *Kinetik der Phasenbildung* (Leipzig, 1939).
[2] Farkas, *Z. physik. Chem.*, 1927, **125**, 236.
[3] Kaischew and Stranski, *Z. physik. Chem. B*, 1934, **26**, 317 ; *Physik. Z.*, 1935, **36**, 393.
[4] Becker and Döring, *Ann. Physik*, 1935, **24**, 719.
[5] Volmer and Schultze, *Z. physik. Chem. A*, 1931, **156**, 1.

* Typical conditions may be taken as such that the vapour pressure lies within a few powers of 10 of 1 mm. Hg : or the temperature between about 0·5 and 0·8 times T_b, the boiling point in ° K.

crystals) is attributed to the presence of dislocations. In the early stages of the development of dislocation theory by Polanyi,[6] Orowan [7] and Taylor,[8] only one aspect of the dislocation was recognized, in which the displacement direction was normal to the dislocation line. In 1939 Burgers [9] drew attention to the " screw " form assumed by the dislocation when the displacement is parallel to the dislocation line, and the developments of dislocation theory by Mott and Nabarro,[10] Frank [11] and others emphasize the fact that dislocation lines can be curved and exhibit any orientation. Fig. 1 shows (*a*) as a continuous deformation of a plane, (*b*) in a block model, the form of a simple cubic crystal when a screw dislocation emerges normally at the

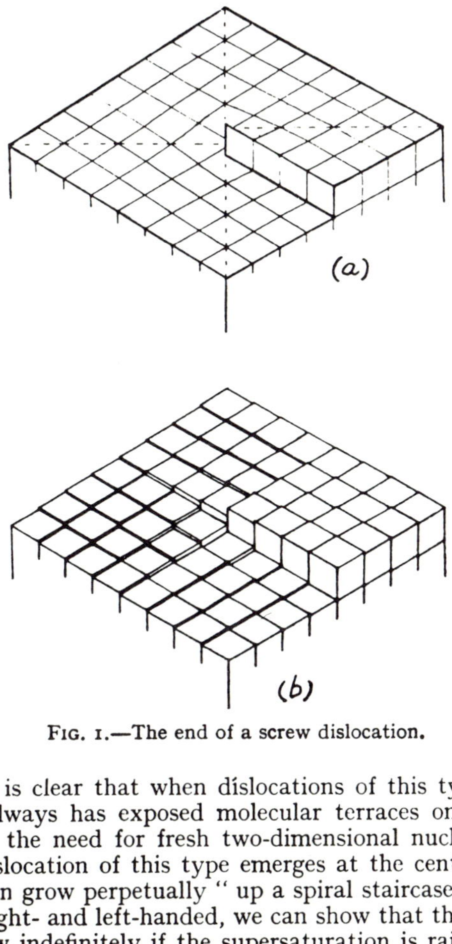

FIG. 1.—The end of a screw dislocation.

cube-face. It is clear that when dislocations of this type are present, the crystal face always has exposed molecular terraces on which growth can continue, and the need for fresh two-dimensional nucleation never arises. If just one dislocation of this type emerges at the centre of the face, that crystal face can grow perpetually " up a spiral staircase." If there are two, respectively right- and left-handed, we can show that the terrace connecting them will grow indefinitely if the supersaturation is raised to such a value that the diameter, l_0, of the critical two-dimensional nucleus is less than the distance between them. More precisely, since the critical nucleus is not

[6] Polanyi, *Z. Physik*, 1934, **89**, 660.
[7] Orowan, *Z. Physik*, 1934, **89**, 634.
[8] Taylor, *Proc. Roy. Soc. A*, 1934, **145**, 362.
[9] Burgers, *Proc. Kon. Ned. Akad. Wet.*, 1939, **42**, 293.
[10] Mott and Nabarro, *The Strength of Solids* (Physical Society, London), 1948, p. 1.
[11] Frank, *ibid.*, p. 46.

circular, we should say that growth occurs when the critical nucleus, correctly oriented, will pass between two points in the positions of the two dislocations. Likewise when a single dislocation is close to the boundary of the face, growth occurs if the critical nucleus will pass between the dislocation and the boundary. If it will not pass, the terrace rests in a position corresponding to a portion of the boundary of the critical nucleus (Fig. 2). For rough calculations we may approximate the boundary as a circle of radius $\frac{1}{2}l_0$, and we have a close analogy to the equilibrium (or non-equilibrium) of a bubble at an orifice.

Measuring in molecular spacings, we have

$$l_0 = \varphi/\boldsymbol{k}T \ln \alpha,$$

where α is the saturation ratio: since the supersaturation is small, $(\alpha - 1)$ may be written in place of $\ln \alpha$. φ is the neighbour-neighbour binding energy of the crystal. For a rule of thumb we may use Trouton's rule and take $\varphi/\boldsymbol{k}T$ as $3{\cdot}5\ T_b/T$. Thus at an absolute temperature of 0·6 of the boiling point of the material, and a supersaturation of 1 %, l_0 is about 600 molecular spacings: or 6000 at 0·1 %.

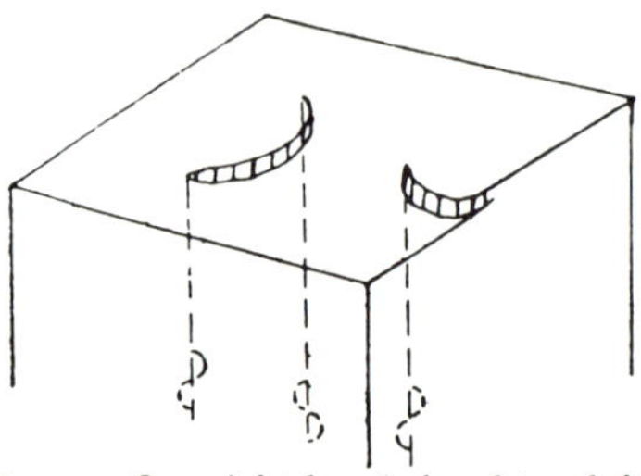

FIG. 2.—One right-handed and two left-handed screw dislocations ending in a crystal face.

In the theory of mechanical deformation, it is commonly estimated that in an annealed metal crystal there are about 10^8 dislocation lines intersecting each square centimetre. This appears to be the limiting perfection attainable in metals, and would be classed, indeed, as high perfection from the point of view of X-ray diffraction. (This limit may well be connected with the impurity content.) This is only an order-of-magnitude estimate, and does not necessarily apply to other materials than metals: but it serves as a guide, and suggests that critical supersaturations of the order of a fraction of 1 % are reasonably to be expected.

We are perforce limited for quantitative discussion to the case of growth from the vapour of homopolar crystals, since the classical nucleation rate theory has not been quantitatively developed for any other case. But experimental indications, and such theoretical guesswork as we can make, suggest that the conditions governing the growth from solution of crystals, including ionic crystals, are substantially similar. It is possible, but less certain, that the same is true of growth from the melt. Perhaps the most distinctively different case is that of the growth, either from vapour or solution, of crystals of highly non-equiaxed organic molecules. There are indications that in such cases growth proceeds through the formation of adsorption films, dense, but differing in molecular orientation from the bulk crystal (e.g., liquid, or liquid-crystalline) within which subsequent rearrangement occurs.

The general importance of dislocations for crystal growth accounts immediately for many observations, such as the individual behaviour of each crystal face, particularly on the microscopic scale, leading sometimes to such unexpected results as the formation of lath-shaped crystals for a lattice of cubic symmetry.

Under steady uniform supersaturation the terrace attached to an isolated growth-promoting dislocation in a crystal otherwise perfect in its neighbourhood will grow outwards in a spiral of which the spacing between turns, and the rate of their advance, will be uniform at a considerable distance from the centre. Near the centre the rate of advance must be less, since the radius of curvature of the spiral terrace-line must remain less than $\frac{1}{2}l_0$. At given supersaturation and distance from other terrace-lines, the rate of advance, v, is an increasing function of the radius of curvature, being 0 when it is $\frac{1}{2}l_0$ and, say, v_∞ when the terrace becomes straight. Supposing it increased very steeply to v_∞ for a very small increase of the radius of curvature above $\frac{1}{2}l_0$, the inner portion of the spiral would be an arc of a circle of this radius, and the spiral would make $v_\infty/\pi l_0$ turns per second. The spacing of turns in the outer region of the spiral would then be πl_0. Actually v will not increase infinitely steeply to v_∞, so that the number of turns per second will be less, and the spacing greater, but still of the same order of magnitude: let us guess $2\pi l_0$. Macroscopically this spirally terraced hill would appear to be a flat cone, with its sides inclined at an angle of $\frac{1}{2}\pi l_0$ radians to the true lattice surface—say, 1 min. for a supersaturation of 1 %, and in other cases pro-

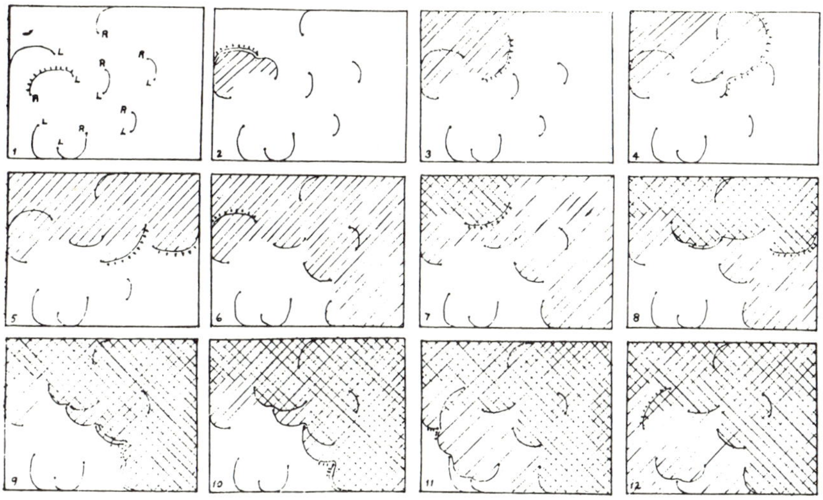

FIG. 3.—The influence of random dislocations on crystal growth.

portional to the supersaturation. We have so far neglected the dependence of the rate of advance of the terrace on its direction. On this account, instead of a flat cone, we shall have a more or less sharply defined flat pyramid having the symmetry appropriate to the crystal face.

When growth spreads from an isolated pair of dislocations, respectively right- and left-handed, the observable result will be practically the same provided that the supersaturation suffices to make l_0 less than their separation. If it is not, no growth occurs at all. If it is, the terrace-line connecting them repeatedly spreads out on one side, wraps round, and meets itself on the other side: thus forming a short connecting terrace-line again, and an outward-growing closed loop. The rate of formation of such loops will be the same in order of magnitude as the rate of formation of turns of the spiral from a single dislocation. The result will be a flat cone or pyramid indistinguishable from the previous case.

In the case that there is a random distribution of dislocations a variety of phenomena occur, exhibited pictorially in Fig. 3 (*1* to *12*). In these pictures there is supposed to be positive supersaturation, so that each terrace makes a curve convex on its cliff side, running from one dislocation to another of opposite sense, or to the boundary of the crystal face. The supersaturation is supposed to be low, and fluctuating, so that in general only one terrace-line—normally the longest—moves at a time. In each picture the terrace due to move next is specially marked. Shading indicates the area covered by fresh growth after picture *1*. Among points to be specially noticed are :

(i) the holding up of one terrace-line behind another (pictures *2*, *8*, *9*, *10*) or at a dislocation connected to another terrace (*9*, *10*, *11*, *12*) ;

(ii) the way in which a close pair of dislocations connecting a terrace facing the advancing terrace break it and so impede its passage (many examples, especially pictures *4*, *5*, *6*). In these two ways, a dislocation pair holding a terrace facing *either* way is an obstruction to the passage of an advancing terrace ;

(iii) the way in which the obstructions of type (i) are broken down (*10*, *11*, *4*) ;

(iv) the relatively impregnable region in the bottom left-hand corner. In *6* growth into this region has ceased by obstruction of type (ii) : a new advancing terrace is held up again (at *9*) by obstruction of type (i) : but this obstruction is more easily broken down, and by *12* all but a small portion of the crystal face has increased in thickness by at least one monolayer. A third advancing terrace, now commencing, will finally overrun this strong-point : however, a continuous line of closely spaced dislocations could be totally impenetrable below a critical supersaturation.

Obstruction of type (i) requires a little more consideration. One way in which type (i) obstruction can break down is the following. Each " pinned " terrace-line holds its equilibrium form by small statistical fluctuations back and forth. When two terraces lie together, the lower cannot fluctuate back under, nor the upper fluctuate forward over the other. Thus they exert a small effective repulsion on each other, and the lower one may be pushed beyond the critical curvature when the upper one arrives. Secondly, there is only a small region of seriously deformed crystal structure and hence of seriously reduced binding energy for new molecules, in the immediate neighbourhood of the dislocation. When the two terraces at such a point of obstruction face each other at an acute angle, as in pictures *9*, *10*, *11*, only a small number of molecules need condense in these unfavourable positions to enable the terrace to link across and pass on. If the obstruction arises from a straight row of dislocations spaced l_1 apart, this angle becomes smaller as l_0 decreases, diminishing rapidly from 120° when l_0 becomes less than $2l_1$. The obstruction will be ineffective when l_0 is significantly smaller than this value. Even when it is greater, a comparatively small number of molecules (say, 6 to 10) in unfavourable positions suffice to overcome the obstruction, so that statistical fluctuations (negligible with regard to obstruction of type (ii)) can be effective. Thus obstruction of type (i) probably only imposes a delay rather than a total prohibition of growth.

In view of this, one might suppose it a particularly important question whether the numbers of right-handed and left-handed dislocation-ends in a face are equal or not : the surplus of one kind might have very long terrace-lines linking them to the boundary. This is not the case for reasonably uniformly or randomly distributed dislocations, unless the relative numbers are very different. If, in an area A, n dislocations of one kind and m of another are formed randomly, like head and tail throws of a coin, the probable excess of one kind, $n - m$, is $1{\cdot}35\ (n + m)^{1/2} \approx 1{\cdot}9\, n^{1/2}$: but if it does not exceed $\pi n^{1/2}$ (where $n > m$) the dislocations linked to the boundary can all be close to it, and the longest necessary terrace inappreciably longer than in the case $n = m$, i.e., still of the order of magnitude $(A/n)^{1/2}$. In a systematically deformed crystal it is possible to have $n \gg m$, and then there is always a terrace-line at least as long as the $(m + 1)$th furthest dislocation of the n from the boundary.

The most important aspect of the type (i) obstruction by a fence of dislocations all of the same kind is that every boundary between two crystal blocks inclined at a small angle to each other constitutes such a fence (see Fig. 4, and Fig. 12 in ref. 9). The distance between the dislocations in the fence, measured in molecular spacings, is equal to the reciprocal angle of rotation between the two blocks, whether for " tilt," " twist " or more general boundaries, so long as the angle is small. A possible cause of the *visible* growth terraces sometimes seen advancing over a crystal face is that a number of molecular terraces have accumulated behind such a fence, and then been set free by a rise in supersaturation : but it is possible to think of alternative causes which can bring about the same " bunching " of molecular growth terraces into visible ones.

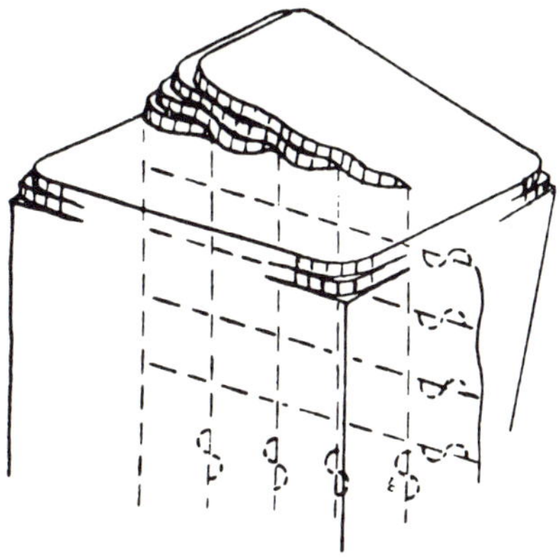

FIG. 4.—" Twist " boundary—a crossed grid of screw dislocations.

It remains to elaborate the concept of dislocations somewhat, to free it from the simplifications introduced by considering only simple cubic crystals, with cube faces. In the general case the important property of a dislocation is its displacement vector (or Burgers vector). If this vector has a component normal to the crystal face on which the dislocation line ends, there will be an associated molecular terrace in the face, promoting crystal growth. But there is also an important distinction between perfect dislocations, whose displacement vectors are lattice vectors, and imperfect dislocations whose displacement vectors lead in general from a lattice position to a twin-lattice position. Such is the dislocation with displacement vector $\left(-\frac{1}{3}, \frac{1}{3}, -\frac{1}{3}\right)$, produced in the close-packed cubic crystal by omission of part of a (111) close-packed plane of atoms. This dislocation cannot glide,[12] but must lie in its (111) plane : ending in a $(11\bar{1})$ or similar surface, it attaches a molecular growth terrace in the usual way—but every time this growth terrace reaches the trace in the surface of the missing plane of atoms, the terrace must pause unless there is a definite supersaturation. At this boundary the lattice is not continuous but has a translation-twin relationship. Instead of the usual 6 contacts per added atom, characteristic of cubic close-packing, one row of atoms added at this boundary make 7 contacts each, and the next row only 5. The latter loosely bound row will only be formed at a definite supersaturation, or by statistical fluctuation after delay.

[12] Frank, *Proc. Physic. Soc. A*, 1949, **62**, 202.

We must give a brief account of the origin of the dislocations which, it is suggested, dominate crystal growth. The chief origins which have been thought of so far are:

(i) Surface nucleation of layers in improper (e.g., twin) positions and proper positions simultaneously on the same face. Where these meet there is a dislocation. It must be remembered that the initial nucleation of the crystal always takes place at high supersaturation, much more than adequate for the Becker-Döring condition for surface nucleation on a perfect face.

(ii) Formation of one-dimensional dislocations in the edge row of the growing terrace (cf., van der Merwe's contribution to this Discussion). Such one-dimensional dislocations have an energy similar to the latent heat of evaporation of a molecule, and consequently exist in thermal equilibrium. During rapid growth at high supersaturation they can be trapped in an edge row, developing into two-dimensional and thence into three-dimensional dislocations.

(iii) The development of curvature in the growing crystal owing to the presence of impurities (a subject to be treated at length elsewhere). This ultimately leads to stress in the surface which demands a certain supersaturation for further growth, which can then continue if, and only if, dislocations are formed.

(iv) When *systems* of dislocations are present (particularly sub-grain boundaries) it is probable that the stress they would cause in perfect crystal compels the formation of further members of the system in growth: i.e., sub-grain *boundaries* are propagated in lineage structure.

(v) Aggregation of molecular vacancies into flat collapsed cavities (the edges of which are dislocation loops) whenever the temperature of a crystal is lowered. This is very likely the process responsible for the intensification of X-ray reflection (the so-called establishment of mosaic structure) when an organic crystal is plunged in liquid air.

(vi) Plastic yield under mechanical stress: this is believed not to create dislocations *ab initio* but to multiply those already present.[11]

These various considerations indicate that the initial dislocations necessary for growth are formed inevitably in the conditions needed for nucleation: and that to secure the best attainable perfection thereafter we must ensure small supersaturation (this involves good stirring, or there will be large supersaturation at the corners when it is small at the centre of a face), high purity of materials, steady temperature and absence of mechanical stresses.

The effect which dislocations have upon crystal growth produces a rather odd natural selection both of imperfection and perfection in crystals. The nucleation stage with high supersaturation makes a population of initial nuclei of varied, mostly rather great, imperfection. If growth is now carried out at low supersaturation, only a few of these seeds, in which dislocations are relatively far apart, will grow. The lower the supersaturation at this stage, the fewer and the more perfect the seeds which will actually grow. But the completely perfect crystal will not grow in any circumstances: the conditions which could cause it to grow would also soon make it imperfect.

Part IX

PERIODIC BOND CHAINS

Editor's Comments on Papers 26 Through 28

26 HARTMAN and PERDOK
On the Relations Between Structure and Morphology of Crystals

27 DONNAY and DONNAY
Four Papers: 1961 (Authors' Abstracts and Editor's Review)

28 WOLFF and GUALTIERI
PBC Vector, Critical Bond Energy Ratio and Crystal Equilibrium Form

Hartman and Perdok published a set of three papers in 1955 (Paper 26) that employed the kinetic considerations of Burton, Cabrera, and Frank for the identification of planes of relative stability. These papers introduced the concept of the periodic bond chain (PBC), which was identified in the known structure by chemical considerations. According to whether a plane contained two or more PBC vectors, a single PBC vector, or none, it was designated F, S, or K for flat, stepped, or kinked, F being the most important, S occasional, and K rare. Statistical studies based on Goldschmidt's data after the methods of Niggli (Braun, 1932[1]) were used as the standard against which to measure the success of the method. In these and subsequent papers, attachment energies calculated along the PBC vectors were shown to "represent the morphological importance much better than the mesh areas" of Bravais. Under the title "An Approximate Calculation of Attachment Energies for Ionic Crystals,"[2] Hartman developed equations based on the calculation of electrostatic potentials for chains and layers of chains (method of Madelung and Kleber). There were therefore two distinct approaches to the PBC method. The first a qualitative classification of faces, identification of PBC vectors, and the classification of faces relative to these vectors; the second, a quantitative calculation of energies for ionic crystals from their structures and based on electrostatic potential. Of necessity, such calculation involves the factor of distance—the distances between ions.

In 1963,[3] Hartman added to the hypothesis considerations of the thickness of the growth layer d_{hkl}. Since d_{hkl} is inversely proportional to the mesh area, the PBC method (Hartman's designation) was therefore brought closer to the original Bravais rule and to the inner transition layers of Niggli and Buerger. In 1965[4] the consequences of the occurrence of submultiples of d_{hkl} due to glide planes, screw axes, or pseudosymmetry were incorporated into the method. With Raymond Kern in 1964, Hartman explained the modification of habits by considering the addition or alteration of bond chains with adsorbed impurities.[5]

Donnay and Harker, to the extent that they had confined their attention to true "equipoints" of the space group in their first paper, left out of consideration the approximations to symmetry of actual structure. With attention focused on bond chains by Hartman and Perdok, the symmetry of the bond assemblage was used to account for departures from the predictions of the Donnay-Harker rule. J. D. H. Donnay very kindly provided the articles abstracted in Paper 27 as representative of the development of the theory in the early 1960s.

Major developments in the practical application of the theory of solids have led to considerable interest of an extramineralogical character in the problem of surfaces. As a significant example of the extensive literature treating equilibrium form, Paper 28, by Wolff and Gualtieri, is included. The closest atom-to-atom link is treated as the strongest bond (PBC vector) and the direction of maximum surface energy on the Wulff plot. The surface energies of planes (hkl) for a number of elements are calculated as the component of the maximum normal to the plane. Surface energy components due to second- and third-nearest-neighbor interactions are calculated in the same way. This molecular approach should be compared with the calculations for the Brillouin and Jones zones (Paper 20). For first-nearest-neighbor interactions of elemental solids, for example, the equilibrium polyhedron coincides with the first Brillouin zone.

The authors are grateful to Professor Lacmann for corrections to Table I, Part 2, on p. 572.[6] The table for beta tin should be as follows:

<table>
<tr><th>Structure</th><th>Nearest-Neighbor Interaction</th><th>Spherical Polygon</th><th>σ_{hkl} Multiplied by $ac(h^2 + k^2 + (a/c)^{2/2})^{1/2}$</th><th colspan="4">Equilibrium Planes in Order of Their Apperance</th></tr>
<tr><td>Beta tin</td><td>1st</td><td>(101, 011, $\bar{1}01$, $0\bar{1}1$)</td><td>$2l$</td><td></td><td>1st</td><td>2nd</td><td>3rd</td></tr>
<tr><td></td><td></td><td>(011, 101, $\bar{1}01$, $0\bar{1}1$)</td><td>$2(h + k)$</td><td>1st</td><td>$\underline{011}$</td><td>$\underline{010}$</td><td>011
010</td></tr>
<tr><td></td><td>2nd</td><td>(100, 010, $\bar{1}00$, $0\bar{1}0$)</td><td>$4l$</td><td>2nd</td><td></td><td>none</td><td>010</td></tr>
<tr><td></td><td>3rd</td><td>(001, 011, 121, 332)
(011, 010, 121)
(010, 110, 332, 121)</td><td>$6l$
$2(-h + k + 2l)$
$2(h + k)$</td><td>3rd</td><td></td><td></td><td>011
010
121</td></tr>
</table>

REFERENCES

1. F. Braun, *N. Jb. Min. Geol. Palaont.*, Beil-Bd. A, **65**, 173 (1932).
2. *Acta Cryst.*, **9**., 569–572 (1956).
3. *Z. Krist.*, **119** 65–78.
4. "The Non-uniform Distribution of Faces in a Zone," *Z. Krist.*, **121**, 78–80.
5. "Le Changement de facies par adsorption et la théorie des (PBC)," Note, *Compt. Rend. Acad. Sci. Paris*, **258**, 4591–4593.
6. R. Lacmann, "Surface Energy and Equilibrium Form of Crystals with β-Tin Structure," *J. Cryst. Growth*, **5**, 354–360 (1969).

26

Reprinted from *Acta Cryst.*, **8**, 49–52, 521–529 (1955)

On the Relations Between Structure and Morphology of Crystals. I

By P. Hartman and W. G. Perdok

Kristallografisch Instituut der Rijksuniversiteit, Melkweg 1, Groningen, Netherlands

(*Received 4 October 1954*)

An attempt is made to find relations between crystal structure and crystal morphology on an energy basis. It is concluded that the morphology of a crystal is governed by chains of strong bonds running through the structure. The effective period of such a chain of strong bonds is called a periodic bond chain vector (P.B.C. vector). The faces of a crystal are divided into three classes: (*a*) flat faces or *F*-faces, each of which is parallel to at least two P.B.C. vectors; (*b*) stepped or *S*-faces, each of which is parallel to at least one P.B.C. vector; (*c*) kinked faces or *K*-faces which are not parallel to any P.B.C. vector. *F*-faces are the most important faces; *S*-faces are of medium importance and *K*-faces are very rare or do not occur at all. Two examples are given, concerning the morphology of diamond and urotropine respectively.

Introduction

As soon as crystals became the subject of thorough investigation, attempts were made to correlate morphological development with internal structure. Up to now only geometrical characteristics of the structure had been taken into account (Bravais, Friedel, Fedorov, Niggli, Donnay), but, as crystal growth is a physical process, energy quantities will probably play a more important role than geometrical quantities like row densities and mesh areas.

Criticism of this kind has already been made by Wells (1946*a*, *b*, *c*) and by Buerger (1947), but these authors did not give a method by which the influence of energy factors could be taken into account in a practical way.

In this paper we present some theoretical considerations, which make it possible to correlate crystal morphology and crystal structure, starting from qualitative considerations on bond energies (see also Hartman & Perdok (1952) for a preliminary note and Hartman (1953) for details).

Outline of the main ideas

(*a*) The influence of strong bonds in the structure on the shape of a crystal is strikingly demonstrated by fibrous crystals like asbestos, where chains of strong bonds run parallel to the fibre axis. Niggli (1923), however, showed that even in crystals lacking such extreme properties, a main zone axis is parallel to a *Hauptbindungsrichtung* in the structure, as far as external influences can be excluded. It will be shown, however, that these *Hauptbindungsrichtungen* have to be replaced by the directions of chains of strong bonds.

(*b*) Gibbs (1906) showed thermodynamically that in the case of a crystal in equilibrium with its surroundings the condition that $\Sigma\sigma_i F_i$ is a minimum applies. Here σ_i is the specific surface free energy of the *i*th face and F_i the area of this face.

Wulff (1901), later Liebmann (1914) and more recently von Laue (1943) showed the Gibbs condition to be equivalent to the statement that a point can be located in the crystal (the Wulff point) from which the distances to the different crystal faces are proportional to the specific surface free energies, or, in other words, that the surface area of a crystal face —and consequently its morphological importance— decreases in general with increasing surface free energy.

The Gibbs condition, however, has never been applied to practical morphological problems because, apart from the fact that crystal growth is not a reversible process, no method is known by which the surface free energies of an actual crystal can be measured or calculated. Therefore we have to find an energy quantity associated with the crystal faces that is more accessible.

A considerable simplification of the problem is obtained by replacing the surface free energies by the surface energies, as defined by Born (1923) in his lattice theory; this means that only bond energies are taken into account.

Instead of this surface energy, however, we take the 'attachment energy', which is defined as the bond energy released when one building unit is attached to the surface of a crystal face. Now we assume that the time needed for the formation of a bond decreases with increasing bond energy. Therefore, the displacement velocity of a crystal face increases with increasing attachment energy; consequently the attachment energy will be representative for its morphological importance because a crystal face will disappear the sooner during crystal growth the faster it moves away from the growth centre. Though our assumption cannot be supported by strict thermodynamical and kinetical evidence, we found that the attachment energies, provided that these can be calculated, represent the morphological importance much better than the mesh areas (see Part II to appear later).

(*c*) Consider a two-dimensional crystal bounded by (10) and (01), each unit cell of which contains one building unit (Fig. 1(*a*)). Suppose there are two types

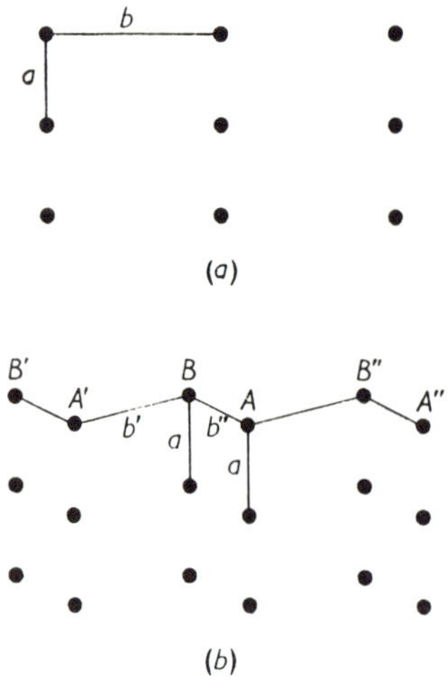

Fig. 1. (*a*) Two-dimensional crystal in which bond *a* is stronger than bond *b*. The crystal will be elongated in the direction of bond *a*.

(*b*) Two-dimensional crystal in which bonds *a* and *b''* are of the same strength, while bond *b'* is much weaker. The strong bond *b''* has no influence on the shape of the crystal, because it is the stronger bond in the chain $-b'-b''-b'-b''-$. The shape of the crystal is determined by the bonds *a* and *b'*. The crystal will be elongated in the direction of bond *a*.

of bonds: *a* in the direction [10] and *b* in the direction [01]; the bond energy of *a* is larger than that of *b*. Now the attachment energy on (10) is *a*, that on (01) is *b*, so that (10) has a greater displacement velocity than (01). Consequently the crystal will be elongated into the direction [10], i.e. the direction of the stronger bond.

Suppose now, that in a second case each cell has two building units, *A* and *B* (Fig. 1(*b*)), and that they are bound by a strong bond, the bond between *A* and *B''* being weak. We denote this weak bond by *b'* and the strong bond between *A* and *B* by *b''*; the bonds *a* between like building units *A* and like building units *B* in the [10] direction are assumed to have both the same strength as *b''*.

The growth in the direction [01] is now governed by two bonds *b'* and *b''*, of which the weaker determines the displacement velocity, so that the crystal will have smaller dimensions in the [01] than in the [10] direction. This means that the influence of the strong bond *b''* is reduced by its combination with *b'*.

From these examples three conclusions can be drawn:

(1) For a crystal to grow in the direction of a strong bond, these bonds must form an uninterrupted chain through the structure.
(2) If a bond chain contains bonds of different type, its influence on the shape of the crystal is determined by the weakest bond present in the chain.
(3) Important zone directions must be parallel to bond chains containing strong bonds only.

Obviously these chains are periodical in pattern: each period connects a building unit to a structurally equivalent one. In complicated structures each period may consist of several links or 'partial bond vectors' (representing the distance between centres of building units in length and direction). In general, the effective zone direction can be found as the direction of the sum vector of all the partial bond vectors; hence it is parallel to the translation relating two equivalent building units. This sum vector we call a *periodic bond chain vector* (P.B.C. vector). Sometimes the length of a P.B.C. vector is a submultiple of the translation distance, owing to pseudo-halvings, pseudo-thirdings etc.

Our working hypothesis is, then, that the P.B.C. vectors give the clue to the understanding of the morphological development of a crystal; important zones are parallel to P.B.C. vectors, and correspond to chains that contain only strong bonds. As soon as a chain of otherwise strong bonds contains one weak bond, it loses its morphological importance. Prominent faces are parallel to at least two high-energy bond chains.

(*d*) The use of the term 'strong bond' needs some further explanation. In general a strong bond is understood as a bond requiring a large amount of energy to be broken, or releasing a large amount of energy when it is formed. But we must be aware of the fact that not every strong bond in a crystal structure has necessarily released its energy in the crystallization process. This is quite obvious in the case of molecular structures; the strong, generally covalent, bonds between the atoms in the molecule have been formed before crystallization could take place and consequently do not have any influence on the morphology of the crystal. A similar effect may be observed in other types of structures; sometimes certain strong bonds may already have been formed in the liquid phase (precondensation), causing a difference in morphological development according to whether the crystal has grown from the melt or from the vapour. A strong bond can also lose its significance through the influence of the solvent. Molecules of the solvent may be adsorbed at the surface of a crystal face so that the formation of a new bond in the structure must be preceded by the breaking of the 'adsorption bond'. In this way the net amount of energy released in the crystallization process may be lowered to such an extent that the effect is discernible in the morphological development.

Therefore, it may be emphasized that we have to define a strong bond as a bond releasing a high amount of energy *in the crystallization process*.

(*e*) To obtain a general morphological picture of a crystal species we have to divide the crystal faces into three classes depending on their positions with respect to the P.B.C. vectors: *class F, or flat faces*, containing two or more coplanar P.B.C. vectors, *class S, or stepped faces*, containing one P.B.C. vector and *class K, or kinked faces*, containing no P.B.C. vector (for the terminology cf. Burton & Cabrera (1949); the name 'flat face' does not imply that all building units should be exactly coplanar: the coplanarity refers only to the P.B.C. vectors).

A glance at Fig. 2 will explain the difference between these classes. Suppose there are three P.B.C. vectors: vector $A||[100]$, vector $B||[010]$, and vector $C||[001]$. These P.B.C. vectors define six *F*-faces, namely (001) and (00$\bar{1}$) defined by *A* and *B*; (010) and (0$\bar{1}$0) defined by *A* and *C*; (100) and ($\bar{1}$00) defined by *B* and *C*. Any face parallel to only one of the vectors *A*, *B* or *C* is an *S*-face. In Fig. 2 three of these are shown: (011), (101) and (110). Faces which are not parallel to any of the vectors are *K*-faces, one of which, namely (111), is shown in Fig. 2.

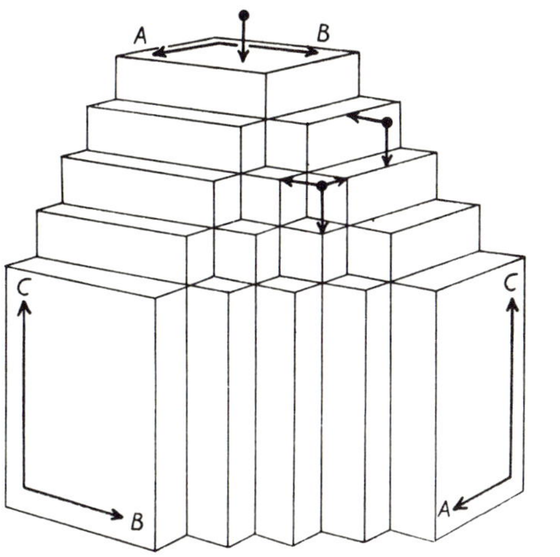

Fig. 2. Hypothetical crystal with three P.B.C. vectors: $A||[100]$, $B||[010]$ and $C||[001]$. The figure shows the *F*-faces (100), (010) and (001); the *S*-faces (110), (101) and (011); and the *K*-face (111).

Another feature can be read off from the figure: the attachment of a building unit to the surface of an *F*-face involves the release of only a relatively small amount of energy, namely the bond energy in forming bonds of those P.B.C. vectors which do not lie in this *F*-face. The attachment of a building unit to the edge of a step of an *S*-face involves the formation of at least one strong bond more than in the case of an *F*-face. The attachment of a building unit into a kink involves the formation of at least one strong bond more than in the case of an *S*-face.

Hence the displacement velocities of *K*-faces are greater than those of *S*-faces, and the displacement velocities of the *F*-faces are the smallest. *Consequently the F-faces will become the larger (final growth) faces, and the K-faces are often not observed or belong to the very rare forms. All the other faces are S-faces.*

(*f*) Up to now we have considered only the morphological importance in connexion with the chance of disappearance of the crystal faces in the growth process. But it can be shown that a chance of appearance has also to be taken into account. For that purpose the very beginning of the crystallization process has to be considered.

When crystal growth is starting from a small germ there is in general a notable degree of supersaturation. The supply of material is then so abundant that any building unit, once attached to the surface of the germ, is already overgrown by a subsequent layer before it can diffuse back into the disordered phase. In this stage all kinds of zones with the directions $mp+nq+$ etc. (p, q, etc. being P.B.C. vectors) grow out with about the same velocity—the attachment energy is not yet the determining factor—so that the germ develops into a 'seed' with an approach to an isometric shape. As time proceeds and the external conditions come closer to 'equilibrium' conditions, the attachment energy begins to direct the growth process; the seed will develop more and more, so that its 'faces' (which might be represented by only one mesh) become real crystal faces with a noticeable extension.

If from now on the conditions of growth remain constant, no new faces can be formed which were not present originally on the seed. It is obvious that a higher index face cannot appear at all, as long as the diameter of the 'seed' is so small that there is no room for one mesh on its surface. Therefore, the greater the dimensions of the mesh corresponding to a crystal face, the smaller is the chance that it will appear, and consequently the less is its morphological importance. At this point the geometry of the lattice enters also into our considerations, which are otherwise confined to energy. The chance of appearance is of interest in comparing the relative importance of *S*-faces within one zone and will be treated more fully in Part II of this paper.

(*g*) Statistical methods sometimes enable us to arrange the crystal forms according to their importance (see, for example, Niggli (1923) and Parker (1923)). In agreement with the conclusions given above, the forms found in the beginning of such a list prove to be *F*-forms, the rest being *S*-forms, while *K*-forms close the list or do not occur at all. This general picture of the morphology was found to be correct in all cases examined so far. In Part II we will treat some minerals in greater detail and show that the theoretical morphological aspect, obtained by calculating the attachment energies, is in very good agreement with the experimental data.

Examples

(1) *Significance of periodic bond chains, illustrated in the case of diamond*

The three minerals diamond, arsenolite and senarmontite crystallize in the same type of structure. In diamond the building units are atoms, linked together by covalent bonds. The building units of the other minerals are molecules and the bonds between them are mainly of the van der Waals type. Both types of bonds decrease very rapidly in strength with increasing distance. Hence only the bond between building units at (0, 0, 0) and ($\frac{1}{4}, \frac{1}{4}, \frac{1}{4}$), and structural equivalents, need to be considered. Two of these bonds form the P.B.C. vector [$\frac{1}{2}, \frac{1}{2}$, 0]. The most prominent form of the three minerals is the octahedron. The faces of this form are, indeed, parallel to three P.B.C. vectors.

Niggli (1941) showed that the explanation on the basis of structure geometry encounters difficulties. Indeed here Niggli's conception of 'row density' cannot be used as a criterion for the importance of a zone. Some rows are covered with matter in such a way that a relatively short distance is followed by a distance three times as long. The question arises whether the shortest distance or the mean distance is determinative for the importance of a zone. According to the first criterion (shortest distance) the zone [111] would be more important than [110] (and presumably the dodecahedron more important than the octahedron). The second criterion (mean distance) indicates the zone [110] as more important than [111], in agreement with observations. These difficulties arise because the bonds between the atoms are not considered. The row [111] contains the shortest distance which corresponds at the same time to the strongest bond, but the three-times longer distance does not correspond to any bond at all.

(2) *Relative importance of F-, S- and K-faces, illustrated in the case of urotropine*

Urotropine crystallizes in a molecular structure in which the centres of the molecules lie in the nodes of a cubic body-centred structure. Two bonds can be distinguished: bond *a* between atoms in (0, 0, 0) and ($\frac{1}{2}, \frac{1}{2}, \frac{1}{2}$) and bond *b* between atoms in (0, 0, 0) and (1, 0, 0). The latter bond is slightly longer so that it will be less strong than bond *a*. Every bond constitutes a P.B.C. vector and hence only the P.B.C. vectors [$\frac{1}{2}, \frac{1}{2}, \frac{1}{2}$] and [1, 0, 0], and their symmetrical equivalents, have to be taken into account. The faces of the forms {110} and {100} are parallel to three and two, respectively, of these P.B.C. vectors, and hence these are *F*-faces. Experiments on the growth of urotropine crystals from their vapour (Stranski & Honigmann, 1950) proved the growth form to be {110}. The equilibrium forms, obtained by tempering at room temperature, are {110}, {100} and {211}, the latter two forms being rather small. According to the theory of Stranski {111} should also be present, when {211} is present. In our terms the observations can be understood because {211} is an *S*-form, whereas {111} is a *K*-form as it is not parallel to any of the P.B.C. vectors. Hence {111} has a greater displacement velocity than {211} and is therefore more apt to disappear.

The authors are very much indebted to Prof. Dr P. Terpstra, head of the Crystallographic Institute, for his encouraging interest and criticism, to Prof. Dr J. D. H. Donnay and Dr G. Donnay for stimulating discussions during the first stage of the work, and to Prof. Dr E. H. Wiebenga for his critical remarks on some thermodynamical questions.

References

Born, M. (1923). *Atomtheorie des festen Zustandes*, 2nd ed., p. 538. Leipzig; Berlin: Teubner.
Buerger, M. J. (1947). *Amer. Min.* **32**, 593.
Burton, W. K. & Cabrera, N. (1949). *Disc. Faraday Soc.* No. 5, 33.
Gibbs, J. W. (1906). *Scientific Papers*, vol. 1.
Hartman, P. (1953). Thesis, Groningen.
Hartman, P. & Perdok, W. G. (1952). *Proc. K. Ned. Akad. Wet.* B, **55**, 134.
Laue, M. von (1943). *Z. Kristallogr.* **105**, 124.
Liebmann, H. (1914). *Z. Kristallogr.* **53**, 171.
Niggli, P. (1923). *Z. Kristallogr.* **58**, 490.
Niggli, P. (1941). *Lehrbuch der Mineralogie und Kristallchemie*, 3rd ed., part 1, p. 355. Berlin: Borntraeger.
Parker, R. L. (1923). *Z. Kristallogr.* **59**, 1.
Stranski, I. N. & Honigmann, B. (1950). *Z. Phys. Chem.* **194**, 180.
Wells, A. F. (1946*a*). *Phil. Mag.* (7), **37**, 184.
Wells, A. F. (1946*b*). *Phil. Mag.* (7), **37**, 217.
Wells, A. F. (1946*c*). *Phil. Mag.* (7), **37**, 605.
Wulff, G. (1901). *Z. Kristallogr.* **34**, 449.

On the Relations Between Structure and Morphology of Crystals. II

BY P. HARTMAN AND W. G. PERDOK

Kristallografisch Instituut der Rijksuniversiteit, Melkweg 1, Groningen, Netherlands

(*Received 27 January 1955*)

The characteristic features of *S*-faces (stepped faces, i.e. faces parallel to one periodic bond chain only) are further developed. After giving some definitions, a discussion follows of the factors determining the morphological importance of *S*-faces. The conclusion is drawn that the development of *S*-faces in a particular zone can be represented by an addition scheme of indices and that an *S*-face is the more important, the greater the relative strength of the bonds in the periodic bond chain to which it is parallel.

An outline is given of the way in which the morphology of an actual crystal can be derived from the structure. This is illustrated in the case of cubic closest packing. The result is that the octahedron is the most important form, followed by the cube, provided crystallization takes place from the vapour. Crystallization from the melt involves the consideration of pre-condensation effects. In this case the influence of the strongest bond on the morphology diminishes and weaker bonds become more important. This leads to a preponderance of the cube, a relation actually observed with copper.

Introduction

(*a*) In Part I (Hartman & Perdok, 1955) the forms of a crystal were divided into three classes with decreasing morphological importance: *F*-, *S*- and *K*-forms.* Niggli (1923) introduced a method of obtaining quantitative data on the morphological importance. He defined the *Fundortpersistenz* (*F*-values) and the *Kombinationspersistenz* (*P*-values), as the percentage of 'localities in which', or 'combinations on which' the form has been observed. The drawback of this method is that crystals showing rare forms enter with the same weight into the statistics as crystals showing common forms only, in spite of the fact that the latter are far more abundant. As a consequence, the lower *F*- and *P*-values are much less reliable than the higher values. Therefore special attention is here paid to the forms with high *F*- and *P*-values, these forms being *F*-forms in most cases. The relative importance of *F*-forms is determined by the attachment energy (see Part I). With the *S*-forms several influences have to be taken into account. Before discussing these influences some definitions have to be given.

(*b*) Let us consider *S*-faces in one particular zone. In our view the surface of an *S*-face must consist of parts of adjacent *F*-faces. In Fig. 1 two *F*-faces (F_1 and F_2) and an *S*-face (BB'') are projected parallel to their zone axis. The actual profile of the *S*-face is $BAB'A'B''$; one step BAB' consists of two flat parts BA and AB', parallel to F_2 and F_1 respectively. Therefore we can say that an *S*-face consists of alternating strips of adjacent *F*-faces. The smallest possible strip

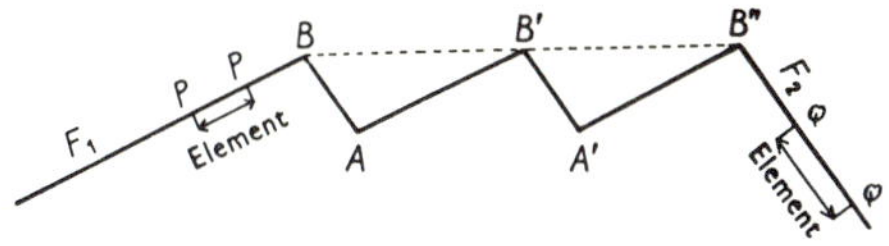

Fig. 1. Profile of two *F*-faces and one *S*-face. F_1 and F_2 are *F*-faces; their zone axis $[F_1F_2]$ is normal to the plane of the drawing. BB'' is an *S*-face showing actually the profile $BAB'A'B''$. AB' and AB are parts of the adjacent *F*-faces F_1 and F_2, respectively; they are called *flat parts*. A is the *step edge*. $BB' = B'B''$ is the *profile period*. The flat part AB consists of one element of the face F_2 and the flat part AB' consists of three elements of the face F_1. PP is the *element period* of F_1 in the zone $[F_1F_2]$ and QQ is the *element period* of F_2 in the zone $[F_1F_2]$.

of an *F*-face we call an 'element of the *F*-face'. In Fig. 1 the flat part AB represents one element of the *F*-face F_2 and the flat part AB' represents three elements of the *F*-face F_1. The distance BB' we call the 'profile period' of the *S*-face. The distances PP and QQ we call the 'element periods' of the *F*-faces.

Morphology of S-faces

The factors determining the importance of an *S*-face are related to its chance of appearance and its chance of disappearance. (We do not use the word 'chance' in its exact thermodynamical sense.)

(1) As pointed out in Part I, a face can appear only in the first stage of crystal growth. It must occupy an area of at least one mesh on the seed, so that the chance of appearance of an *S*-face decreases with increasing mesh area. For *S*-faces of one and the same zone the chance of appearance decreases with increasing profile period. Hence, if the element periods

* *F*: Flat faces, containing two or more coplanar periodic bond chain (P.B.C.) vectors.
S: Stepped faces, containing one P.B.C. vector.
K: Kinked faces, containing no P.B.C. vector.

of the *F*-faces do not differ much in length, an addition scheme of indices like that of Goldschmidt will be determinative for the chance of appearance (cf. Niggli, 1941, p. 541).

As an example we consider the zone [001] of a crystal with the *F*-faces (100) and (010). Suppose that one element of (100) and of (010) each, determine (110) as an *S*-face. A following *S*-face can be found by taking two elements of an *F*-face and one element of the other *F*-face. In this way the faces (210) and (120) are obtained. The profile periods of these faces are in general longer than the profile periods of (110), so that it is more probable that (110) appears on the seed, than (210) or (120). The addition scheme of indices reads:

(100)						(010)	..... *F*-faces
			(110)				..... S_1-face
	(210)				(120)		..... S_2-faces
(310)		(320)		(230)		(130)	..... S_3-faces

There are in general 2^{n-1} S_n-faces. For small n we may say that the chance of appearance depends on the level (n) in the addition scheme of indices. If, however, the element periods of the *F*-faces differ much in length, then the profile period of, say, (310) may be smaller than that of (120). In such a case the addition scheme becomes 'asymmetric' as it were, and then it becomes a Baumhauer series, rather than a Goldschmidt normal series (cf. Donnay, 1937).

(2) The chance of disappearance depends on the relative displacement velocities of the faces. If we assume that every building unit landing on the flat part of a step migrates on it, and moves to the edge, where it is captured, then the displacement velocity is determined by the attachment energy of a building unit to the edge of a step. This means that *S*-faces of one and the same zone all have the same displacement velocity. If a building unit is supposed to be captured on the flat part also, then the average attachment energy of a building unit to a step (including flat part and edge) determines the displacement velocity, so that *S*-faces with long flat parts would have the smallest velocities; therefore the S_1-face would be the most unimportant of all.

In a large number of cases we have found that S_1-faces are more important than S_n-faces of higher order n in the same zone. Therefore we are forced to suppose that for *S*-faces in one and the same zone the profile period is the determining factor for their importance.

(3) The relative importance of *S*-faces belonging to different zones depends not only on the profile period but also on the bond energies in the different periodic bond chains. As to this last factor, two possibilities arise, either of which may be responsible for the importance of *S*-faces: (1) the attachment energy of a building unit to the edge of a step; (2) the strength of the weakest bond in the periodic bond chain parallel to the zone.

As yet, our observations do not permit us to distinguish between these two possibilities, because in most cases a higher attachment energy (possibility (1)) is caused by weaker bonds in the chain (possibility (2)). Therefore, in general, we can say that *S*-faces with smaller attachment energies would be more important. In the majority of cases this has been found to be correct. Moreover, it appeared that *S*-faces of second and higher order seldom have a *P*- or *F*-value higher than 10, and, if so, they occur almost exclusively in the zones with the smallest attachment energies for the *S*-faces, which again shows the decisive role played by high-energy periodic bond chains.

Some general rules for practical applications

Up to now we have defined an *F*-face as being a face parallel to at least two P.B.C. vectors. The converse, however, is not true: not every face parallel to two P.B.C. vectors is an *F*-face. For it often occurs that one of the chains is a zigzag chain, and that the face is perpendicular to the plane in which this zigzag chain lies. It may happen, then, that the surface of the so-called *F*-face actually consists of two elements of neighbouring *F*-faces, so that it has to be considered as an *S*-face (an example will be given in Part III). Sometimes it may even become a *K*-face. This phenomenon occurs quite often and reduces the number of genuine *F*-faces considerably. A consequence of this is that the number of periodic bond chains is also limited. When the period of a chain consists of periods of other chains, it may be neglected, because the combination of such a chain with another chain will not give an *F*-face, but an *S*-face. The general way to derive the morphology of a crystal from its structure runs, then, as follows:

(1) The strong bonds are found by means of chemical considerations.

(2) Chains of strong bonds are sought, these chains being submitted to the following conditions: (*a*) it must be possible to divide the whole structure into a fabric of chains with stoichiometric composition which do not have any bond in common; (*b*) the chain should not entirely consist of periods of other chains. The period of a chain is in general the structural period, but sometimes it may be one half of the latter, owing to pseudo-centring.

(3) The *F*-faces can be obtained, for example, by drawing in a stereogram the zone circles corresponding to the P.B.C. vectors; the points of intersection of these zone circles represent stereographic poles of possible *F*-faces. In general not all faces found in this way are genuine *F*-faces; some of them may be *S*- or even *K*-faces. Since an *F*-face grows in layers, the following conditions should be fulfilled: (*a*) it must be possible to divide the whole structure into slices which do not have any atom in common and which have stoichiometric composition; (*b*) the slice should not entirely consist of elements of other *F*-faces, because in this case the face would be an *S*-face.

(4) The S-faces can be obtained by combining the elements of the F-faces. The element periods (PP and QQ in Fig. 1) are determined by the periodicity of the bonds in the slice of the F-face. In practice, however, the elements are not characterized by the element periods but by a P.B.C. vector Z, lying in the element, of which the element period is the projection, i.e. $PP = Z \sin \chi$, where χ is the angle between the vector Z and the zone axis under consideration.

Morphology of cubic closest packing

As an example the metals crystallizing in the cubic closest packing may be taken. As yet little is known quantitatively about the metallic bond, but it is generally assumed that the bond energy is a rapidly decreasing function of the distance.

(1) *Bonds*

For the moment we will confine ourselves to nearest neighbour–neighbour interaction. Then there is only one type of bond, namely that between atoms in $(0, 0, 0)$ and $(\frac{1}{2}, \frac{1}{2}, 0)$, and its symmetrical equivalents.

(2) *P.B.C. vectors*

The bond is periodically repeated and hence there is only one kind of P.B.C. vector, which consists of this one bond. The vector may be denoted by $[\frac{1}{2}, \frac{1}{2}, 0]$ and its corresponding bond energy may be a.

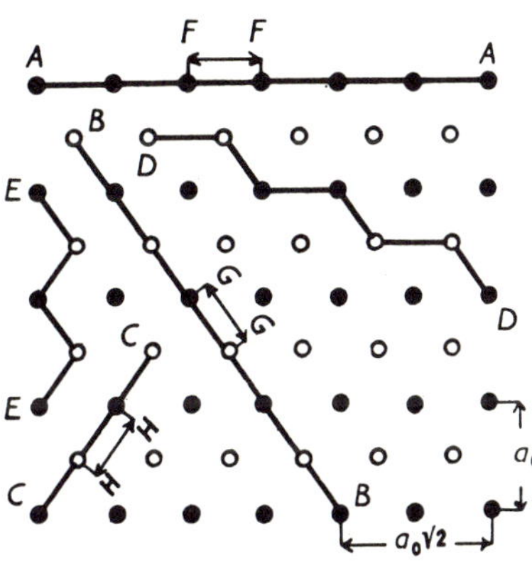

Fig. 2. Projection of the cubic closest packing on $(1\bar{1}0)$. Each small circle and each dot represents a periodic bond chain parallel to $[\bar{1}10]$. A–A is a slice of the F-face (001); B–B is a slice of the F-face (111); C–C is a slice of the F-face $(11\bar{1})$; D–D is the profile of the S_1-face (113) and E–E is the profile of the S_1-face (110). F–F, G–G and H–H are elements of (001), (111) and $(11\bar{1})$, respectively.

●: atoms at heights 0 and 1 above the plane of the paper.
○: atoms at height $\frac{1}{2}$ above the plane of the paper.
—: bond a.

(3) *F-forms*

The P.B.C. vector $[\frac{1}{2}, \frac{1}{2}, 0]$ and its symmetrical equivalents define the F-forms {111} and {100}. In Fig. 2 slices of these faces are shown. An atom attaching to the surface of (111) forms three bonds with it so that the attachment energy is $3a$. An atom attaching to the surface of (100) forms four bonds with it, so that the attachment energy is $4a$. Hence the octahedron is more important than the cube since it has the lower attachment energy. As a matter of fact, Forty & Frank (1953) obtained crystals of silver from the vapour which were bounded by {111} and {100}, the octahedron distinctly predominating.

(4) *S-forms*

In the zone $[1\bar{1}0]$ we find the F-faces (001), (111) and $(11\bar{1})$. Slices of these faces are shown in Fig. 2. The profile of the S_1-face between the two F-faces (001) and (111) consists of alternating elements of these F-faces. An element of (001) is characterized by the P.B.C. vector $[\frac{1}{2}, \frac{1}{2}, 0]$; an element of (111) is characterized by the P.B.C. vector $[\frac{1}{2}, 0, \frac{1}{2}]$. These two vectors have as their sum-vector $[1, \frac{1}{2}, \frac{1}{2}]$, which defines the face (113) in the zone $[1\bar{1}0]$ under consideration: hence (113) is the S_1-face. Therefore the addition scheme of indices becomes (cf. Niggli, 1941, p. 541):

(002)				(111)	F-faces
		(113)		.....	S_1-face
	(115)		(224)	.....	S_2-faces

In this scheme the indices of the face (001) have to be doubled in order to obtain (113) as the first S-face.

In the same way it can be shown that the S_1-face between (111) and $(11\bar{1})$ is (110). The corresponding addition scheme of indices is:

(111)				$(11\bar{1})$	F-faces
		(220)		.....	S_1-face
	(331)		$(33\bar{1})$	.....	S_2-faces

The profile period of (110) (a_0) is smaller than the profile period of (113) ($\frac{1}{4}a_0\sqrt{22}$) and therefore the form {110} is more important. For the order of importance we find now: {111}, {100}, {110}, {113}, ..., which is the same as the order predicted by the law of Donnay & Harker (1937).

(5) *More detailed picture*

In order to get an idea of the behaviour of other faces, we suppose that the interaction energy of two copper atoms depends on their distance according to the formula $E_r = E(r/r_0)^{-6}$. E is the interaction energy between the atoms in $(0, 0, 0)$ and $(\frac{1}{2}, \frac{1}{2}, 0)$ and r_0 their distance. If bonds with an energy smaller than $0{\cdot}01\,E$ are neglected, then we find the bonds given in Table 1.

The bonds b and c are much weaker than a, hence they have a minor influence on the morphology, except in the case that bond a is less significant for the crystallization, e.g. when the bond has been formed

Table 1. *Bonds in the structure of copper*

Bond	Between atoms in	r (A)	Energy
a	(0, 0, 0) and ($\frac{1}{2}$, $\frac{1}{2}$, 0)	2·55	1·00 E
b	(0, 0, 0) and (1, 0, 0)	3·61	0·12 E
c	(0, 0, 0) and (1, $\frac{1}{2}$, $\frac{1}{2}$)	4·42	0·04 E

already in the disordered state (cf. Part I, § (d)). In molten copper near the solidifying point a considerable short-range order can be assumed. The strong bonds a are then already formed to a great extent (pre-condensation) and hence the process of crystallization is mainly determined by the weaker bonds. For the morphology this means that the F-forms defined by these become more prominent. To describe this process the bonds a should be credited with a smaller energy.

The bonds b and c constitute each a complete P.B.C. vector. Hence they introduce the F-forms {110}, {113}, {210} and {531}. Table 2 shows the

Table 2. *Attachment energies of F- and S-forms*

Form	Bonds formed when one atom attaches to the surface	Energy of attachment	Energy of attachment, when bond a is neglected
{111}	$3a+3b+9c$	3·72	0·72
{100}	$4a+b+12c$	4·60	0·60
{110}	$5a+2b+10c$	5·64	0·64
{113}	$5a+3b+10c$	5·76	0·76
{210}	$6a+2b+10c$	6·64	0·64
{531}	$6a+3b+10c$	6·76	0·76
S-face in zone [110]	$5a+3b+12c$	5·84	0·84
S-face in zone [100]	$6a+2b+12c$	6·72	0·72
S-face in zone [211]	$6a+3b+11c$	6·80	0·80

attachment energies in two extreme cases: crystallization from the vapour in which no pre-condensation occurs, and crystallization under such circumstances that the influence of bond a can be neglected (as presumably in the melt).

If we compare the order of importance for the various cases with that predicted by the law of Donnay & Harker and with the equilibrium form predicted by Stranski & Kaischew (1931), we obtain the results in Table 3. It appears that with increasing pre-condensation the cube becomes more important than the octahedron, which is even preceded by {110} and {210}. Now in fact it has been observed that copper crystallizes from its melt in cubes, while from the vapour octahedra are generated (Groth, 1906–19). The metals silver and gold crystallize from their melts in octahedra. This may be due to the greater screening capacity of these atoms, so that the bonds b and c are still weaker (cf. the similar case of zinc and cadmium, Stranski, 1949).

Table 3. *Forms, listed in the order of decreasing morphological importance, according to various assumptions*

Attachment energy with bond a	Attachment energy without bond a	Equilibrium form, Stranski & Kaischew	Law of Donnay & Harker
{111}	{100}	{111}	{111}
{100}	{110}	{100}	{100}
{110}	{210}	{110}	{110}
{113}	{111}	{113}	{113}
S-forms in zone [110]	S-forms in zone [100]	{210}	{331}
{210}	{113}	{531}	{210}
S-forms in zone [100]	{531}		{112}
{531}	S-forms in zone [211]		etc.
S-forms in zone [211]	S-forms in zone [110]		

Recently Shafranovskii & Malkova (1950) found that forms {$hk0$} have an unexpectedly high frequency on naturally occurring crystals of copper, silver and gold. Some of them are more frequent than {113}. It follows from Table 2 that this can easily occur when the bond a is ineffective in the crystallization process.

(6) *Conclusion*

The morphology, as deduced from the structure, is in agreement with observations on crystals when grown from the vapour. The prominence of the octahedron and the cube can be traced back to the influence of nearest neighbour interaction energy.

Unlike purely geometrical theories, the one presented here affords an explanation of the effect of pre-condensation on morphology.

References

Donnay, J. D. H. (1937). *Amer. Min.* **22**, 209.

Donnay, J. D. H. & Harker, D. (1937). *Amer. Min.* **22**, 446.

Forty, A. J. & Frank, F. C. (1953). *Proc. Roy. Soc.* A, **217**, 262.

Groth, P. (1906–1919). *Chemische Kristallographie*, vol. 1, p. 3. Leipzig: Engelmann.

Hartman, P. & Perdok, W. G. (1955). *Acta Cryst.* **8**, 49.

Niggli, P. (1923). *Z. Kristallogr.* **58**, 490.

Niggli, P. (1941). *Lehrbuch der Mineralogie und Kristallchemie*, 3rd ed., part 1. Berlin: Borntraeger.

Shafranovskii, I. I. & Malkova, K. M. (1950). *Mém. Soc. Russe Minéral.* **79**, 294. (*Chem. Abstracts* (1951), p. 4177*d*.)

Stranski, I. N. (1949). *Disc. Faraday Soc.* (No. 5), 13.

Stranski, I. N. & Kaischew, R. (1931). *Z. Kristallogr.* **78**, 373.

On the Relations Between Structure and Morphology of Crystals. III

By P. Hartman and W. G. Perdok

Kristallografisch Instituut der Rijksuniversiteit, Melkweg 1, Groningen, Netherlands

(*Received 27 January 1955*)

The morphology of barite is deduced from the structure. It is found that there are six *F*-forms, namely {001}, {010}, {211}, {210}, {101} and {100}. An estimate of the relative importance of these faces is obtained by approximate calculations of the attachment energies. The results are in satisfactory agreement with observations.

The form {011} is found to be a *K*-face. Its relatively high morphological importance is explained in terms of dehydration and impurity effects.

All other forms of some importance appear to be *S*-forms. Crystals of some compounds isotype with barite, grown from pure aqueous solutions, show only the six *F*-forms, and {011} due to the dehydration effect. This suggests the assumption that the occurrence of *S*- and *K*-forms on minerals is due to a period of dissolution followed by growth in an impure environment.

1. Introduction

In Parts I and II (Hartman & Perdok, 1955*a*, *b*) it was shown, by means of theoretical considerations, how the morphology of crystals can be deduced from the structure. In the present article the results obtained are applied to the structure of barite.

Barite has space group *Pnma* with $a = 8{\cdot}8625$, $b = 5{\cdot}4412$, $c = 7{\cdot}1401$ Å. The positions of the barium and sulfur atoms are given in Table 1. In the following the sulfate ions as well as the barium ions are considered as point charges.

Table 1. *Positions of barium and sulfur atoms in the structure of barite*

Atom	x	y	z	Atom	x	y	z
Ba(1)	0·182	$\frac{1}{4}$	0·161	S(1)	0·068	$\frac{1}{4}$	0·695
Ba(2)	0·318	$\frac{3}{4}$	0·661	S(2)	0·432	$\frac{3}{4}$	0·195
Ba(3)	0·682	$\frac{1}{4}$	0·339	S(3)	0·568	$\frac{1}{4}$	0·805
Ba(4)	0·818	$\frac{3}{4}$	0·839	S(4)	0·932	$\frac{3}{4}$	0·305

2. Bonds and P.B.C. vectors

High-energy bonds are found between Ba^{++} ions and SO_4^{--} ions belonging to the first coordination sphere. Five types of bonds can be distinguished: every ion is surrounded by seven ions of the opposite sign, making with it two bonds *a*, two bonds *a'* and one of *b*, *c* and *d* each (see Fig. 1 and Table 2). The energy of these bonds, expressed in arbitrary units, is $1/r$, where *r* is the distance in Å between two ions.

Table 2. *High-energy bonds in the barite structure*

Bond	Between ions*	'Energy', $1/r$ (Å^{-1})
a	S(1) and Ba(2)	0·284
a'	S(1) and Ba(4,$\bar{1}$00)	0·273
b	S(1) and Ba(1,001)	0·287
c	S(1) and Ba(1)	0·254
d	S(1) and Ba(3,$\bar{1}$00)	0·235

* The notation Ba(4,$\bar{1}$00) is used to denote the barium atom at a distance [$\bar{1}$00] from the barium atom Ba(4).

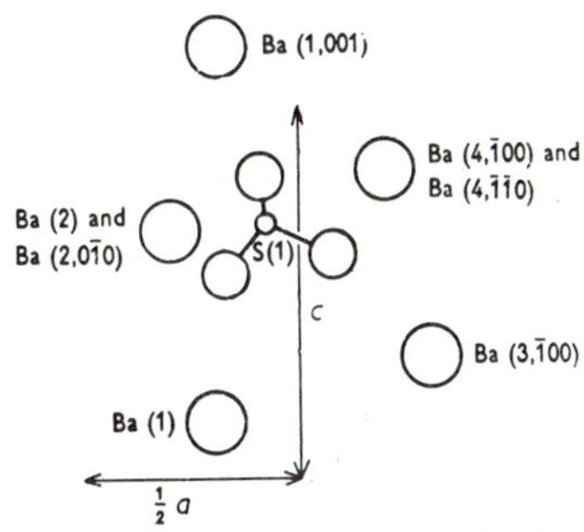

Fig. 1. Sulfate ion surrounded by seven barium ions in the structure of barite. The figure represents a projection on (010). Circles of intermediate size represent oxygen atoms, which are not shown in the other figures (cf. also Table 2 and Fig. 3(*b*)).

By combining these bonds, periodic bond chains can be constructed; for example, combination of bond

Table 3. P.B.C. *vectors of the barite structure*

P.B.C. vector		Atoms that form part of one period of a chain	Bonds per period
A	[$\frac{1}{2}$, 1, 0]	S(1)–Ba(2)–S(3,010)	$a+a'$
B	[0, 1, 0]	S(1)–Ba(2)–S(1,010)	$a+a$
C	[$\frac{1}{2}$, 0, 0]	S(1)–Ba(2)–S(3)	$a+a'$
D	[0, 0, 1]	Ba(1)–S(1)–Ba(1,001)	$c+b$
E	[$\bar{1}$, 0, 1]	Ba(1,100)–S(1,001)–Ba(3)–S(3)–Ba(1,001)	$c+d+b+d$
F	[$\bar{1}$, 1, 1]	Ba(1,100)–S(4)–Ba(4)–S(3,010)–Ba(1,011)	$a'+c+a+d$
G	[1, 1, 0]	S(1)–Ba(1,001)–S(2,001)–Ba(4)–S(1,110)	$b+a+d+a'$
H	[0, $\frac{1}{2}$, $\frac{1}{2}$]	S(4)–Ba(3)–S(3) or S(4)–Ba(4)–S(3)	$a+b$ $c+a$

b with bond c gives the P.B.C. vector [001]. In Table 3 eight P.B.C. vectors are given.

The periods of the vectors do not always agree with the lattice periods. This is demonstrated clearly by the bond chain parallel to [120]. This chain is composed of bonds a and a' and runs

S(1)–Ba(2)–S(3,010)–Ba(4,010)–S($\bar{1}$,120)
(cf. Fig. 3(a)).

This part of the chain covers a structural period. Now the second half of the chain (from S(3,010) to S($\bar{1}$,120)) can be obtained from the first half by the action of the glide plane a, parallel to (001) at height $\frac{3}{4}$. Therefore the period referred to bond energy is half the structural period. In the same way the energy period of the vector C is [$\frac{1}{2}$, 0, 0] instead of [100]. The energy period of the vector H is half the structural period, owing to the action of the glide plane n, parallel to (100).

3. *F*-forms

The eight P.B.C. vectors define twelve possible F-forms; seven of them appear to have F-character, the others being S-forms. The F-forms are {001}, {011}, {010}, {211}, {210}, {101} and {100} (cf. Fig. 2). Because F-forms contain two or more P.B.C. vectors, it must be possible to divide the whole structure into slices with stoichiometric composition parallel to the F-faces. For the F-faces (100), (210) and (010) slices are shown in Fig. 3(a), which represents a projection of the structure on (001). The elements of the F-faces are indicated in the figure; the period of an element can be characterized by a P.B.C. vector lying in it. One element of (210) can be characterized by the vector [$\frac{1}{2}$, $\bar{1}$, 0] and one element of (010) can be characterized by the vector [$\frac{1}{2}$, 0, 0]. In the figure also a slice of the face (110) is shown. It is seen that the profile of this slice consists actually of alternating elements of (210) and (010). Therefore (110) is not an F-face but an S_1-face in the zone [001]. With the aid of a structure

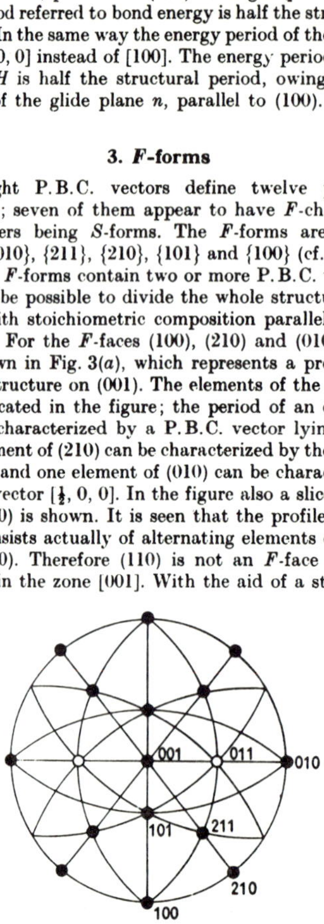

Fig. 2. P.B.C. vectors and F-forms of barite in stereographic projection. The P.B.C. vectors are represented by the corresponding zone circles. The faces of the particular K-form {011} are indicated by open circles.

model it was found that also the forms {121}, {111}, {213} and {212} are S-forms.

The relative importance of F-forms is determined by the attachment energies. These energies have been calculated, taking into account only the bond energies given in Table 2. The results are summarized in Table 4, which also gives the attachment energies for each separate place of attachment.

As an example of the manner in which these quantities have been calculated {001} and {101} may be taken. In Fig. 3(b) two slices of (001) are shown. Suppose that ions of the upper slice attach to the lower one. The ion Ba(1,003) forms bonds $b+d$ with the sulfate ions of the lower slice. Ba(3,003) forms one bond c; S(2,003) forms bonds $c+d$, while S(4,003) forms one bond b. Thus, the number of bonds formed per ion is: $(2b+2c+2d)/4$. The energy of a bond is taken (in very rough approximation) as $4e^2/r$, where r is the distance between two ions. These 'energies' are given in Table 2. Substitution of these energies in the expression $(2b+2c+2d)/4$ gives $0{\cdot}388 \times 4\ e^2.\text{Å}^{-1}$ as the mean attachment energy per ion.

In the same way this quantity is calculated for {101}.

Table 4. *Attachment energies of F-faces*

Face	Bonds formed when an ion attaches to the surface	Energy	Mean attachment energy per ion in units 4 $e^2.\text{Å}^{-1}$
(001)	$b+d$	0·522	
	$c+d$	0·489	
	b	0·287	
	c	0·254	$(2b+2c+2d)/4$ = 0·388
(210)	a	0·284	
	a'	0·273	
	$a+d$	0·519	
	$a'+d$	0·508	$(2a+2a'+2d)/4$ = 0·396
(101)	b	0·287	
	$2a'$	0·546	
	$2a'+c$	0·800	
	c	0·254	
	b	0·287	$(4a'+2b+2c)/5$ = 0·435
(211)	a	0·284	
	a'	0·273	
	$a+d$	0·519	
	$a'+d$	0·508	
	$a+b$	0·571	
	$a+c$	0·538	
	$a'+b$	0·560	
	$a'+c$	0·527	$(4a+4a'+2b+2c+2d)/8$ = 0·472
(100)	$2a'$	0·546	
	$2a'+d$	0·781	
	d	0·235	$(4a'+2d)/3$ = 0·521
(010)	$a+a'$	0·557	$a+a'$ = 0·557
(011)	$a+a'+b$	0·844	
	$a+a'+c$	0·811	
	$a+a'+b+d$	1·079	
	$a+a'+c+d$	1·046	$(4a+4a'+2b+2c+2d)/4$ = 0·945

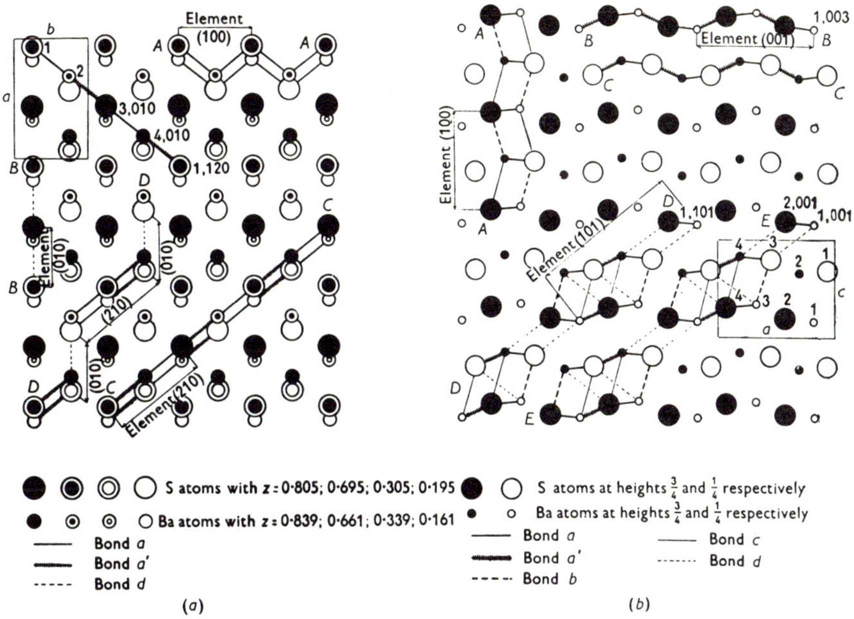

Fig. 3. (*a*) Projection of the structure of barite on (001). *A–A* is a slice of (100); *B–B* is a slice of (010); *C–C* is a slice of (210); *D–D* is a slice of (110), showing its *S*-character. The unit cell is outlined.
(*b*) Projection of the structure of barite on (010). *A–A* is a slice of (100); *B–B* and *C–C* are slices of (001); *D–D* and *E–E* are slices of (101). The unit cell is outlined.

Suppose again that ions of the upper slice of (101) in Fig. 3(*b*) attach to the lower slice. Ba(1,101) forms bond *b* with the sulfate ions of the lower slice. Ba(3,100) forms bonds $c+2a'$; S(2,101) forms one bond *c*; S(3,100) forms $2a'$; S(4,100) forms *b*. The ions Ba(2,200), Ba(4,100) and S(1,200) do not form bonds in the first coordination sphere with the ions of the lower slice, so that they cannot be attached to it. Hence the number of bonds formed per ion is $(4a'+2b+2c)/5$, which gives $0{\cdot}435\times 4\ e^2.\text{Å}^{-1}$ as the mean attachment energy per ion.

The attachment energies of the other faces have been calculated in the same way (Table 4). In Table 5 the order of the *F*-faces, obtained in this way, is compared with the observed order (*P*- and *F*-values have been taken from Braun, 1932), and with the order given by the law of Donnay & Harker (1937).

For the *F*-faces in the zones [001] and [120], the attachment energies have been calculated more precisely. To obtain these, the electrostatic potential of an ion with respect to a periodic bond chain was calculated for various chains parallel to the zones [001] and [120], by means of the method of Kleber (1940). The results are given in the following scheme:

Zone [001]; electrostatic potential of the sulfate ion S(1) with respect to chains of ions going through the ions indicated

		S(3,$\bar{1}$00) 0·217		
	Ba(4,$\bar{1}\bar{1}$0) −0·054		Ba(4,$\bar{1}$00) −0·054	
S(1,0$\bar{1}$0) −0·038		S(1) —		S(1,010) −0·038
	Ba(2,0$\bar{1}$0) 0·302		Ba(2) 0·302	
		S(3) −0·250		

Zone [001]; electrostatic potential of the barium ion Ba(1) with respect to chains of ions going through the ions indicated

		Ba(3,$\bar{1}$00) −0·238		
	S(4,$\bar{1}\bar{1}$0) 0·264		S(4,$\bar{1}$00) 0·264	
Ba(1,0$\bar{1}$0) −0·038		Ba(1) —		Ba(1,010) −0·038
	S(2,0$\bar{1}$0) 0·001		S(2) 0·001	
		Ba(3) 0·228		

Table 5. *Comparison between observed and calculated order of importance*

Face	P	F	$(P+F)/2$	Attachment energy (Table 4)	Attachment energy (electrostatic potential)	Donnay–Harker relative order
(001)	97·8	94·8	96·3	0·388	0·058	4
(210)	91·9	91·3	91·6	0·396	0·079; 0·072	5
(101)	86·4	88·1	87·2	0·435	—	1
(011)	85·9	82·5	84·2	0·945	—	3
(211)	69·2	59·9	64·5	0·472	0·134	6
(010)	70·8	54·0	62·4	0·557	0·218	7
(100)	61·1	50·0	55·5	0·521	0·261	2

Zone [$\bar{1}$20]; electrostatic potential of the sulfate ion S(2) with respect to chains of ions going through the ions indicated

S(1)	Ba(2)	S(3,010)
−0·020	0·218	−0·036
Ba(1)	S(2)	Ba(3,010)
0·084	—	0·126
S(1,00$\bar{1}$)	Ba(2,00$\bar{1}$)	S(3,01$\bar{1}$)
−0·033	−0·028	0·000

Zone [$\bar{1}$20]; electrostatic potential of the barium ion Ba(1) with respect to chains of ions going through the ions indicated

Ba(4,$\bar{1}\bar{1}$0)	S(1)	Ba(2)
−0·088	0·093	0·008
S(4,$\bar{1}\bar{1}$0)	Ba(1)	S(2)
0·140	—	0·106
Ba(4,$\bar{1}\bar{1}\bar{1}$)	S(1,00$\bar{1}$)	Ba(2,00$\bar{1}$)
0·060	0·117	−0·060

In this scheme the chains are grouped around the central ion in such a way that they form a projection, as it were, on (001) and on (2$\bar{1}$0). The potential energies are expressed in units e.^{2}Å^{-1}.

The potential energies of the chains going through the central ions were also calculated and found to be 3·079 for the zone [$\bar{1}$20] and 2·945 for the zone [001], both per molecule $BaSO_4$. The lattice energies then become 3·766 and 3·776 respectively, based on the calculations starting with the zones [$\bar{1}$20] and [001] respectively.

The attachment energies for the F-faces in the two zones can be calculated by taking the mean of the potentials at various places of attachment. In the case of (001), for example, the attachment energy is the mean of (−0·088+0·093+0·008), (0·060+0·117−0·060), (−0·020+0·218−0·036) and (−0·033−0·028+0·000). The face (210) lies in both zones; in the zone [001] the attachment energy is 0·079, while in the zone [$\bar{1}$20] it is 0·072. The agreement is satisfactory.

From Table 5 it appears that the order given by the attachment energies agrees well with the experimental order. The only serious deviation is the face (011). The attachment energy is half the total energy $(2a+2a'+b+c+d)$, so that the face (011) must be regarded as a kinked face. The outermost layer consists of ions of the same sign only and the face can therefore be compared with the octahedron face of rock salt. According to Stranski (1928) the surface of the octahedron consists of steps of cube faces. With barite the surface of (011) must then be built up from flat parts of (001), (210) and ($\bar{2}$10); that means in our terminology that it is a K-face (see also Buckley, 1951, p. 199). The question arises why a K-face can have such high F- and P-values. The answer must be sought in the influence of impurities, which enhance (011). Buckley (1935) found that the habit of $KClO_4$ crystals, which have the same structure as barite, is modified by inorganic and organic impurities, the faces enhanced being (101) and (011) in most cases. From the experiments of Kern (1953) on rock salt it may be expected, moreover, that $KClO_4$ crystals grown slowly will not show (011), whereas this face will appear when growth is rapid.

4. S-forms

The development of the S-forms in the zone [001] can be read off from Fig. 3(a). The first S-face between (100) and (210) is obtained by taking one element of each F-face. An element of (100) is characterized by the P.B.C. vector [010] and an element of (210) is characterized by the P.B.C. vector [$\bar{\frac{1}{2}}$, 1, 0]; the sum-vector is [$\bar{\frac{1}{2}}$, 2, 0], so that the S_1-face is (410). In the same way it is found that the first S-face between (210) and (010) is (110). The addition scheme of indices is:

Zone [001]

```
200        210   210         010..... F-faces
     410               220        ..... S1-faces
  610   620         430   230     ..... S2-faces
```

The development of the other zones was studied with the aid of a structure model. The addition schemes of indices are:

Zone [010]

```
100        101   101         001
     201               102
  301   302         203   103
```

Zone [$\bar{1}$20]

```
001        211   211         210
     212               421
  213   423         632   631
```

Zone $[\bar{1}01]$

101 010
111
212 121

Zone $[\bar{1}11]$

101 211
312
413 523

Zone $[01\bar{1}]$

211 100
311
522 411

Zone [110] contains one *F*-face only and hence no *S*-faces can occur in this zone.

Table 6. *Comparison between observed morphological importance of barite forms, their F-, S- or K-character and the order according to the law of Donnay & Harker*

Form	*P* value	Character	Donnay–Harker
{001}	97·8	*F*	6
{210}	91·9	*F*	7
{101}	86·4	*F*	1
{011}	85·9	*K*	3
{010}	70·8	*F*	12
{211}	69·2	*F*	9
{100}	61·1	*F*	2
{410}	38·9	S_1 [001]	26
{102}	37·3	S_1 [010]	8
{111}	26·5	S_1 [101]	4
{213}	25·9	S_2 [120]	28
{230}	25·9	S_2 [001]	>29
{201}	24·4	S_1 [010]	5
{214}	16·7	S_3 [120]	>29
{212}	15·7	S_1 [120]	13
{310}	14·6	S_2 [001]	>29
{110}	11·4	S_1 [001]	16

5. Survey

In Table 6 all forms of barite with a *P* value higher than 10 have been tabulated (Braun, 1932). The dominant position of the *F*-forms is clear. The anomaly of {011} can be explained by the particular surface structure. As a consequence of {011} being a *K*-form, all forms {0*kl*} will also have *K* character.

S-forms of higher order occur only in the zones [001] and [120], the most important zones of barite. The potential energy of the periodic bond chains corresponding with these zones must be higher than that of the other zones, because these chains deviate less from a straight line than the other chains (cf., for example, the potential energies of the zones [100] and [110] for NaCl, which are respectively 2·773 and 2·126 units, according to Kleber, 1940).

Outside the naturally occurring crystals there are a number of laboratory products that are isotype with barite. The morphology of some of these (the perchlorates of K, Rb, Cs, Tl and NH_4 and potassium permanganate) is described by Groth (1906–1919, vol. 2, pp. 167 *et seq.*). The forms observed on these compounds appear to be the six *F*-forms and the *K*-form {011}. Their frequencies are:

Observed on 6 compounds: {001}, {210} and {101}.
Observed on 5 compounds: {211}.
Observed on 4 compounds: {011} and {010}.
Observed on 3 compounds: {100}.

The three forms of barite with the highest *P*-values have also here the highest frequencies.

There is a striking difference between the mineral barite and the isotype laboratory products. The latter show only *F*-forms (and in this case the particular *K*-form {011}), while barite shows also a large number of *S*- and *K*-forms. This difference in number of forms between a mineral and a laboratory product appears to be rather general (unpublished investigations). The occurrence of *S*- and *K*-forms requires a big seed (Hartman & Perdok, 1955*a*) which might be formed during a period of dissolution. During renewed growth the *S*- and *K*-forms are then eventually stabilized through the influence of impurities, which are always present in the liquid from which a mineral crystallizes. Thus the assumption is made that the occurrence of *S*- and *K*-forms on minerals is due to alternating periods of dissolution and growth in an impure environment.

It is evident from the last column of Table 6 that the morphology of barite does not obey the law of Donnay & Harker at all. We were unable to find an appropriate pseudo-structure which should give {001}, {210} and {101} as the most important forms. The question why the law of Donnay & Harker is not obeyed will be discussed elsewhere.

References

Braun, F. (1932). *N. Jb. Min. Geol. Paläont.* Beil.-Bd. A, **65**, 173.

Buckley, H. E. (1935). *Z. Kristallogr.* **91**, 375.

Buckley, H. E. (1951). *Crystal Growth.* London: Chapman and Hall.

Donnay, J. D. H. & Harker, D. (1937). *Amer. Min.* **22**, 446.

Groth, P. (1906–1919). *Chemische Kristallographie.* Leipzig: Engelmann.

Hartman, P. & Perdok, W. G. (1955*a*). *Acta Cryst.* **8**, 49.

Hartman, P. & Perdok, W. G. (1955*b*). *Acta Cryst.* **8**, 521.

Kern, R. (1953). *Bull. Soc. franç. Minér. Crist.* **76**, 325.

Kleber, W. (1940). *N. Jb. Min. Geol. Paläont.* Beil.-Bd. A, **75**, 72.

Stranski, I. N. (1928). *Z. phys. Chem.* **136**, 259.

27

FOUR PAPERS: 1961 (Authors' Abstracts and Editor's Review)

J. D. H. Donnay and G. Donnay

This review of the four Donnay and Donnay papers was prepared expressly for this Benchmark volume by Cecil J. Schneer, from "'Assemblage liaisons' et structure cristalline," Compt. Rend., **252**, *908–909 (1961); "Ébauche d'une structure cristalline par analyse morphologique,"* Compt. Rend., **252**, *1982–1983 (1961); "Ponctualisation des charges dans les structures cristallines du type ionique,"* Compt. Rend. , **253**, *291–292 (1961); "Centers of Charges Inferred from Barite Morphology,"* Soviet Phys. Cryst., **6**, *679–684 (1962)*

The PBC concept of Hartman and Perdok (Paper 26) seemed to provide an explanation for the remaining discrepancies between the Donnay–Harker "generalization of the law of Bravais" (the terminology is that of G. Donnay and J. D. H. Donnay) and experience. Friedel had spoken of two cells, one revealed by morphology and one revealed by diffraction, for example the cubic face-centered structural cell of rock salt with edge $\bar{a}$ and the primitive cubic cell of edge one-half $\bar{a}$, derived from morphology.

In 1938, J. D. H. Donnay had extended the Donnay–Harker "generalization of the law of Bravais" (the terminology is Donnay's) by considering the space group obtained from morphological data alone.[1] In four papers written in 1961 (summarized below), J. D. H. Donnay and G. Donnay introduced further "generalization of the law of Bravais" by drawing the analogy between systematic absences of faces in special zones and the accidental extinctions of X-ray diffraction when all atoms are in special positions. Systematic absences of faces in special zones might locate atoms in positions that are special with respect to the morphological subcell.[2] By considering centers of ionic charges independently of their sign,[3] the PBC concept of Hartman and Perdok was joined with Bravais theory.

The 1961 Papers

1. "'Assemblage liaisons' et structure cristalline," *Compt. Rend. Acad. Sci. Paris,* **252**, 908–909 (1961).

Accidental extinctions juxtaposed to the systematic extinctions of the morphological group, for the faces of a zone normal to a plane of symmetry, indicate in projection on the plane of symmetry, the existence of a submultiple cell, a pseudocell of the structure but a true cell of the bond assemblage.

2. "Ébauche d'une structure cristalline par analyse morphologique," *Compt. Rend. Acad. Sci. Paris,* **252**, 1982–1983 (1961).

One can establish the principal traits of the structure of columbite starting from the morphological data, remarkably abundant in this particular case. The results obtained agree very well with the X-ray structure of Sturdivant (1930).

3. "Ponctualisation des charges dans les structures cristallines du type ionique," *Compt. Rend. Acad. Sci. Paris,* **253**, 291–292 (1961).

The centers of nearest neighbor ionic charges function independently of their sign as homologous points of their bond assemblage. The structure of barite gives examples in two dimensions (projections on *yz* and *zx*) and in one dimension (projection on *x*).

4. "Centers of Charges Inferred from Barite Morphology," *Soviet Phys. Cryst.,* **6**, 679–684 (1962).

A comparison of the systematic extinction criteria of the structural space group *Pnma* with the criteria that actually govern the morphology of barite enables one to predict, in the crystal structure, certain halvings that give information on the bond assemblage in space, in plane projections, and in line projections. The predicted features are indeed found on the known barite structure. Centers of charges, whether positive or negative, act as equivalent points: not only individual ions (Ba^{2+}, SO_4^{2-}), but also pairs of neighboring cations or anions are thus punctualized in the various projections.

NOTES AND REFERENCES

1. *Am. Mineralogist*, **24**, 184 (1938).
2. Donnay and Donnay, Paper 1 above. "It is not surprising that there are two space groups for these two groups represent the *symmetry of position* of different periodic assemblages representing distinct properties: atomic assemblage (X-rays), assemblage of points of attachment of periodic bond chains 'bond assemblage' (morphological development): to different properties, different symmetries."
3. P. Hartman, "Proposition for the Interpretation of the Generalized Law of Bravais (law of Donnay and Harker)," Note, *Compt. Rend. Acad. Sci. Paris,* **252**, 3288–3289 (1961).

28

Reprinted from *Am. Mineralogist,* **47**, 562–584 (1962)

PBC VECTOR, CRITICAL BOND ENERGY RATIO AND CRYSTAL EQUILIBRIUM FORM

G. A. WOLFF[1] AND J. G. GUALTIERI, *U. S. Army Signal Research and Development Agency Fort Monmouth, New Jersey.*

ABSTRACT

The various methods of deriving crystal equilibrium forms are reviewed. Crystal habit and microcleavage are presented in terms of specific surface energy, PBC vectors and two-dimensional nucleation. The influence of hemihedrism on the specific surface energy is discussed. For compounds or elements having bonds of different strengths, the concept of a critical bonding ratio is introduced, and its value shown to be readily determined. The role of adsorption and of departures from ideal surface structure is demonstrated.

INTRODUCTION

The idea that crystal habit is closely related to the internal structure of the crystal was first expressed by A. Bravais, who emphasized the importance of crystal planes of high reticular density. Along this line, Donnay and Harker (1937) later demonstrated the influence of screw axes and glide planes on crystal habit. Wells (1946) and Buerger (1947) also dealt with this subject. The significance of the atomic array in determining crystal habit was then shown by Kossel (1928), Stranski (1928), Hartman (1958), Kern *et al.* (1956) and Kleber (1937). Stranski and others later estimated the influence of departures from the ideal surface structure (Stranski and Honigmann, 1950) and of adsorption (Stranski, 1956; Kleber, 1957; Hartman, 1959) on crystal morphology. There also have been several attempts at correlating crystal habit with surface free energy and its plot in spherical coordinates (Wulff's plot or "γ plot" (Wulff, 1901; Yamada, 1924; Herring, 1950)). This approach appears desirable to the authors since it permits generalizations and predictions to be made with regard to the theoretical growth, or cleavage habit. It is demonstrated herein that differences between experimental and derived habits yield specific information with regard to the nature of adsorption, surface lattice deformation, and the character of bonding, both in the bulk and at the surface of crystals. The special role of microcleavage habit (Wolff and Broder, 1959) is also discussed. A correlation of Wulff's plot, PBC (Periodic Bond Chain) vectors (Hartman and Perdok, 1955) crystal habit, microcleavage and activation energy of two-dimensional nucleation will be presented. In addition, the relation of Wulff's plot to Stranski's method (Stranski, 1928) of determining equilibrium forms is shown. The development of crystal habits differing from the theoretically derived equilibrium forms can be explained through the establishment of

[1] Present address: The Harshaw Chemical Company, Solid State Research Laboratory, 2240 Prospect Avenue, Cleveland 15, Ohio.

new PBC vectors; the latter result from atomic displacements (Stranski and Honigmann, 1950; Wolff and Broder, 1959; Wolff, 1962) in the surface or through the adsorption of foreign atoms.

The utilization of the PBC vector as introduced by Hartman and Perdok (1955) greatly facilitates the understanding of the equilibrium form of crystals and its derivation from the crystal structure. A periodic bond chain in a crystal plane is an array of atoms or crystal building blocks which are held together by bonds. In each array every atom, or building block, is bonded more strongly to the crystal than the atom or building block in the equilibrium position on the crystal surface; or, expressed in a different way, the detachment energy of each atom, or building block, is greater than the average or mean detachment energy of the atoms or building blocks of the surface layer when they are detached from the crystal one by one. In this way, the periodic bond chains are easily determined for simple crystal structures. For complicated structures an unambiguous and easy way for their determination is by the calculation of Wulff's surface energy plot as shown in this paper. The periodic bond chains are identical to the directions of the "surface energy valleys" of this plot. The PBC vectors denote the direction and magnitude of the full translation distance of the periodic bond chains.

Formulae of Gibbs and Wulff

Gibbs (1928) found that a crystal is in equilibrium with its own vapor phase when for constant volume, its total surface energy is a minimum.

$$\sum_i \sigma_i A_i = \sigma_1 A_1 + \sigma_2 A_2 + \cdots \sigma_n A_n = \text{minimum} \tag{1}$$

σ_i = specific free surface energy of the ith crystal plane.
A_i = area of the ith crystal plane.

Wulff (1901) showed that this statement is equivalent to the equation

$$\frac{\sigma_i}{h_1} = \frac{\sigma_2}{h_2} = \frac{\sigma_3}{h_3} \cdots \frac{\sigma_n}{h_n} = \frac{\sigma_i}{h_i} = \text{constant} \tag{2}$$

where h_i = central distance of the ith crystal plane.

It is implicit in Gibbs' original formula (not given here) that a crystal will assume the equilibrium form only when it is of microscopic dimensions. Polyhedral macroscopic crystals are, in all probability, not equilibrium forms but growth forms. Whereas the boundary planes of the former are determined by their specific surface free energies, the boundary planes of the latter result from slow growth in the normal direction which, in turn, is induced by the presence of a two-dimensional nucleation energy. According to Stranski and Kaishev, (1935) these planes are the same as the planes that belong to the equilibrium form. This does not mean, however, that the growth form is identical with the equilibrium form, since the relative sizes of the various planes usually differ in the two forms. In a limited way, this consideration holds true also for imperfect crystals which grow via screw dislocations. Their growth rate is largely determined by the activation energy of the one-dimensional nucleation.

Practical Derivation of Equilibrium Forms

A crystal equilibrium form may be derived in two ways:

1. The first method consists of varying the central distance of all crystal planes until they are proportional to the specific surface free energy (Wulff's method) (Stranski, 1956). This construction follows from eq. (2). In a first approximation, the specific surface free energy may be replaced by the specific surface free energy at 0° K, *i.e.*, the specific surface energy.
2. In the second method, one removes from a crystal model all "building blocks" that are bonded less strongly than those in the semicrystalline position, *i.e.*, growth position (Stranski, 1928). As an example, should the evaporation probability of a "building block" occupying a corner position be greater than that in a semicrystalline position, the corner "building block" will evaporate and leave the position vacant. In a like manner, neighboring loosely bound "building blocks" will evaporate and a new plane will appear. The resulting form, the equilibrium form, will contain "building blocks" whose evaporation probability is less than or equal to that of the semicrystalline position.

The specific surface energy σ_i is defined as being half the energy per unit area required to separate the crystal along the ith plane. This results from the fact that two surfaces are created and only half the energy of separation per surface need to be considered. This is correct, however, only when both surfaces are identical. For Van der Waals, metallic and covalent bonding where the interaction energy decreases with interatomic distance; the separation energy increases with the number of atoms at first, second or higher nearest neighbor distances which are separated from their respective neighbors during the process. Consequently, σ_i will be equal to half the sum of the products $\Sigma_j N_j E_j$ of the number of different neighbors (N_1, N_2, N_3 . . . etc.) involved in the separation process and their energy of interaction (E_1, E_2, E_3 . . . etc.) per unit area of the ith plane.

In two dimensions, Wulff's construction is a polar diagram of the specific surface energies of the various planes which belong to a particular zone. (Fig. 5 and 6). The crystal form which corresponds to the lowest surface energy may be obtained by selecting the planes enclosing the smallest area. The respective planes are orthogonal to lines drawn from the origin and intersect these lines at distances proportional to the specific surface energy of the specific plane.

In a three-dimensional representation, the specific surface energies σ_i of all planes are plotted in spherical coordinates (Fig. 1). Equations for σ_i may then be derived in terms of crystalllgraphic indices hkl. Once these equations are determined for a sufficient number of planes, the areas can be found in spherical space where the equations hold for all planes. These areas are bounded by crystallographic zones of low surface energy which form spherical polygons. Each of the areas has one maximum (σ_{max}) representing the antipole of a sphere with the pole passing through the origin (Wulff's point, or center) of the crystal. The surface energy diagram of the boundary zones, or "valleys," are intersections of two such spheres. The intersections of "valleys" are known as "cusps." Hence, "cusps," "valleys," and "maxima" in a Wulff's diagram correspond respectively to "planes," "edges," and "corners" in an equilibrium form (Fig. 1).

For any type of bonding, such as nearest neighbor bonding, it can be shown that the ratio

$$\sigma_i/\sigma_{max} = \cos\delta$$

where δ is the angle between the ith plane vector and the maximum plane vector. This is equivalent to: (See Figure 2)

$$\vec{\sigma}_i(\vec{\sigma}_i - \vec{\sigma}_{max}) = 0 \tag{3}$$

or

$$\sigma_i(\sigma_i - \sigma_{max}\cos(\vec{\sigma}_i, \vec{\sigma}_{max})) = 0 \tag{4}$$

with

$$\vec{\sigma}_i = u_i\vec{a} + v_i\vec{b} + w_i\vec{c} \tag{5}$$

and

$$\vec{\sigma}_{max} = u_{max}\vec{a} + v_{max}\vec{b} + w_{max}\vec{c} \tag{6}$$

and since

$$\cos(\vec{\sigma}_i, \vec{\sigma}_{max}) = \cos(\vec{\sigma}_i{}^*, \vec{\sigma}_{max}) = \frac{\vec{\sigma}_{max} \cdot \vec{\sigma}_i{}^*}{\sigma_{max} \cdot \sigma_i{}^*}, \tag{7}$$

substitution yields

$$\sigma_i = \frac{\vec{\sigma}_i{}^* \cdot \vec{\sigma}_{max}}{\sigma_i{}^*}. \tag{8}$$

Defining

$$\sigma_i = \sigma_{hkl}$$

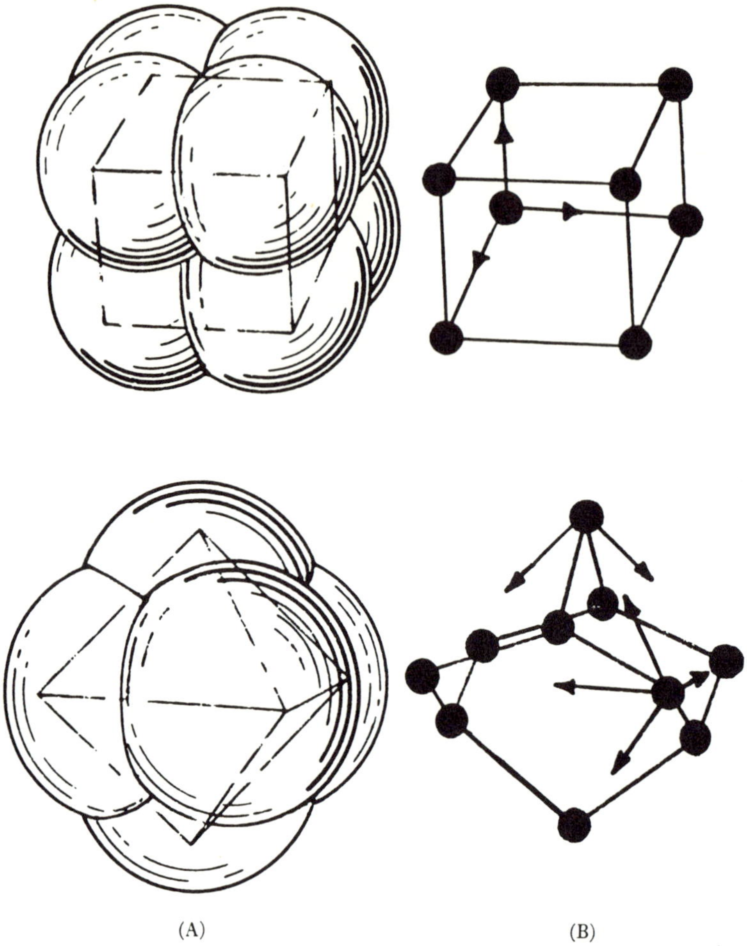

FIG. 1 A) Surface free energy plot of a crystal of simple cubic structure in spherical coordinates. Only nearest neighbor interaction is taken into account. Each cusp corresponds to an equilibrium plane (001); each valley connecting the cusps correspondingly represents an edge. This edge is [100] and is the common boundary to two adjacent (001) planes. The valleys are "complete," *i.e.* they can be traced 360° from cusp to cusp around the plot along one zone. A complete valley corresponds to a straight array of atoms. The equilibrium form (001) is inscribed.

B) Surface free energy plot for a crystal of diamond structure in spherical coordinates. Only nearest neighbor interaction is taken into account. The cusps correspond to (111) planes and the valleys connecting them correspond to [011] edges which are common to two neighboring (111) planes. The valleys are "incomplete," *i.e.* they cannot be traced 360° around the plot. An incomplete valley corresponds to a zig-zag array. The equilibrium form (111) is inscribed. The corresponding PBC vectors are given at the right.

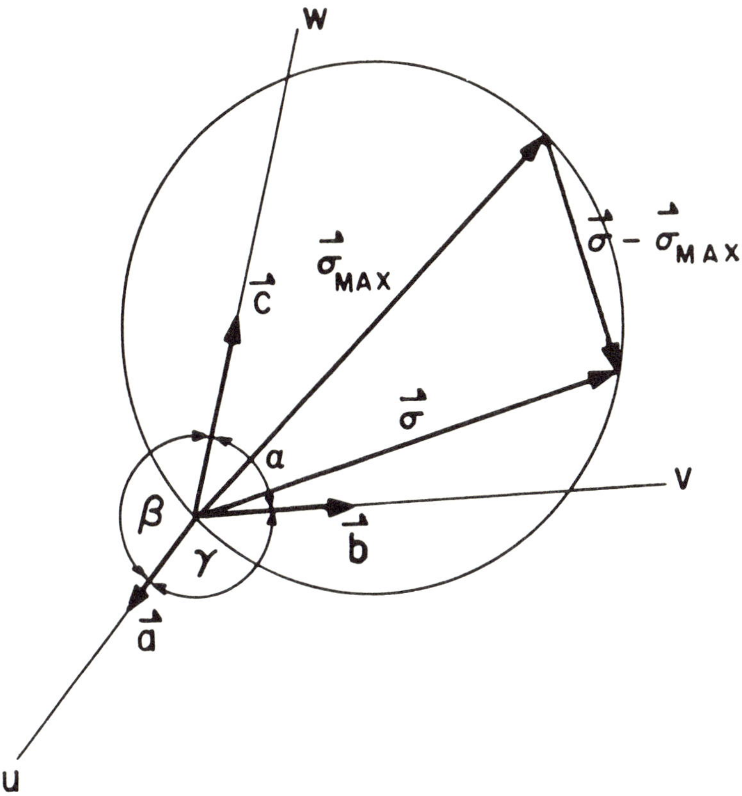

FIG. 2. Vector representation of surface energy. $\vec{\sigma}$max = maximum point in Wulff's plot. $\vec{\sigma}$ = surface energy of any plane.

and its respective vector in reciprocal space

$$\vec{\sigma}_i^* = h_i\vec{a}^* + k_i\vec{b}^* + l_i\vec{c}^*,$$

we then have

$$\sigma_{hkl} = (u_{max}h_i + v_{max}k_i + w_{max}l_i)/(h^2b^2c^2 \sin^2\alpha + k^2c^2a^2 \sin^2\beta + l^2a^2b^2 \sin^2\gamma + 2klbca^2(\cos\beta\cos\gamma - \cos\alpha) + 2lhcab^2(\cos\gamma\cos\alpha - \cos\beta) + 2hkabc^2(\cos\alpha\cos\beta - \cos\gamma))^{1/2}. \quad (9)$$

σ_{max} can be derived by simply summing all PBC vectors of the unit cell which form an angle of less than 90° with any vector within a spherical polygon. For example, [001]+[010]+[100]=[111] for the spherical triangle $h, k, l \geq 0$ in simple cubic structure. This holds for structures having linear atomic arrays. For structures having zig zag arrays, only so-called

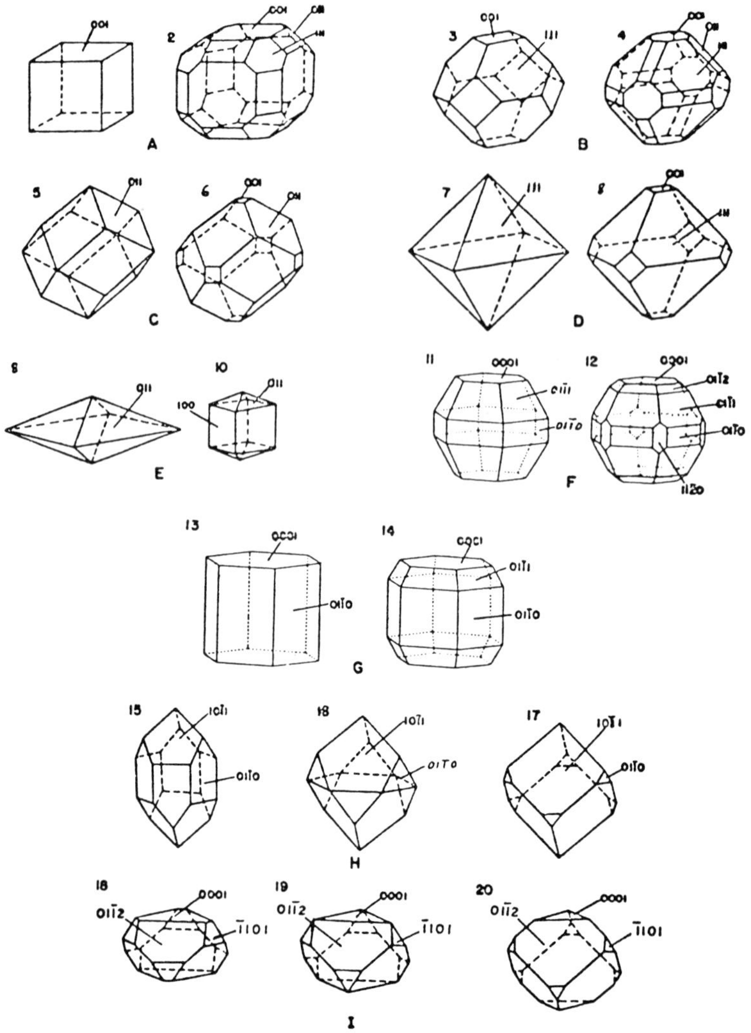

FIG. 3. Nos. 1–20 Three-dimensional drawings of the equilibrium forms in clinographic projection.

1) s.c. as a result of 1st NN interaction only. 2) s.c. 1st and 2nd NN only. 3) f.c.c. 1st. 4) f.c.c. 1st and 2nd. 5) b.c.c. 1st. 6) b.c.c. 1st and 2nd. 7) diamond 1st. 8) diamond 1st and 2nd. 9) β-Sn 1st. 10) β-Sn 1st and 2nd. 11) h.c.p. 1st. 12) h.c.p. 1st and 2nd. 13) wurtzite 1st. 14) wurtzite 1st and 2nd. 15) Se, where $E_\beta < \frac{1}{2}E_\alpha$. 16) Se, $E_\beta = \frac{1}{2}E_\alpha$. 17) Se, $E_\beta > \frac{1}{2}E_\alpha$. 18) As, $E_\beta < \frac{1}{3}E_\alpha$. 19) As, $E_\beta = \frac{1}{3}E_\alpha$. 20) As, $E_\beta > \frac{1}{3}E_\alpha$. (s.c.—simple cubic; f.c. —face centered cubic; b.c.c.—body centered cubic; h.c.p.—hexagonal close-packing; NN—nearest neighbor.)

"effective" PBC vectors alone are added. For example, in diamond for the spherical square $h, k \leq l$, σ_{max} is given by

$$2[011] + 2[0\bar{1}1] \quad \text{or} \quad 2[\bar{1}01] + 2[101] = [004]$$

In this case, the prospective vectors [110], [1$\bar{1}$0], [$\bar{1}\bar{1}$0], and [$\bar{1}$10] are not effective PBC vectors since after a separation parallel to these vectors, every atom or molecule becomes separated from as many neighbors as there are in the equilibrium position; that is, from one atom or molecule too many.

As a result, the following three conditions hold: 1) Any plane which truncates a corner of the equilibrium form contains atoms or molecules in the equilibrium position. These may be attached or detached without any activation energy of one- or two-dimensional nucleation "A_0"; *i.e.* one- or two-dimensional nucleation in this plane is not necessary. 2) Any plane parallel to and truncating an edge of the equilibrium form is parallel to one PBC vector and hence has atoms or molecules bonded to one neighbor more than the atom or molecule in the equilibrium position. A one-dimensional nucleation energy "A_1" is a requisite for the growth of such planes[1]. 3) A plane which is parallel to a plane of the equilibrium form is parallel to at least two PBC vectors, and thus contains atoms or molecules that are bound to at least two more neighbors than those atoms or molecules in the equilibrium position. In this type of plane there exists a two-dimensional nucleation energy "A_2"; *i.e.*, these planes grow via two-dimensional nucleation. This holds true mainly for perfect crystals. As a consequence, there exists in crystal equilibrium forms and growth or solution forms only flat A_2 planes of the type mentioned under 3), striated A_1 planes of the type mentioned under 2), or roughened A_0 planes as mentioned under 1). (These terms have been used by Honigmann (1959), and the planes correspond to "K," "S" and "F" planes (kinked, stepped and flat) of Burton and Cabrera (1949).)

It was mentioned earlier that the atoms or molecules in the corner positions of the equilibrium form are attached to as many neighbors as those in the equilibrium position. Consequently these same corner atoms or molecules will be left untouched in Stranski's derivation of the equilibrium form.

The formulae for specific surface energy of the planes of some of the more common structures are listed in Table I. The equilibrium forms which result when only first nearest neighbor interaction is taken into account, or when interaction is considered of first-plus-second nearest neighbors, have been drawn in Fig. 3 in three-dimensional perspective by standard crystallographic techniques. Stereographic plots of the equi-

[1] This result holds true for perfect and imperfect crystals (ideal and real crystals) as well.

TABLE I

Specific energies of common elemental structures. The number of free bonds ("dangling bonds") at the surface are given for 1st, 2nd, 3rd and sometimes 4th nearest neighbor interaction only. For example, for the plane (314) of the simple cubic structure, the corresponding value is

$$\frac{(h+k+l)n_1 + 2(k+2l)n_2 + 4ln_3}{a_0^2(h^2+k^2+l^2)^{1/2}} = \frac{8n_1 + 18n_2 + 16n_3}{a_0^2 26^{1/2}}$$

or

$$\frac{(h+k+l)n_1 + 2(k+2l)n_2 + 2(h+k+l)n_3}{a_0^2(h^2+k^2+l^2)^{1/2}} = \frac{8n + 18n_2 + 16n_3}{a_0^2 26^{1/2}}$$

n_1, n_2, n_3 represent the free bond energies for 1st, 2nd and 3rd nearest neighbor interaction. Since (314) lies at the boundary of the spherical triangle and quadrangle bounded by the planes (011), (101), (110) and (011), (101), (0$\bar{1}$1), ($\bar{1}$01) respectively, both formulae hold true for this plane for 3rd nearest neighbor interaction. It should be noted that "hexagonal diamond structure" is related to wurtzite in the same way as is the (cubic) diamond structure related to sphalerite, *i.e.* if identical atoms are assumed to occupy the sulfur and zinc positions in wurtzite, this results in a "hexagonal diamond" structure.

TABLE I, PART 1

Structure	Nearest Neighbor Interaction	Spherical Polygon	σ_{hkl} Multiplied by $a^2(h^2+k^2+l^2)^{1/2}$	Equilibrium Planes in Order of Their Appearance			
				•	1st	2nd	3rd
simple cubic	1st	(001, 010, 100)	$h+k+l$	1st	$\underline{001}$	$\underline{011}$ 001	011
	2nd	(001, 111, $\bar{1}$11)	$2(k+2l)$				
	3rd	(011, 101, 0$\bar{1}$1, $\bar{1}$01)	$4l$	2nd		$\underline{111}$ 001	$\underline{112}$ 011
		(101, 110, 011)	$2(h+k+l)$	3rd			011

• Intersects; *e.g.*, in simple cubic structure—(1st intersects 2nd) means valleys due to 1st NN interaction intersecting valleys due to 2nd NN interaction yield cusps $\underline{(011)}$ and (001). Underlined index refers to the first time this index has appeared.

T..BLE I, PART 1 (*continued*)

<table>
<tr><th>Structure</th><th>Nearest Neighbor Interaction</th><th>Spherical Polyg-n</th><th>σ_{hkl} Multiplied by $a^2(h^2+k^2+l^2)^{1/2}$</th><th>Equilibrium Planes in Order of Their Appearance</th></tr>
<tr><td>face centered cubic</td><td>1st
2nd
3rd</td><td>(001, 111, $\bar{1}$11)
(001, 010, 100)
(001, 012, 113, 102)
(113, 135, 111, 315)
(102, 315, 101, 3$\bar{1}$5)
(315, 111, 101)
(102, 113, 315)</td><td>$4(k+2l)$
$4(h+k+l)$
321
$12(h+k+2l)$
$8(2h+3l)$
$4(5h+2k+5l)$
$4(2h+k+7l)$</td><td><table><tr><th>•</th><th>1st</th><th>2nd</th><th>3rd</th></tr><tr><td>1st</td><td>$\underline{001}$
$\underline{111}$</td><td>$\underline{011}$
001</td><td>$\underline{113}$ 011
111</td></tr><tr><td>2nd</td><td></td><td>001</td><td>$\underline{012}$ 111
011</td></tr><tr><td>3rd</td><td></td><td></td><td>$\underline{135}$ 111 011
113 012</td></tr></table></td></tr>
<tr><td>body centered cubic</td><td>1st

2nd
3rd</td><td>(101, 011, $\bar{1}$01, 0$\bar{1}$1)
(101, 011, 110)
(001, 010, 100)
(001, 111, $\bar{1}$11)</td><td>$4l$
$2(h+k+l)$
$2(h+k+l)$
$4(k+2l)$</td><td><table><tr><th>•</th><th>1st</th><th>2nd</th><th>3rd</th></tr><tr><td>1st</td><td>$\underline{011}$</td><td>011</td><td>$\underline{112}$
011</td></tr><tr><td>2nd</td><td></td><td>$\underline{001}$</td><td>001
011</td></tr><tr><td>3rd</td><td></td><td></td><td>$\underline{111}$
001</td></tr></table></td></tr>
<tr><td>cubic diamond</td><td>1st
2nd
3rd</td><td>(111, $\bar{1}$11, $\bar{1}\bar{1}$1, 1$\bar{1}$1)
(001, 111, $\bar{1}$11)
(113, $\bar{1}$13, $\bar{1}\bar{1}$3, 1$\bar{1}$3)
(113, 1$\bar{1}$3, 101)
(101, 113, 111)</td><td>$4l$
$8(k+2l)$
$20l$
$4(3h+4l)$
$4(4h+2k+3l)$</td><td><table><tr><th>•</th><th>1st</th><th>2nd</th><th>3rd</th></tr><tr><td>1st</td><td>$\underline{111}$</td><td>111</td><td>$\underline{011}$
111</td></tr><tr><td>2nd</td><td></td><td>$\underline{0C1}$
111</td><td>$\underline{113}$ 111
011</td></tr><tr><td>3rd</td><td></td><td></td><td>113 111
011</td></tr></table></td></tr>
</table>

TABLE I, PART 2

Structure	Nearest Neighbor Interaction	Spherical Polygon	σ_{hkl} Multiplied by $ac(h^2+k^2+(a/c)^2l^2)^{1/2}$	Equilibrium Planes in Order of Their Appearance
beta tin	1st	(101, 011, $\bar{1}01$, $0\bar{1}1$) (011, 101, $10\bar{1}$, $01\bar{1}$)	$2l$ $2(h+k)$	see beta tin sub-table below
	2nd	(100, 010, $\bar{1}00$, $0\bar{1}0$)	$4l$	
	3rd	(101, 011, $\bar{1}01$, $0\bar{1}1$) (101, 011, 010, 100)	$6l$ $2(h+k+2l)$	
			σ_{hkl} Multiplied by $ac(h^2+k^2+hk+\frac{3}{4}(a/c)^2l^2)^{1/2}$†	
hexagonal close packed	1st	(0001, $10\bar{1}1$, $01\bar{1}1$) ($10\bar{1}1$, $01\bar{1}1$, $01\bar{1}0$, $10\bar{1}0$)	$4h+4k+3l$ $6h+6k+l$	see hexagonal close packed sub-table below
	2nd	($10\bar{1}2$, $01\bar{1}2$, $\bar{1}102$, $\bar{1}012$, $0\bar{1}12$, $1\bar{1}02$) ($10\bar{1}2$, $01\bar{1}2$, $01\bar{1}1$, $11\bar{2}0$, $10\bar{1}1$) ($10\bar{1}1$, $11\bar{2}0$, $10\bar{1}0$)	$3l$ $4h+4k+l$ $6h+2k-l$	
	3rd	($l\geq 0$)	$2l$	

Equilibrium Planes in Order of Their Appearance — beta tin:

•	1st	2nd	3rd
1st	011	010	011- 010
2nd		none	010
3rd			011 010

Equilibrium Planes in Order of Their Appearance — hexagonal close packed:

•	1st	2nd	3rd
1st	0001 $10\bar{1}1$ $10\bar{1}0$	$10\bar{1}2$ $11\bar{2}0$ $10\bar{1}1$ $10\bar{1}0$	1010
2nd		$10\bar{1}2$ $10\bar{1}1$ $11\bar{2}0$	$10\bar{1}0$ $11\bar{2}0$
3rd			none

TABLE I, PART 2 (*continued*)

Structure	Nearest Neighbor Interaction	Spherical Polygon	σ_{hki} Multiplied by $ac(h^2+k^2+hk+\frac{3}{4}(a/c)^2l^2)^{1/2}$ †	Equilibrium Planes in Order of Their Appearance
"hexagonal diamond" (wurtzite type)	1st	$(0001, 10\bar{1}0, 01\bar{1}0)$	$2h+2k+l$	(see below)
	2nd	$(0001, 10\bar{1}1, 01\bar{1}1)$ $(10\bar{1}1, 10\bar{1}0, 01\bar{1}0, 01\bar{1}1)$	$2(4h+4k+3l)$ $2(6h+6k+l)$	
	3rd	$(10\bar{1}2, 11\bar{2}0, 01\bar{1}2, \bar{1}2\bar{1}0$, etc.) $(10\bar{1}2, 11\bar{2}0, 10\bar{1}1)$ $(10\bar{1}1, 11\bar{2}0, 2\bar{1}\bar{1}0)$	l $2(h-k)$ $2l$	

Equilibrium Planes in Order of Their Appearance:

*	1st	2nd	3rd
1st	$\underline{0001}$ $\underline{10\bar{1}0}$	$10\bar{1}1$ 0001 $10\bar{1}0$	$10\bar{1}2$ $11\bar{2}0$ $10\bar{1}1$
2nd		0001 $10\bar{1}0$ $10\bar{1}1$	$10\bar{1}2$ $11\bar{2}0$ $10\bar{1}1$
3rd			$11\bar{2}0$

*** Intersects; *i.e.*, in simple cubic structure—(1st intersects 2nd) means valleys due to 1st NN interaction intersecting valleys due to 2nd NN interaction yield cusps (011) and (001). Underlined index refers to the first time this index has appeared.**

† Since the energies of the nearest neighbor bonds within the (0001) layer and normal to it are assumed to be identical, the values given apply only to a value of $c/a=(8/3)^{1/2}$.

If the values of the energies of these bonds within the (0001) layer, E_a, and normal to it, E_c, are not equal, then $c/a \neq (8/3)^{1/2}$ and the specific surface energy values for the various planes are changed. This results in a modification of the crystal forms as a function of $P=E_c/E_a$. It is to be expected that $P<1$, and E_a = NN interaction energy. For hexagonal close packing, for example, for any given c/a ratio, the surface energy values of the spherical triangle and quadrangle change from the above expressions to $4(h+k)+3Pl$ and $2(2+P)(h+k)+Pl$, respectively. Here, with varying P, the crystal form changes only its ratio of axial to radial dimension; for hexagonal diamond structure, however, the crystal form abruptly changes with the appearance and disappearance of specific planes at the critical values $P=1$ and $P=3$.

TABLE I, PART 3

Structure Type	Bond Energy Ratio	Nearest Neighbor Interaction	Spherical Polygon	σ_{hkl} Multiplied by $ac(h^2+k^2+hk+3/4(a/c)^2l^2)^{1/2}$	Equilibrium Planes in Order of Their Appearance
selenium	$R \leq \frac{1}{3}$	1st & 2nd	$(10\bar{1}1, \bar{1}101, 0\bar{1}11)$ $(10\bar{1}1, 01\bar{1}0, \bar{1}101)$ $(10\bar{1}1, 10\bar{1}0, 01\bar{1}0)$	$(1+2R)l$ $2Rh+4Rk+l$ $4Rh+4Rk+(1-2R)l$	see Se; R < ⅓
	$R \geq \frac{1}{3}$	1st & 2nd	$(10\bar{1}1, \bar{1}101, 0\bar{1}11)$ $(10\bar{1}1, 01\bar{1}0, \bar{1}101)$ $(10\bar{1}1, 01\bar{1}\bar{1}, 10\bar{1}0)$	$(1+2R)l$ $2Rh+4Rk+l$ $4Rh+2k+(1-2R)l$	see Se; R > ⅓
	$R \leq \frac{1}{3}$†	3rd Se; 4th Te	$(l \geq 0)$	$2l$	
	$R \geq \frac{1}{3}$†	3rd Se; 4th Te	$(0001, 10\bar{1}1, 01\bar{1}0, \bar{1}101, 0001)$ $(10\bar{1}1, 11\bar{2}0, 01\bar{1}0)$ $(10\bar{1}1, 10\bar{1}0, 11\bar{2}0)$	$2l$ $2h$ $2(k+l)$	
	$R > 0$	4th Se; 3rd Te	$(0001, 10\bar{1}0, 000\bar{1}, 01\bar{1}0, 0001)$	$6(h+k)$	

Se; R < ⅓

*	1st	1st & 2nd	3rd	4th
1st	none	$\underline{10\bar{1}0}$	none	$10\bar{1}0$
1st & 2nd		$\underline{10\bar{1}1}$ $10\bar{1}0$	$10\bar{1}0$	$10\bar{1}1$ $10\bar{1}0$ $\underline{10\bar{1}2}$
3rd			none	$10\bar{1}0$
4th				$\underline{00C1}$

Se; R > ⅓

*	1st & 2nd	3rd	4th
1st & 2nd	$\underline{10\bar{1}1}$ $\underline{10\bar{1}0}$	$10\bar{1}1$ $10\bar{1}0$ $\underline{11\bar{2}0}$	$10\bar{1}1$ $10\bar{1}0$ $\underline{10\bar{1}2}$
3rd		$10\bar{1}1$ $10\bar{1}0$	$10\bar{1}1$ $10\bar{1}0$
4th			$\underline{0001}$

* Intersects; e.g., in simple cubic structure—(1st intersects 2nd) means valleys due to 1st NN interaction intersecting valleys due to 2nd NN interaction yield cusps $(\underline{C11})$ and (001). Underlined index refers to the first time this index has appeared.

† It is assumed that 3rd and 4th nearest neighbor interactions are negligible as compared to 1st and 2nd nearest neighbor interactions and that, therefore, the surface structure of the various planes as derived from 1st and 2nd nearest neighbor interaction remains unchanged.

TABLE I, PART 3 (*continued*)

Structure Type	Bond Energy Ratio	Nearest Neighbor Interaction	Spherical Polygon	σ_{hkl} Multiplied by $ac(h^2+k^2+hk+3/4(a/c)^2l^2)^{1/2}$	Equilibrium Planes in Order of Their Appearance
arsenic	$R \le \frac{1}{3}$	1st & 2nd	$(0001, 01\bar{1}2, \bar{1}101)$ $(01\bar{1}2, \bar{1}101, \bar{1}10\bar{2})$	$3(-2Rh+(1-R)k+Rl)$ $-4Rh+(R+3)k+Rl$	As; $R \le \frac{1}{3}$ (see below)
	$R \ge \frac{1}{3}$	1st & 2nd	$(0001, 01\bar{1}2, \bar{1}012)$ $(01\bar{1}2, \bar{1}101, \bar{1}012)$ $(01\bar{1}2, \bar{1}101, \bar{1}10\bar{2})$	$3((R-1)h+(1-R)k+Rl)$ $-(3R+1)h+(3R+1)k+l$ $-4Rh+(R+3)k+Rl$	As; $R > \frac{1}{3}$ (see below)
	$R > 0$	3rd	$(0001, 01\bar{1}0, 000\bar{1}, \bar{1}100, 0001)$	$12k$	

As; $R \le \frac{1}{3}$

•	1st	1st & 2nd	3rd
1st	0001	$01\bar{1}2$ $01\bar{1}\bar{1}$	0001
1st & 2nd		$01\bar{1}2$ $01\bar{1}\bar{1}$	$01\bar{1}2$ $01\bar{1}\bar{1}$
3rd			0001

As; $R > \frac{1}{3}$

•	1st	1st & 2nd	3rd
1st	0001	$01\bar{1}2$ $01\bar{1}\bar{1}$	0001
1st & 2nd		$01\bar{1}2$ $01\bar{1}\bar{1}$	$01\bar{1}2$ $01\bar{1}\bar{1}$ $01\bar{1}4$
3rd			0001

$R = \frac{\text{2nd nearest neighbor bond energy}}{\text{1st nearest neighbor bond energy}}$

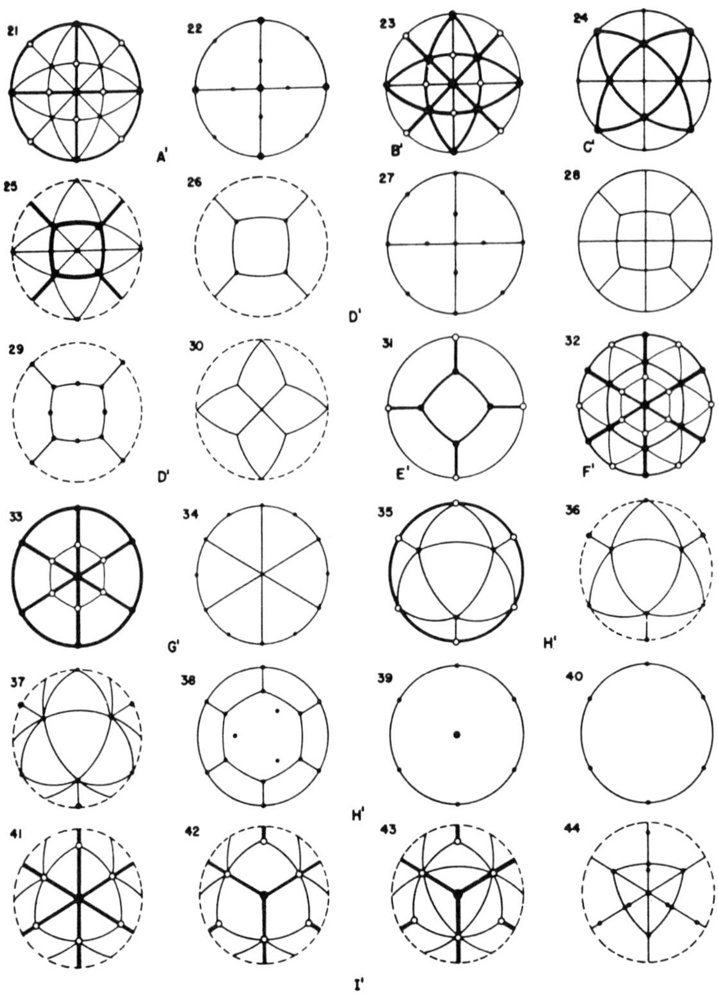

FIG. 4. Nos. 21–44. Stereograms of equilibrium forms, microcleavage patterns, and thermal etch patterns. In stereograms of equilibrium forms, heavy lines represent crystal zones (or crystal edges) as a result of 1st NN interaction only. Light lines correspond to zones as a result of 2nd NN interaction only. Large filled circles represent cusps due to 1st NN valleys intersecting each other. Small filled circles represent cusps due to 2nd NN valleys intersecting each other; and large open circles represent cusps due to 1st NN valleys intersecting 2nd NN valleys.

Filled circles in all other stereograms represent planes with their degree of development indicated by the size of the circles. Lines represent zones, and dashed lines are used only to librium planes for all structures investigated are to be found in Fig. 4. Figure 1 gives two examples of three-dimensional Wulff's plots with their corresponding PBC vectors.

Discussion of Wulff's Plots, Habit, and PBC Vectors, Microcleavage and Activation Energy of Nucleation

In evaluating Wulff's plots it is useful to correlate the crystallographic zones to the corresponding strong bonding arrays or PBC vectors (Hartman and Perdok, 1955). These zones may be found by a microcleavage (Wolff and Broder, 1959; Wolff, 1962) or habit study. The three-dimensional Wulff's plots (Fig. 1) exhibit maxima, valleys and cusps. Where the valleys (Kleber, 1959) include an angle of 360° (Fig. 1) the atoms lie either along one straight array ([001] in s.c.) or are composed of double or triple zig zag arrays ([001] in rutile and marcasite structures, or [001] in wurtzite and h.c.p., respectively). Any other zig zag array or PBC vector of this type results in a valley and includes an angle of less than 360°. This is also true for a PBC vector composed of two vectors of different strengths (E_1, E_2), *e.g.*, in As and Se. In cases such as these, at some critical ratio of the strengths of the two vectors ($E_2/E_1 = \frac{1}{3}$ for As and $E_2/E_1 = \frac{1}{2}$ for Se), one type of valley disappears to be replaced by another type. A very good approximation of the ratio of the bond strengths is arrived at by comparing these theoretically derived habits with those observed on natural, synthetic or microcleaved crystals. For example, the three forms of this type are shown for As, Sb and Bi in Fig. 3, Nos. 18, 19 and 20, respectively, and for Se and Te in Fig. 3, No. 15 (Dana, 1920, 1951).

There are certain cases where experimentally observed and theoretically calculated crystal forms are not identical. In this case, valleys in the Wulff's plot of these crystals exist for the experimental form but do not exist in the derived form and, in some cases, valleys do not exist for the experimental form where they do exist for the derived form. These valleys

---

indicate the outline of the stereograms.

21) s.c. equilibrium form. 22) NaCl microcleavage. 23) f.c.c. equilibrium form. 24) b.c.c. equilibrium form. 25) diamond equilibrium form. 26) diamond microcleavage. 27) HgTe microcleavage. 28) $AuGa_2$ microcleavage. 29) InSb microcleavage. 30) $\alpha-$Sn microcleavage. 31) $\beta-$Sn equilibrium form. 32) h.c.p. equilibrium form. 33) wurtzite equilibrium form. 34) CdS microcleavage. 35) Se, $E_\beta < \frac{1}{2} E_\alpha$ equilibrium form. 36) Se, $E_\beta = \frac{1}{2} E_\alpha$, equilibrium form. 37) Se, $E_\beta > \frac{1}{2} E_\alpha$, equilibrium form. 38) Te thermally etched at 350° for three hours. 39) Te microcleavage. 40) HgS (cinnabar) microcleavage. 41) As, $E_\beta < \frac{1}{3} E_\alpha$, equilibrium form. 42) As, $E_\beta = \frac{1}{3} E_\alpha$ equilibrium form. 43) As, $E_\beta > \frac{1}{3} E_\alpha$ equilibrium form. 44) Sb thermally etched at 600° C. for six hours. Attention is called to the zones (lines) connecting two planes (filled circles) which correspond to the edge joining the two planes. The lines also correspond to PBC vectors. Light lines are not shown wherever heavy lines appear.

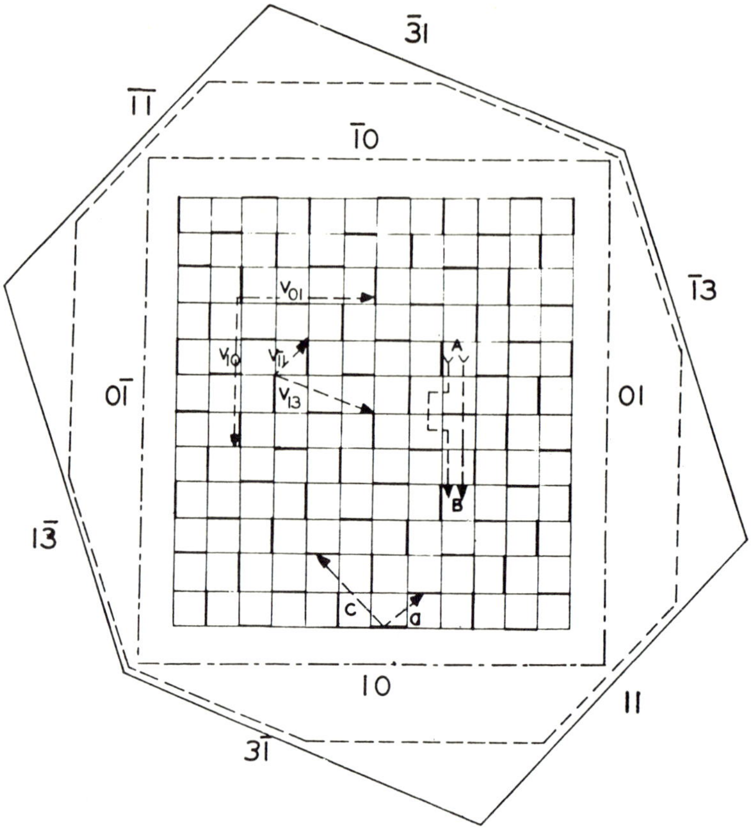

FIG. 5. A two-dimensional representation of the I_2 or Br_2 type structure in the b or (010) plane. The direction follows c and a as indicated. Strong bonds are represented as heavy lines; weak bonds are represented as light lines. The bonds above and below the plane are not considered. They correspond to close-packing. It should be noted that the straight vector AB represents the least energetic mode of bond separation when the heavily drawn bonds are as strong as the light bonds. In this case, a square results as an equilibrium form in two dimensions. The identical vectors V_{01} and V_{10} then prevail. When the heavily drawn bonds are more than three times stronger than the weak bonds, the two vectors V_{11} and V_{13} then prevail. It is obvious that a vector of the type V_{10} in the direction of AB will circumvent the strong bond and cut the three neighboring weak bonds which, together, are weaker than the strong bond. Thus, in this case V_{11} and V_{13} holds for $R<\frac{1}{3}$. R represents the bond strength ratio E_β/E_α. For $R<\frac{1}{3}$ ($\bar{1}3$) and (11) prevail; for $1>R>\frac{1}{3}$ ($\bar{1}3$), (11), (01), and (10) are observed; and for $R=1$ there are only (01) and (10) outlines. In this discussion, bondings above and below the plane are not considered.

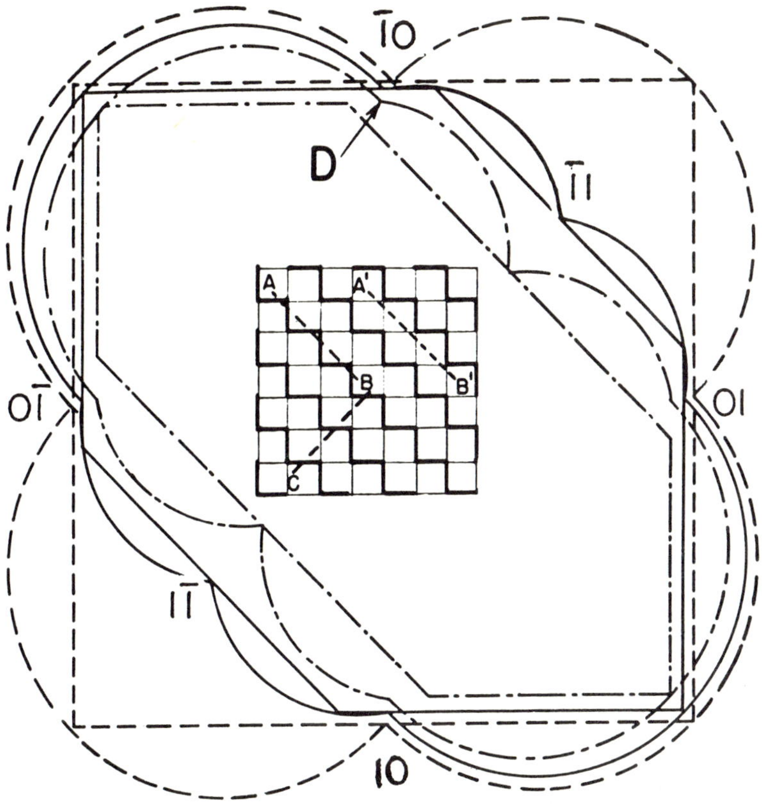

FIG. 6. As type structure; (0112) plane. The bonds above and below the plane of projection are not considered. Two types of bonds are present, strong bonds (heavy lines) and weak bonds (light lines). Two typical types of cuts, parallel to (1̄1), AB or A′B′ are possible. AB cuts only strong bonds, When $E_\beta = E_\alpha$, *i.c.* the strengths of the bonds are equal, it would make no difference cutting BC, AB or A′B′. The resulting habit in this case is a square. For $E_\beta < E_\alpha$ a truncation of type (1̄1) appears.

(and their associated PBC vectors) result from the absorption of foreign atoms onto the crystal surface and/or from the displacement of surface atoms into energetically more favorable positions. Such changes, although quite negligible for close packed metals and Van der Waals type crystals, are rather common in covalent type crystals.

It is possible from a habit study to determine fairly precisely the positions of the displaced and/or absorbed atoms. The diamond structure, for instance, presents a good example. In this structure the surface atoms lack either one or two neighbor atoms. This results in "dangling" bonds. For the planes $l \geq h \geq k$, the number of single "dangling" bonds (denoted by s) per unit area as a function of h, k and l is $4h$ and that of the double "dangling" bonds (denoted by d) is $-4h+4l$. Bivalent foreign atoms absorbing to the surface will generally bond only to surface atoms with two "dangling" bonds. Upon absorption the surface energy per bond E_s (referring now to what were previously "dangling" bonds) will be lowered to a new value E_d. This is equivalent to $\alpha=(E_s-E_d)/E_s$ where α is the relative amount by which the surface energy per bond was lowered. The specific surface energy in this case will be given by $\sigma_{hkl}=sE_s+dE_d$ per planar cell area or by $(4[\alpha h+(1-\alpha)l]E_s)/a_0^2(h^2+k^2+l^2)^{\frac{1}{2}}$. For small values of $\alpha(\alpha<\frac{1}{2})$ (001) appears in addition to (111) and two PBC vectors, [110] and [$1\bar{1}0$], (both associated with the appearance of (001)) come into existence.

Critical Bonding Ratio

An investigation has been made of crystal habit of the elements Se, Te, Po, Sn, As, Sb, Bi, Br_2 and I_2. All exhibit a bonding which could be described as mixed sp^3 and p^3 bonding. The atoms of Vb and VIb elements are in a nearly octahedral arrangement, surrounded respectively by 3+3 and 2+4 nearest and next-nearest neighbors. In simple cubic Po, which has a 6-fold coordination, there is no difference between nearest and next-nearest neighbors. In arsenic type structures the theoretical habit is shown in Fig. 3, Nos. 18, 19, 20. Here, the interaction between next-nearest neighbors is less than $\frac{1}{3}$, equal to $\frac{1}{3}$ and greater than $\frac{1}{3}$, respectively, of the interaction between nearest neighbors. In this case the habit changes from tabular to pseudo-cubic (For comparison with natural habit, see Dana, 1951). The theoretical habit for the selenium type structure is given in Nos. 15, 16, and 17 of Fig. 3. The interaction between next-nearest neighbors is less than $\frac{1}{2}$, equal to $\frac{1}{2}$, and greater than $\frac{1}{2}$ of that between nearest neighbors, respectively. As a result, the habit changes from hexagonal prismatic to pseudocubic.

To illustrate this principle of critical bonding ratio, consider a crystal of only two dimensions as shown in Fig. 5. It is seen that (11) and ($\bar{1}3$) are the lines (two-dimensional "planes") with the lowest "specific outline free energy," *i.e.* the two dimensional "specific surface free energy." (Wulff's two-dimensional plots and crystals are enclosed.) The crystal habit corresponding to the outermost "habit" outline is observed only when the bond energy between next-nearest neighbors E_2 is less than or equal to one-third the bond energy between nearest neighbors E_1. When the ratio E_2/E_1 is greater than the critical value $\frac{1}{3}$, it is obvious that the habit changes. In Case 1 with $E_2/E_1 \leq \frac{1}{3}$, only the PBC vectors $V_{\bar{1}1}$ and

V_{13} exist (outermost "habit" outline); in Case 2, with $1 > E_2/E_1 > \frac{1}{3}$, the two equivalent vectors V_{01} and V_{10} also exist (intermediate "habit" outline). In Case 3, with $E_2/E_1 = 1$, only V_{01} and V_{10} exist (innermost square "habit" outline). In each case, the following habits are derived:

Case 1) $\{11\}$ and $\{1\bar{3}\}$ are present. Vectors $V_{\bar{1}1}$ and V_{13} are parallel to the corresponding outline of the two-dimensional crystal.

Case 2) $\{11\}$, $\{1\bar{3}\}$ and $\{01\}$ are present. With increasing E_2/E_1, $\{01\}$ increases in size as compared to $\{11\}$ and $\{13\}$.

Case 3) The last two outlines disappear when $E_1 = E_2$ (or $E_2/E_1 = 1$) and only $\{01\}$ is found. The symmetry obtained is now quadratic and shows (Fig. 5) the cross section (100) of a crystal with simple cubic structure.

In this connection it should be noted that the crystal illustrated in Fig. 5 can be cut in two ways between A and B. Either six bonds between next-nearest neighbors (vectors V_2) are cut or three bonds between next-nearest neighbors and one bond (vector V_1) between nearest neighbors are cut. If the former is true ($E_2/E_1 \leq \frac{1}{3}$), the strong V_1 bond is circumvented by cutting the three weak V_2 bonds next to V_1. Should the latter be true, $1 \geq E_2/E_1 > \frac{1}{3}$. Then less energy is expended when V_1 is cut in the process. Consequently, V_{10} is established as the new PBC vector.

There are two more examples, enargite and chalcopyrite, which may be used to demonstrate the estimation of the bond strength ratio which is obtained by comparing habits which have been theoretically and experimentally derived. The experimentally determined microcleavage habit is used in these examples. The advantage of this method of determining habit is that a unique form results, that is, there is only one such habit for each material (Wolff and Broder, 1960). While there is a disadvantage to this method in the neglect of hemihedral properties, conclusions drawn concerning bond strengths within the bulk of the crystal are unaffected.

For enargite a bond ratio estimate of three between the As—S bond and the Cu—S bond results, giving good agreement between the theoretical and microcleavage habit. For chalcopyrite the bond strength ratio between Fe—S and Cu—S (or vice versa) is at least two (or $\frac{1}{2}$). For symmetry reasons it is not possible to decide which bond is stronger.

Discussion of Hemihedrism

Most experimental or theoretical determinations of σ_{hkl} consider only the energy of separation parallel to the plane (*hkl*). What is found is the sum of surface energies of the two new planes. The specific surface energy, however, of each plane remains obscure. The usual procedure is to divide the total surface energy by two so as to assign a surface energy to each plane; but, as previously mentioned, this holds only when the new planes possess identical structure. That this is, indeed, true may be drawn from conclusions made from habit studies. For example, the tetrahedral habit of sphalerite (ZnS) shows the negative $(\bar{1}\bar{1}\bar{1})$ plane to be more developed than its positive (111) counterpart. This indicates that the negative $(\bar{1}\bar{1}\bar{1})$ plane is the more stable plane. It, therefore, in accordance with Wulff's law has a lower specific surface energy.

The differences in the two planes can be described in terms of the differences in rearrangement of atoms, chemical bonds, and bonding orbitals of the atoms in their surfaces. For example, in sphalerite the two new planes would have zinc atoms on (111) and sulfur atoms on $(\bar{1}\bar{1}\bar{1})$. Since in the bulk sulfur has **ca.** 7 electrons and zinc **ca.** 1 (Wolff and Broder, 1959), it is very probable that the sulfur $(\bar{1}\bar{1}\bar{1})$ electronic configuration would be more stable by virtue of its extra electrons with respect to tetrahedral bonding. The conclusion arrived at is that for materials lacking a center of symmetry the "anionic" plane is more stable. Habit studies, in confirmation of this, indicate a more highly developed "anionic" plane. A difference in the two planes appears also in the III-V compounds and in etching experiments.

Assuming a surface energy assigned to the different surface atoms as denoted by $(A)_s$, $(A)_d$, $(C)_s$, and $(C)_d$ for anionic and cationic single and double "dangling" bonds, then the σ_{hkl} formula for sphalerite structure materials is

$$\sigma_{hkl} = \frac{2[(A)_s - 2(A)_d + (C)_s]h - 2[(A)_s - (C)_s]k + 4(A)_d l}{a_0^2(h^2 + k^2 + l^2)^{1/2}}$$

When $(A)_s = (A)_d \equiv (A)$, $(C)_s \equiv (C)$, and $(C)/(A) = \beta \geq 3$,

$$\sigma_{hkl} = \frac{[2(\beta - 1)(h + k) + 4l](A)}{a_0^2(h^2 + k^2 + l^2)^{1/2}} = \frac{4(\bar{h} + \bar{k} + \bar{l})}{a_0^2(h^2 + k^2 + l^2)^{1/2}}$$

and a tetrahedron with only tetrahedral "anionic" planes appears. For many materials of this structure type (GaP, ZnS and others) (111), $(\bar{1}\bar{1}\bar{1})$ and (001) are observed. This can easily be explained when $(A)_d/(A)_s < 1$. This is illustrated in Fig. 7. In this way, therefore, it is possible to investigate the bonding structure of the surface atoms.

Ionic Crystals

Most arguments which have been set forth thus far may correctly be applied only to elements, *i.e.*, to materials with Van der Waals, metallic or covalent bonding. In this case, the bonding and electron distribution between the atoms is symmetrical and the interaction energy between the atoms falls off rapidly with interatomic distance. As a rule, however, compounds differ from elements in that the electron distribution of the bonding is asymmetrical. This is due to the partial ionic character of the

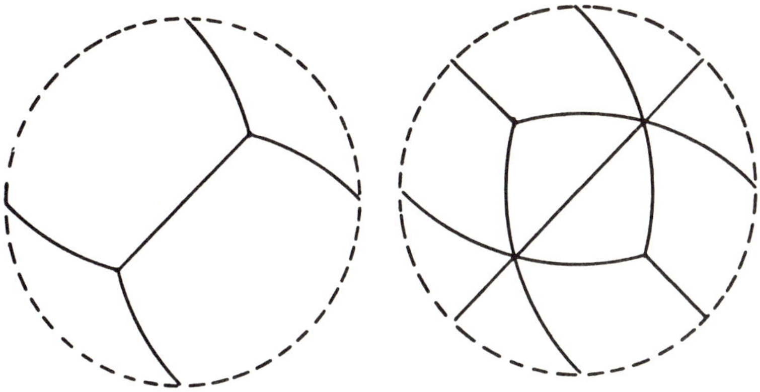

A B

FIG. 7. Stereograms representing equilibrium forms of crystals having sphalerite structure. A) a negative tetrahedron ($1\bar{1}1$) exists when $\sigma_{111}/\sigma_{1\bar{1}1}=\beta\geq 3$; B) a negative tetrahedron ($1\bar{1}1$) with minor development of the positive tetrahedron (111) for $3>\beta>1$. C) major ($1\bar{1}1$), minor (111), (001); not shown here. This form is quite common and appears when the atoms with two "dangling" bonds move to more favorable surface positions or rearrange their electronic structure.

bonding in which there is an attractive force exerted by first nearest and other neighbor atoms as well as a repulsive force exerted by still other neighbors.

For crystals with Van der Waals, metallic and covalent bonding the interaction energy between atoms decreases with r^{-n} where $n=6$ and r is the interatomic distance. For electrostatic bonding, on the other hand, $n=1$. In this type of bonding it appears that the interaction energy of more distant atoms should still be considered while, for all practical purposes, in the former type only up to next-nearest neighbors need to be considered. This renders the calculation of the surface energy and the equilibrium form for ionic crystals extremely difficult by the method outlined here. Sufficient experimental agreement has been found, however, between the calculated equilibrium forms and observed microcleavage forms to justify, in a first approximation, the validity and usefullness of the principles cited herein. This agreement is probably due to the partial ionic character of most compounds and also probably to the screening effect of nearer ions on the interaction on more distant ions.

To take into account the ionic nature of compounds, the specific surface free energy may be approximated by a formula of the type

$$\sigma_{hkl} = (\sigma_{I} + a\sigma_{II} + b\sigma_{III} + c\sigma_{IV})hkl$$

where a, b, and c are constants and I, II, III, IV refer to 1st, 2nd, 3rd, and 4th nearest neighbor interactions.

Acknowledgments

The authors wish to express their thanks to Drs. A. A. Giardini, J. A. Kohn, F. W. Leonhard and to Mr. J. Hietanen for interesting discussions and constructive criticism of the manuscript.

References

Buerger, M. J. (1947) The relative importance of the several faces of a crystal, *Am. Mineral.*, **32**, 593–606.

Burton, W. K. and Cabrera (1949) Crystal growth and surface structure, *Disc. Faraday Soc.* **5**, 33–39.

Dana, J. D. (1951) The System of Mineralogy, John Wiley and Sons, New York.

Donnay, J. D. H. and D. Harker (1937) A new law of crystal morphology extending the law of Bravais, *Am. Mineral.*, **22**, 446–467.

Gibbs, J. W. (1928) Collected Works, Longman Green and Co., New York.

Hartman, P. (1959) The effect of adsorption on the face development of the equilibrium form, *Acta Cryst.*, **12**, 429–439.

——— (1958) The equilibrium forms of crystals, *Acta Cryst.*, **11**, 459–464.

——— and W. G. Perdok, (1955) On the relations between structure and morphology of crystals, *Acta Cryst.*, **8**, 49–52, 521–529.

Herring, C. (1950) Some simple theorems on the free energies of crystal surfaces, *Phys. Rev.*, **75**, 344–350.

Honigmann, B. (1959) Gleichgewichts- und Wachstumformen von Kristallen, *Fort. phys. Chemie*, Dr. Dietrich Steinkopff Verlag, Darmstadt.

Kern, R. and J. C. Monier (1956) Interprétation des formes caractéristiques de cristaux appartenant aux mériédries non centrées, *Bull. Soc. Franç. Min.* **78**, 461.

Kleber, W. (1937) Die Struktur theoretische Diskussion kristall-morphologischer Fragen, *Fort. Min.*, **21**, 169–224.

——— (1957) Über flächenspezifische Adsorption und Solvation, *Zeit. phys. Chem.*, **206**, 327–338.

Kossel, W. (1928) Die molekularen Vorgänge beim Kristallwachstum, Leipziger Vortraege, Hirzel, Leipzig, pp. 1–46.

Stranski, I. N. (1956) Propriétés des surfaces des cristaux, *Bull. Soc. Franç. Min.* **79**, 359–383.

——— (1928) Zur Theorie des Kristallwachstums, *Zeit. phys. Chem.*, (A) **136**, 259–278.

——— and B. Honigmann (1950) Die spontane Einstellung von Gleichgewichtsformen an Hexamethylentetraminkristallen, *Zeit. phys. Chem.*, **194**, 180–198.

——— and R. Kaishev (1935) Crystal growth and formation of nuclei, *Phys. Zeit.* **36**, 393–403.

Wells, A. F. (1946) Crystal habit and internal structure, *Phil. Mag.*, **37**, 184–199; 217–236; 605–630.

Wolff, G. A. (1962) Zum Problem der Gleichgewichtsformen von Kristallen. *Zeit. phys. Chem.* (N.F.) **31**, 1–22.

——— and J. D. Broder (1959) Microcleavage, bonding character and surface structure in materials with tetrahedral coordination, *Acta Cryst.*, **12**, 313–323.

——— (1960) Cleavage and the identification of minerals. *Am. Mineral.* **45**, 1230–1242.

Wulff, G. (1901) Zur Frage der Geschwindigkeit des Wachstums und der Auflösung der Kristallflächen, *Zeit. Krist.*, **34**, 449–530.

Yamada, M. (1924) Über die Kristallformen und die Kristallgitter, *Phys. Zeit.*, **25**, 289–296.

Manuscript received, May 20, 1960.

Part X

SURFACE ENERGY AS A FUNCTION OF STRUCTURE

Editor's Comments on Papers 29 Through 33

It was argued by Friedel in his critique of Wulff's theorem that a solid, unlike a liquid crystal, was incapable of adjusting its surface form to the demands of equilibrium.[a] A snowflake surface is hardly an example of an equilibrium form, yet in its remarkably complex symmetry it affords a clear example of a coupling between distinct facets during growth of a solid, a twelvefold coupling at that (dihexagonal). Paper 29 on the snowflake undertakes to propose an explanation. It is a topic that began with Kepler and Descartes. Although this paper considers only the snowflake, the problem for two symmetrically related facets on any crystal is the same.

The development of detailed mathematical methods for the calculation of X-ray diffraction intensities and the solutions of the structures of the major part of the mineral kingdom made the exact calculation of vectorial growth forces possible. The Donnay-Harker rule, as Buerger had observed, was not applicable to configurations not related by symmetry operations (e.g., the Na to the Cl in rock salt). In passing from geometry to physics, and from

[a] *Leçons de cristallographie*, p. 260.

space groups to atomic configurations, McLachlan in 1952 (Paper 32) considered first the departure from special positions of the atoms of a structure and second the species of atoms under consideration. The parallels that had already been observed between morphological aspect and X-ray intensities now reappeared in the equations for *laminarity*, a function expressing the dominance of the crystal faces. Regrettably, the work was largely overlooked, because it was published only in the *Bulletin of the University of Utah Experimental Station*. With the addendum written expressly for this volume, it becomes the most general rigorous treatment in print of the problem of the calculation of vectorial forces and therefore surface energies.

It is the editor's view that the several other methods of calculation from the structure approach equivalence to McLachlan's as they take into consideration more and more of the complexity of the structure. See Mott and Jones's use of the geometric structure factor for the diamond configuration[b] as well as the Bravais lattices, and also Papers 26, 28, 30, and 31.

[b] See footnote on p. 203

29

Reprinted from *Proc. Natl. Acad. Sci.*, **43**(1), 143–151 (1957)

THE SYMMETRY OF DENDRITIC SNOW CRYSTALS

By Dan McLachlan, Jr.

STANFORD RESEARCH INSTITUTE, MENLO PARK, CALIFORNIA

Communicated by Sterling Hendricks, November 1, 1956

Snow crystals are of interest in two major respects: (*a*) meteorologically and (*b*) artistically. The meteorological interest in snow crystals was greatly enhanced by the work of Langmuir[1, 2] and Schaefer,[3–6] who have done such extensive pioneering work on the seeding of clouds with material having a structure similar to that of ice, with the object of gaining some control over rainfall throughout the country. Their most effective seeding materials are frozen carbon dioxide and silver iodide. From the artistic standpoint the interest in the beauty of snow crystals goes back to the most primitive times, but we know that as early as 1555 Olaus Magnus, Archbishop of Upsala, published a woodcut of a snow crystal in a book on the general subject of "Natural Phenomena." The most authentic early drawings of snow crystals were made by Scoresby[7] in 1820. One of the first to use photography for the study of shapes of snow crystals was Hellmann[8] in 1893. The most extensive work on the photography of snow crystals was carried out by W. A. Bentley, of Jericho, Vermont, who apparently spent a good portion of his life at this task, judging from his publications between 1901 and 1927.[9] In 1931 W. J. Humphreys assisted Bentley in compiling about 2,400 of Bentley's most interesting photographs into a beautifully arranged book.[10] All recent work on the study of snow crystals is contained in a well-illustrated book by Nakaya.[11] This book lists eighty-seven references and is outstanding for its complete accounts of the experiments of Nakaya and Hanajima in their artificial production of snow crystals to duplicate almost every form produced naturally.

The present discussion deals with the artistic aspects of snow crystals, but it has a scientific interest because it tries to explain the causes of the interesting shapes of snow crystals and, furthermore, because these explanations may be carried over into other fields wherein dendritic growth and branching are evident.

Three outstanding characteristics of dendritic snow crystals are immediately apparent: (*a*) they exhibit a hexagonal pattern; (*b*) the crystals are all different in shape, as though no simple law (such as governs "normal" growth) operated to cause a relationship between the crystals; and (*c*) the branches of the six trees on the dendritic crystals form six almost identical sets. The hexagonal character of snowflakes is in perfect accord with the crystal structure or atomic arrangement of the water atoms in the ice crystal. Bernal and Fowler have worked out the structure of ordinary ice, or ice I,[12] while the structures of ice formed at high pressures have been worked out by Megaw,[13] Burton and Oliver,[14] and McFarlan.[15] The fact that no two snowflakes are alike seems to surprise no one, since experience teaches that the probability of a difference existing between two objects increases with the complexity of detail in the objects. However, the fact that the six trees (see Fig. 6) branching from a common center in a snow crystal are almost identical in spite of their intricacy is a thing that excites the wonder of everyone.

It is the great similarity between the six parts of a dendritic branching snow crystal that this paper is written to explain. The problem has often been expressed by the question, "How does one branch of the crystal know what the other branches are doing during growth?" Such regularity and symmetry are not so uncommon among the plants, especially in the flowers and blossoms, and among the animals, particularly in animals of the sea, where hormones and nerves provide a means of co-ordinating the development and activities of living organisms (Thompson[16]); but for an inanimate organism to duplicate its development in growth in such minute detail is not easily explained. The explanation of the co-ordination in growth among the six branches of a snow crystal is based (in this discussion) upon the theory that it is caused by standing waves of the thermal and acoustical type within the crystal. We will first discuss the nature of the standing waves[17] within bodies of various shapes and then discuss the kinetics of the deposition of molecules of H_2O on the crystal.

The Nature of Standing Waves.—When an elastic substance is disturbed by applying some force momentarily to some portion of it so as to displace that portion from its equilibrium position, that portion of the substance will (as soon as the force is removed) return to its initial position. But, owing to the momentum of that portion, upon its arrival at the initial position, it will "coast" beyond equilibrium. Again in a position of nonequilibrium due to "overshooting," the body again finds itself elastically deformed and returns through the equilibrium position with momentum which carries it to the position of original displacement. This process may be repeated many times, depending upon how rapidly the energy is dissipated by friction, either internal or external. The number of times which this cycle is repeated per second is called the frequency ν and depends upon the restoring force F and the mass M of the displaced material; for simple systems this may be expressed as

$$\nu = \frac{1}{2\pi}\sqrt{\frac{F}{M}}, \tag{1}$$

where ν is the frequency. At the same time vibrations may be carried to other portions of the body. The number of centimeters per second with which the disturbance is transmitted from one portion of the body to another is called the "velocity" of the wave.

If the body is not infinite in extent but is limited to some defined shape, the waves will be reflected at the boundaries, and the reflected waves will mix with other oncoming waves. There are certain frequencies of vibration whose values ν are such that there is a sort of coherence between all reflected waves and the oncoming waves. For these frequencies, which are called "characteristic frequencies," so-called "standing waves" result. The standing waves appear not to be moving. Their apparent stationary condition results from the fact that the oncoming and reflected waves have equal amplitude and equal frequencies and their velocities are equal but in opposite directions. The frequencies of standing waves depend upon the initial conditions of displacement and upon the shape and dimensions of the body. These conditions are called "boundary conditions." From the boundary conditions the characteristic frequencies may be solved for, provided that the boundary conditions are sufficiently simple.[18, 19] The shapes for which solutions may be obtained are (*a*) linear systems, parallelograms, and parallelepipeds, to which one-dimensional, two-dimensional, and three-dimensional Fourier series apply, respectively; (*b*) circular systems and cylindrical systems, to which Bessel functions apply; and (*c*) spherical systems, to which Legendre polynomials apply.

It may be assumed that a long, unbranched spine of a dendritic crystal of snow complies very nearly to a one-dimensional system. When in transverse vibration, the displacement $\phi(x, t)$ of a point at distance x along its length at any instant of time t can be expressed by the Fourier series

$$\phi(x, t) = \sum_m A_m \cos\frac{2\pi m x}{a}\cos 2\pi\nu_m t, \tag{2}$$

where a is the length of the spine, m is an integer, and A_m is the amplitude of the mth wave. In this respect the spine is analogous to a vibrating string free at both ends. As is known to all musicians, the relative values of the amplitudes A_m for each m value (or harmonic) is influenced (1) by the position x_p at which the string is plucked, struck, or bowed and (2) by the position x_d at which the string is arrested by the finger. The point of greatest interest to us is that, in a string of length a which is arrested at the center, for example, all wave lengths a/m' are reduced in amplitude, where m' is an odd integer; or, in general, if the string is damped at some distance from the end expressed by a/g where g is any chosen integer, then the string cannot vibrate at any wave lengths except those which are expressed by $\lambda_p = 2a/gm''$, where m'' is any integer.

If it is assumed that the arresting of a vibrating string by placing a finger on that point is equivalent to loading the string at that point with a mass which is infinite in comparison to the density of the string, then it becomes of interest to know what may happen if the added mass is finite. We shall not go into this matter, but fortunately this problem has been worked out,[17] and we shall merely state the results: (1) a mass of weight m attached to a string reduces the frequency of vibration of all standing waves except those for which the point of attachment is a node, and (2) the mass reduces the amplitude of vibration at the point of attachment. As the mass is increased, the amplitudes and frequencies decrease, until at infinite mass fully developed nodes are induced.

The two-dimensional standing waves, as observed on a rectangular drumhead,[18] are very convenient to study, because when such a drumhead is set into vibration by striking, sawdust scattered upon the membrane is observed to accumulate upon the nodal points or nodal lines. When such a drumhead is struck and then arrested at some point, other points are observed to be induced nodes, as evidenced by the accumulation of sawdust in small piles at these positions. Or, if the drumhead is arrested by placing a straight edge one-third of the way along the a dimension, then an induced nodal line is formed at $2a/3$.

Examples of Branching in Snow Crystals.—The theory which we wish to present is that snow crystals grow from the deposition of water molecules upon small nuclei of ice which are under thermal vibration corresponding to a temperature of 250°–273° K.; the molecules strike and bounce off the nuclei, with relatively few sticking; the number of those which strike and stay to contribute to the size of the growing crystal is greatly influenced by the frequency and amplitude of vibration at any point on the crystal; if a portion of a spine or branch of a dendrite becomes more heavily loaded than the surrounding portions, the vibrations are somewhat restricted, as is a string with a mass attached; these loaded portions become more loaded with further growth; therefore, the growth from the loaded point becomes a runaway process, and thus branching occurs; finally, the loaded portion simulates a node, and the induced nodes sponsor growth at other portions according to a defined pattern.

To clarify the process described in the above paragraph, let us discuss a number of the pictures from the book by Bentley and Humphreys.[10] First let us imagine that Figure 1 is an enlarged photograph of a very small nucleus which in the later stages of its growth has experienced "normal" growth for a sufficiently long time to have established a "normal" hexagonal shape. If conditions were to change so as to be conducive to dendritic growth,[20] then the six corners would find themselves receiving more molecules and losing more heat of crystallization than the flat portions. If this situation prevailed, the corners would "sprout" dendrites, as shown in Figure 2.

After a very much longer time the dendritic "sprouts" will have grown to quite extended lengths, as shown in Figure 3. Up to this point, the growth of the dendrites can be explained on the basis of the known theories.[20] To account for the branching, however, one must consider the standing-wave phenomena.

FIGS. 1–6.—Some typical snow crystals photographed by W. A. Bentley. (Reproduced by permission from *Snow Crystals*, by Bentley and Humphreys. Copyright 1931, McGraw-Hill Book Co.)

Let us consider that each of the six spines in Figure 3 is an array of molecules[21] arranged in the order required by the structure of ice and that all the molecules are in thermal agitation in a manner demanded by the theories of black-body radiation and Planck's distribution law, but that the distribution of energies between the modes of vibration is influenced by the boundary conditions. If one of the spines becomes loaded at some point, say one-fourth the distance from their point of attachment, then nodes are induced at equal intervals along the spine. As shown in Figure 4, these nodes will "sprout" dendritic branches that are equally spaced. Figure 5 shows the equal spacing of the branches on a dendrite where the interval is one-eleventh of the spine length, and the nuclear size is extremely minute.

The question may arise as to how the waves which stand in one of the six branches are coupled with those in the other five branches. Experimentally one may build a model of six such spines from thin drill-rod material and weld them together, then place them on a thin dental rubber dam sheet. Sawdust may be used to study the wave forms produced by vibrating them. The torque through the point of intersection transmits the same frequencies and induces the same nodes in all branches in an identical manner.

Figure 6 is an interesting example of a crystal which changed its vibrational modes. This crystal grew dendritically from the center O to the point S according to some system of loading, then grew a slender spine from S to T, and then re-branched systematically again. Also, one may notice secondary branches such as the one marked B. Many of the secondary and tertiary branches have been inhibited by interference or shielding by their neighbors.

Returning to Figure 1, which was chosen for its simplicity, we see markings which indicate that it has formed complete hexagons at several stages in its development. Each time after (or during) the filling-in process, changing conditions produced further dendritic growth. Similar "filling-in" scars can be seen in Figure 2 at G'.

These markings on the interior of the crystals are almost as interesting as the branching characteristics of snow crystals. The complicated markings such as are shown in Figure 4 are a sort of a written record of the atmospheric conditions through which the crystal has gone in its relatively long experience of having been carried up and down repeatedly by upcurrents and gravity between elevations at which the temperature, humidity, and pressure differ extremely with altitude. A study of these markings should eventually become valuable to meteorologists. Nakaya[11] and his co-workers have established, in their experiments on the synthetic production of snow crystals, the conditions of ambient temperature, humidity, and supercooling under which each of the various types of snow crystals is formed. If it can be assumed that natural snow crystals of a given type can be used as evidence of the conditions in the air, then meteorology has a new tool.

In this discussion there is one point that has not been clarified. This point has to do with the question, "What is the mechanism by which some point on a dendritic spine becomes more heavily loaded with molecules than other points?" We shall consider this point statistically.

Some Statistics of Snow Formation.—Let us consider some facts which might enable us to imagine what goes on in the atmosphere during the growth of a snow crystal.

At 0° C., saturated air contains 4.835 gm. of water per cubic meter, according to the Smithsonian tables. At 20° C. the air is saturated when there is only 0.892 gm.

per cubic meter. This means that if saturated air at 0° C. is chilled to −20° C., about four-fifths of the suspended water molecules will precipitate as ice as fast as these molecules can establish (or find) nuclei upon which to deposit. Expressed in partial pressures, the 4.835 gm. of H_2O per cubic meter is about 0.006 atmosphere, or 4.56 mm. of mercury. This does not appear to be a very high pressure, but it represents 1.6×10^{17} molecules of H_2O per cubic centimeter mixed with about 200 times this many molecules of air. At this congestion, the water molecules are only 0.2×10^{-3} millimeter apart. They do not have far to go to associate with one another, and they should not require very long to accomplish much, since the H_2O molecules are traveling at a mean velocity of 62,700 cm/sec. For example, if an ice crystal is suspended in the saturated atmosphere at 0° C., it is bombarded with water molecules at the rate of 2.5×10^{21} molecules per square centimeter per second, that is, about $^1/_{250}$ mole/sec. This means that every 28 seconds there will be about 79 calories of heat deposited per square centimeter if every molecule sticks (this is assuming a heat of sublimation of 675 cal/mole). This accumulation of heat would raise the temperature and melt the crystal (79 cal/mole is the heat of fusion of ice).

To be more precise, let us consider an actual snow crystal of over-all dimensions of 100 μ or about $^1/_{10}$ mm. Assume that it is similar to the one pictured in Figure 3 having six spines, each 20 times as long as wide and about one-tenth as thick as wide. Such a crystal has an area of 10^{-6} square centimeter. Such an area is bombarded with 2.5×10^{15} water molecules per second. Using the figure 1.4×10^{15} and the volume occupied by water molecules in ice, one concludes that the crystal should grow at a rate of about one centimeter per second in all directions. This velocity of growth is much more rapid than observed rates,[6] since observed rates are approximately 100 μ in 30 seconds or 10^{-4} cm/sec. We are forced to conclude that even at fast rates of growth only about four out of every thousand molecules which strike the crystal surface stick; the other 996 bounce off.

As pointed out by private communication from Sterling B. Hendricks, the present argument applies not only to the sublimation processes but also to the surface migration of water molecules on ice. Murphy[22] has shown that in the bipedal random walk of water at 0° C. there are 10^5 skips along the surface for each jump off the surface, and at −100° C. the ratio is 10^8.

Now let us consider the vibrations within the ice crystal. The longest thermal vibration in the crystal has a wave-length twice the length of the spine or about $^1/_{10}$ mm, that is, in the far infrared, not far from the black-body maximum at 0° C.; and the shortest possible wave is twice the interatomic distance, or about 3 A, that is, in the X-ray region and negligibly weak in intensity. Assuming that the velocity of sound in ice is about 1,500 m/sec, we may calculate the lowest frequency of vibration of a dendritic spine to be 1.5×10^7 and for the highest frequency 6.6×10^{12} cycles/sec. This means that even for the lowest possible frequency of oscillation of the spine of an ice crystal only about 10^{-7} second is required to complete a wave cycle. Using the structure of ice, we may calculate the area of a unit cell of the crystal to be about 33×10^{-16} square centimeter. Following, in our imagination, the oscillation of a unit cell back and forth once every 10^{-7} second, and an exposed area of only 33×10^{-16} square centimeter, in an atmosphere where it is being bombarded at a rate of 2.5×10^{22} times per second per square centimeter with water molecules, we can calculate the number of bombardments per cycle per unit cell to be approximately 8. Since only one of these out of each thousand sticks, we may conclude that, on the average, we must consider an expanse of about 300 unit cells to see one captured molecule per cycle at the slowest possible vibration rate. For higher frequencies the times are shorter and the expanse (over which one may expect an average of one capture per cycle) is greater.

This means that, while we may calculate the average number of molecules captured by a given area per second, we have no means of saying for certain that this really happens. Watching a single area in which we expect one molecule to be captured and confining our observations to the short time of one cycle, we have a good probability of seeing no molecules captured, or we have a fair probability of seeing two captured. According to our earlier study of vibrating strings, if two are captured, the system is loaded more heavily in this region than the average, and further captures are more likely on the next cycle. The runaway process has begun!

Fortunately, the probability of capturing any portion of the expected number of molecules is a problem in a field of statistics which has been worked out.[23] In calculating the deviations from expectation, Poisson developed an equation which has proved reliable in many problems such as production control, plant breeding, and other fields where data can be collected:

$$P(n) = \frac{\epsilon^n e^{-\epsilon}}{n!}, \tag{3}$$

where $P(n)$ represents the probability of n molecules falling on an area where ϵ molecules were expected to fall. This equation can be used to calculate the probability that any chosen area of a crystal may become abnormally overloaded with molecules during one vibration cycle. From a table of Poisson probability values[23] (p. 458), we see that, for $\epsilon = 1$, the probability of getting what we expect is about the same as the probability of getting none. The probability of getting 2 is about one-half that of getting 1, and the probability of getting 3 is about one-sixth. The maximum probability is always on the expected number; for example, when $\epsilon = 5$ the maximum probability is .175 for $n = 5$, and when $\epsilon = 10$ the maximum probability is .125 in the neighborhood of $n = 10$. The tables teach us that if we expect 1 molecule per unit area for a given cycle, we have a probability of about 1 out of 3 of getting it; if we expect 5 molecules, we have a probability of 1 out of 5 of getting it, and if we expected 10, our probability is 1 out of 8. The total probability is unity in any set, but the bulk of the probability is scattered over the many unexpected groupings of events.

This all means that, according to statistics, an uneven loading is very likely in a single vibration of a dendritic spine, and branching is easily initiated.

An interesting experiment would be to grow a piezoelectric crystal while vibrating, to see whether fins could be included at the nodes of the vibrations.

The author is indebted to the Stanford Research Institute and the Engineering Experiment Station of the University of Utah for facilities in carrying out this work. The author is also indebted to the McGraw-Hill Book Company, Inc., for permission to use selected figures from the book by Bentley and Humphreys.[10]

[1] I. Langmuir, *J. Meteorol.*, **5**, 175, 1948.
[2] I. Langmuir, *Proc. Am. Phil. Soc.*, **92**, No. 3, 167, 1948.
[3] V. J. Schaefer, *J. Meteorol.*, **6**, 283, 1949.
[4] V. J. Schaefer, *Chem. Revs.*, **44**, 291, 1949.
[5] V. J. Schaefer, General Electric Research Laboratory Occasional Report No. 11 (Schenectady, N.Y., 1948).
[6] V. J. Schaefer, *J. Appl. Math. and Phys.*, **1**, 153, 1950.
[7] William Scoresby, *An Account of the Arctic Regions, with a History and Description of the Northern Whale-Fishery* (Edinburgh, 1820).
[8] G. Hellmann, *Schneekrystalle* (Berlin, 1893).
[9] W. A. Bentley, *Monthly Weather Rev.*, **29**, 212, 1901; **30**, 607, 1902; **35**, 348, 397, 439, 512, 584, 1907; **46**, 359, 1918; **52**, 530, 1924; **55**, 358, 1927.
[10] W. A. Bentley and W. J. Humphreys, *Snow Crystals* (New York and London: McGraw-Hill Book Co., 1931).
[11] Ukichiro Nakaya, *Snow Crystals* (Cambridge, Mass.: Harvard University Press, 1954).
[12] J. D. Bernal and R. H. Fowler, *J. Chem. Phys.*, **1**, 515, 1933.
[13] H. D. Megaw, *Nature*, **134**, 900, 1934.
[14] E. F. Burton and W. F. Oliver, *Nature*, **135**, 505, 1935; *Proc. Roy. Soc. London*, *A*, **153**, 166, 1935.
[15] R. L. McFarlan, *J. Chem. Phys.*, **4**, 160, 253, 1936; *Phys. Rev.*, **49**, 199, 644, 1936.
[16] D'Arcy W. Thompson, *Growth and Form* (New York: Cambridge University Press, 1948).
[17] C. A. Coulson, *Waves* (London and New York: Interscience Publishers, 1947; 6th ed., 1952).
[18] William Elwood Byerly, *Fourier Series and Spherical Harmonics* (Boston: Ginn & Co., 1893), pp. 126–134.
[19] H. Morgenau and G. M. Murphy, *The Mathematics of Physics and Chemistry* (New York: D. Van Nostrand Co., Inc., 1943).
[20] The formation of spines reaching from a crystalline mass (or a single crystal) into the solvent has been explained (D. McLachlan, Allan Carlson, C. J. Christensen, and Alan King, University of Utah Engineering Experiment Station Bull. 57, 1952) on the assumption that under certain conditions the rate of deposition of molecules on the growing crystal is controlled by the rate at which any one of three things can occur: (1) the arrival of solute molecules to the various points on the surface (or the departure of solvent from the surface), (2) the escape of heat of crystallization from the surface, or (3) the escape of electric charges from the surface. The last of these processes is markedly exhibited by the needles formed in the fast electrodeposition of silver. It has been shown (*ibid.*) that all these processes can be expressed by a conformal mapping of the lines of flow and the equiconcentration contours about bodies of various shapes, and that protrusions on the crystal are favorably shaped to enhance any of the three processes and consequently develop spines.
[21] Leon Brillouin, *Wave Propagation in Periodic Structures* (New York: McGraw-Hill Book Co., 1946; 2d ed., New York: Dover Publications, 1953).
[22] E. J. Murphy, *J. Chem .Phys.*, **21**, 1831, 1953.
[23] Thornton C. Fry, *Probability and Its Engineering Uses* (New York: D. Van Nostrand Co., Inc., 1928), pp. 214–240, 458–462.

30

Reprinted from *Am. Mineralogist*, **55**, 1466–1488 (1970)

MORPHOLOGICAL BASIS FOR THE RETICULAR HYPOTHESIS

CECIL J. SCHNEER, *The University of New Hampshire, Durham, New Hampshire 03824.*

ABSTRACT

The persistence of the forms of a mineral species through the environmental variation represented in geological space and time (P_{hkl}') is used to derive average specific surface energies σ_{hkl} as a function of the multiplicity, m, and the interplanar spacing, d; $\sigma_{hkl} = -C [\ln P'/m]/d^2$ with $-C$ a constant proportional to the absolute temperature. Relative σ are of the order of magnitude of 0.5 kcal/mole at room temperature. The equation is one transform from an empirical morphological intensity space to an incomplete structure factor space. It is confirmed by Fourier transforms to the electron and vector densities, and by the comparision with calculated structure factors with residuals of 0.187, 0.274, and 0.360 for anglesite, barite, and celestite.

The ensemble of the forms of a mineral species should approach equilibrium (a statistical distribution invariant with time) throughout geological time and space. As it approaches equilibrium the observed distribution becomes a function of the structure and the structure is a Fourier transform of the distribution.

INTRODUCTION

The aim of this research is to establish the functional relationship between the forms and the structure of crystals. It is based on the concept that the polyhedral envelope of a well-developed crystal results from a vectorial 'quantized' distribution of growth rates; and second, that these growth rates are a function of the potential energies to be associated with each point of the reciprocal lattice.

Some of the terms that enter into the calculation of these energies are dependent upon the geometric distribution of the particles making up the crystal structure and this paper is concerned with these structure-dependent terms. Other terms are either independent of the structure or dependent upon bulk properties. Such scalar terms do not effect the proportionate frequencies of occurrence or relative sizes of the individual faces of a crystal. The crystal or 'lattice' energy is an example of such a scalar term since it influences growth rate but is invariant with respect to the points of the reciprocal lattice. Curie (1885, p. 146) has considered the change in energy for a change in the shape of a body, the volume being held constant, as proportionate to the variation of the area. He defines specific surface energy as the energy required to increase the surface by one unit, the volume being held constant. The definition does not distinguish between a collective of many crystals and an individual crystal.

THE MORPHOLOGICAL STRUCTURE FACTOR SPACE

Definition. The distribution of the forms of a mineral species in geological time and space is the outcome of a world-experiment establishing the probability of occurrence of each particular form. These probabilities may be approximated by the P_{hkl}'—or persistence-values of Niggli (1941, 504, 523). Forms which persist through a wide range of conditions and time ought to be characterized by lower specific surface energy than forms which are rarely observed. Using the statistical-mechanical relationship between energy and probability (*cf.* equation 6 below) it is possible to derive relative average specific surface energies from the geological persistence of the crystal forms of a mineral species. These energies are the magnitudes associated with points of the reciprocal lattice.

The geometry and the proportions of the direct lattice and of the corresponding reciprocal lattice can be derived directly from the morphology (Friedel, 1905, p. 109). The absolute dimensions can be derived from the molecular weight and the specific gravity to within an integer multiple (number of formula molecules per unit cell). Therefore, a reciprocal lattice with points weighted by empirically determined energies may be derived from the crystal morphology.

If each point of the reciprocal lattice is assigned a weight proportional to an observed magnitude (as for example the intensity of the X-radiation reflected in the point), the reciprocal periodic space so defined may be termed the intensity space (*cf.* James, 1954, p. 349). The intensity space varies with the experimental procedure for weighting the lattice points, as for example, neutron diffraction, or X-ray diffraction (Ewald, 1921, p. 154) which give radically different sets of observed magnitudes. To obtain a space reciprocal to the direct crystal structure it is necessary to convert the intensity space obtained by a particular experimental technic to a reciprocal lattice with points weighted by empirical "structure factors" alone. This is the structure factor space. To be complete, the structure factor space should be very large (containing many points) and the weight of each point should include the phase as well as the magnitude of the structure factor. In practice, no single experimental method provides weights and phases for all conceivable points and therefore any structure factor space derived from experiment is more or less incomplete. If the values P_{hkl} are assigned to the appropriate points of the reciprocal lattice, a limited morphological intensity space is defined. A conversion from this morphological intensity space to the structure factor space is outlined below (equation 10).

The Transform from Morphology to Structure. A number of transforms between morphology and structure have been proposed in the past (Bravais, 1866; Barlow, 1883; Pope and Barlow, 1907; Niggli, 1920; Buerger, 1947; McLachlan, 1952, among others). The rule of Bravais

TABLE 1. PERSISTENCE VALUES AND $\sigma \cdot d$ VALUES USED FOR FOURIER TRANSFORMS

hkl (Niggli)	*hkl* (cell)	Anglesite P'	Anglesite $\sigma \cdot d$	Barite P'	Barite $\sigma \cdot d$	Celestite P'	Celestite $\sigma \cdot d$
102	101	73	66.87	86	67.54	85	70.65
011	011	73	84.33	86	86.70	94	91.19
100	200	38	85.77	61	92.66	37	86.37
122	111	52	67.30	27	49.00	17	38.39
001	002	67	120.84	98	128.20	67	122.44
110	210	81	110.98	92	111.23	68	107.02
104	102	28	81.98	37	87.96	25	79.51
111	211	65	92.70	69	91.82	43	79.94
010	020	32	128.37	71	156.27	8	77.69
112	212	21	68.87	16	55.90	9	34.11
120	220	25	110.90	11	73.36	11	75.62
144	122	2	0.0	2	0.0	14	61.20
210	410	2	0.0	39	144.55	2	0.0
113	213	1	0.0	26	95.50	6	21.57
012	024	6	75.12	4	46.35	6	76.02

All values were scaled to σ_{000}=no. of electrons in unit cell for the maps (Figures 1–4).

(1866) provided that the morphological importance of faces would be in the order of the reticular densities of planes of the direct lattice. The rule of Donnay and Harker (1937) substituted the reticular density of "equipoints" *i.e.*, points added by space group operations as well as true lattice points, for the reticular densities of Bravais. The reticular densities are proportional to the interplanar spacings $d(hkl)$. In the example of Donnay and Harker, the basal plane of quartz $\{0001\}$ should be the most prominent according to the Bravais rule, but each lattice point is replaced by three "equipoints" as a result of the 3_1 or 3_2 screw operation. In the same way, the spacing $d(0001)$ is divided by 3 and the plane under consideration is $\{0003\}$, rather than $\{0001\}$. But for this argument to have physical significance, the density under consideration ought to be related to the density of centers of force or material particles rather than geometric points. The electron density is a possible physical substitute for the densities of ideal points.

The standard method of representing the structure is by mapping the electron density. Given the electron density the complete structure factor space can be calculated directly. The inverse process from the structure factor space to the electron density by Fourier transform is also well known. This transform, shares with the structure factor space, the property of complete independence of any experimental technique. If the transform from the morphologically-derived structure factor space should yield an approximation to the electron density, the morphology would be a function of the electron density to the degree of the approximation, and the function would be to the same degree the inverse of the Fourier transform.

Ordinarily the detail and accuracy of the Fourier electron density map depends upon the inclusiveness and accuracy of the structure factor space. A space derived from diffraction is markedly more accurate as well as inclusive than the space derived from morphology. The morphologically-derived structure factor space is determined for only a single point of the lattice in any rational direction (a single order n for any nh nk nl). In place of hundreds or even thousands of diffraction intensities, meaningful values for morphological frequencies may be two orders of magnitude fewer and of only approximate accuracy.

Results. The first attempt at a Fourier transform from a morphologically-derived structure factor space to an electron density, *i.e.* starting from the external forms of crystals and finishing with the crystal structure was shown at the 1966 meetings of the International Mineralogical Association in Cambridge, England (Schneer and Silva, 1966B).

In order to obtain a quantitative evaluation of the results, error factors were calculated as residuals in the comparison of morphologically derived specific surface energies with structure factors. The structure factors were calculated by weighting known atomic positions with 1) electron scattering factors; 2) neutron form factors (atomic numbers); 3) atomic weights; and 4) by using atomic numbers for metal and sulfur alone, omitting oxygens. Residuals varied from less than 0.20 to 0.40. The omission of oxygen atoms and the use of constant atomic number (neutron form factor) in place of electron scattering factors overweighted the metal ion and brought about the lowest residuals. The lowering of the residuals confirms the testimony of the maps in emphasizing the morphological role of the metal ion.

Although these calculated error factors are at least twice as large as those for the corresponding diffraction data, and relatively unamenable to the refinement procedures of diffraction analysis, the *a priori* probability of matching 15 structure factors with a residual of 0.2 or less is vanishingly small. The relationship cannot be coincidence.

Fourier maps, Figures 1 and 2, constructed from the morphologically-derived structure factor space, are crude but show clearly the location and relative magnitude of the metal ion in the barite group structures. They are only slightly less detailed than the Fourier maps made from X-ray data for the same sharply restricted Fourier series. They are greatly superior to the Fourier maps of the same selected terms with

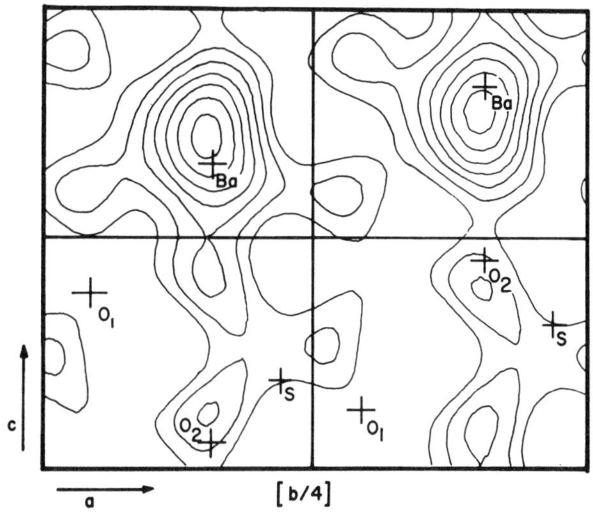

FIG. 1. Fourier transform (electron density map) of the $y(1/4)$ plane of barite, $BaSO_4$, using the values of Table 1 as coefficients. All computations and graphs by computer. Positions of atoms located on the plane by diffraction procedures are marked for comparison. The *signs* of the Fourier coefficients were determined from the structure (diffraction); Schneer (1968).

random magnitudes. Using observed magnitudes (in the form of squared morphologically-derived structure factors) without phase or symmetry or other *a priori* additions to the reciprocal lattice, a preliminary Patterson-Harker (Buerger, 1959) transform from the barite data of Niggli, (Fig. 3a) may be compared with the equivalent map (Fig. 3b) prepared from the known structure data (Wyckoff, 1963). Figure 4 compares the morphologically derived Zn-Zn and S-S vector of sphalerite with the F-F and Ca-Ca vectors of fluorite. It underlines the physical significance of the morphological frequency data.

Wherever parallelism between morphological importance and diffraction intensities is observed, parallel analytic procedures should lead to the same structure. Such parallelism was remarked by Wells (1946) *cf.* Figure 5, and a correlation between Brillouin zones and morphology was pointed out by Schneer (1959). Working backwards, morphological significance should be predictable and calculable from the known structure. The detail of the Fourier map may reflect the relative importance of the different atoms in determining the morphology, as for example in locating the metal ion and ignoring the approximately close-packed anions. For polar crystals the absolute orientation of the structure with respect to the morphology should be obtainable (suggested by J. D. H.

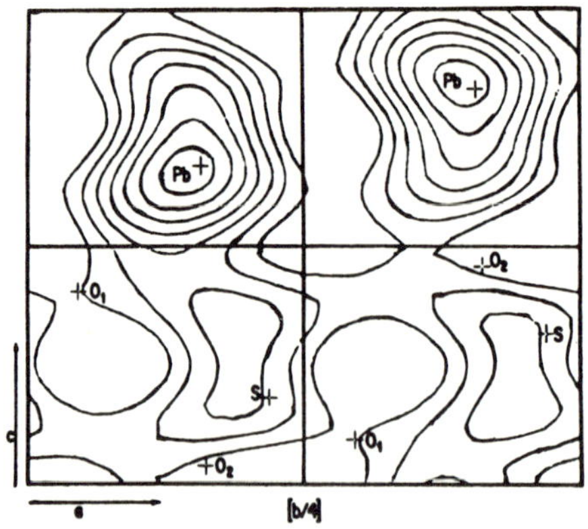

FIG. 2. Fourier transform as in figure 1, but for anglesite $PbSO_4$; data of Table 1, after Schneer (1970).

Donnay). The morphological effects of diadochy, incorporated impurities, defect structures and stacking faults, as well as the complete morphological frequency distribution in geological time and space should be calculable *a priori*.

The Relationship Between Macroscopic Growth and the Fundamental Laws of Crystallography

Introduction. If a crystal is grown uniformly from solution it is often morphologically a better crystal than if it is grown at arbitrarily variant rates. Not only is the body of the crystal clearer, but the external faces of the polyhedral surface approach ideal planes more closely and the edges and corners are more sharply defined, the more slowly the crystal is grown. These are observations made in our laboratory but also reported widely (Buckley, 1951, p. 129) and it is commonly accepted by mineralogists that well-developed single crystals are evidence of uniform growth under conditions permitting the barest minimum of excess saturation. It has also been reported that under these uniform conditions crystal growth is observed to proceed by mechanisms other than that of spiral growth (Egli, in Doremus *et al.*, 408, 409).

Regardless of the particular microscopic mechanism of growth, if the developing crystal is observed macroscopically at intervals of time as for example in time-lapse photography (Schneer and Silva, 1966A), and ifthe Wulff normals on each face[1] are measured as a function of time, then the

[1] Wulff, 1901, proposed that the face-normals from a particular point within the crystal are proportional to the velocities of growth. Such a velocity would be an average for the length of the normal over the total time of growth.

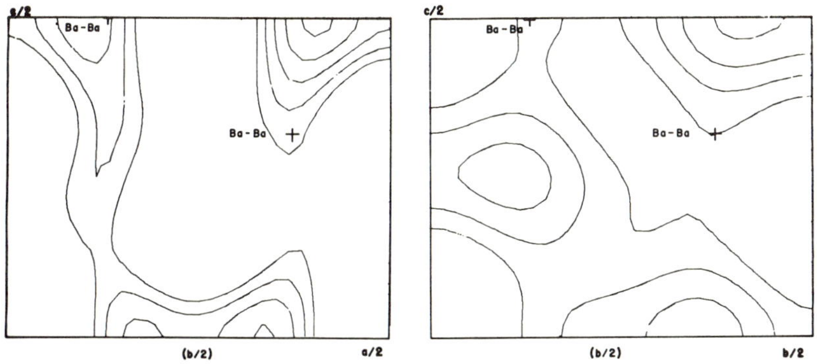

FIG. 3a. Fourier transform (Patterson-Harker map) showing the interatomic vector density for the $y(1/2)$ plane of barite using the same values of Table 1 as in Figure 1 but squared (*without a priori* phase information or symmetry information). The positions of the Ba–Ba interatomic vectors calculated from the diffraction structure are marked (+) for comparison 2.

FIG. 3b. Fourier transform (Patterson-Harker map) for barite exactly as in figure 3a but computed from the diffraction determined structure with coefficients the squares of the calculated X-ray structure factors. The terms of the series were limited as in Figures 1–3a to the list in Table 1.

increments in length, Δl, of these normals, divided by the time difference of the observation, Δt, have the dimensions of velocity and approximate the instantaneous growth rate. We are concerned here solely with those *macroscopic* growth rates in directions perpendicular to the crystal faces $\{hkl\}$, in full awareness that this growth may in most if not all cases, proceed through microscopic processes such as the spiral, with orientation and geometry essentially independent of the crystallography. The crystal is considered to be a black box with a bounding polyhedron which is evidence of its relative growth rates in different directions averaged over the crystal's life-path.

In summary, the macroscopic growth of crystals may be referred to a set of vectors in the directions of the face-normals and identified by the face indices. They have the magnitudes of the rates of extension of the face-normals. These are the "Wulff vectors."

If crystals grew in all directions rather than in the directions of a discrete set of vectors, and if the rates of growth were equal, the resulting surfaces would approach spheres, *cf.* Curie, 1885, p. 147. If crystals grew in all directions and not solely in the directions of the vectors joining the origin to points of the reciprocal lattice, the resulting polyhedral surfaces (*cf.* Gibbs, 1961, p. 316) would contradict the laws of the constancy of interfacial angles and of rational intercepts. Crystals in general do not

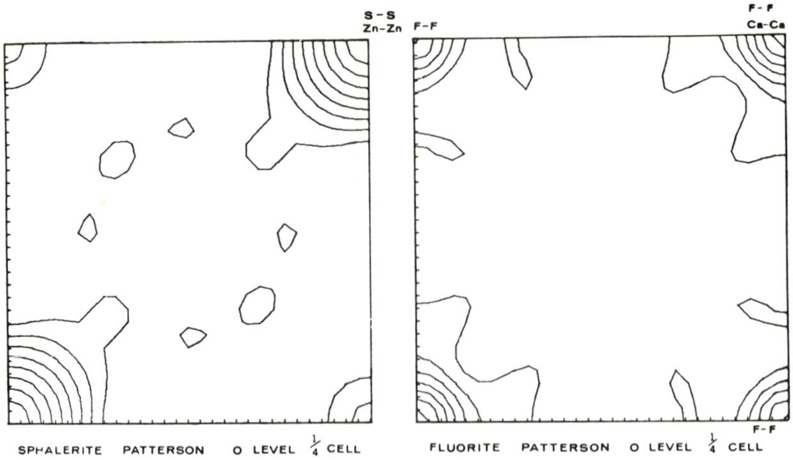

FIG. 4a. Patterson for Sphalerite from the morphological data of Gebhardt (1933). Interatomic vectors at the corners are labeled.
FIG. 4b. Patterson for fluorite from persistences measured on figures in Goldschmidt's *Atlas* (1923).

exhibit spherical surfaces, and do obey the two fundamental laws of crystallography. Therefore we conclude that *the macroscopic growth of a crystal may be represented by vectors in directions from the origin to the points of the reciprocal lattice, with the magnitudes of the rates of extension of the face-normals.*

Environmental Influences on Growth. Any environmental influence on the growth of crystals may be considered in two parts, the first random and the second crystallographically controlled. Random influences affect the persistence of one face as much as another. A property of the fluid medium in which the crystal is growing, such as viscosity, would be a random influence. If a crystal is growing with one face exposed to a current of supersaturated fluid, there is no reason to suppose that another crystal will grow in the same orientation with respect to the current—unless there is a crystallographic control.

Only if the differences in the vectorial growth rates are structurally determined should we expect statistical consistency in the forms and habits of crystals. The forms and habits of crystals may reflect microscopic mechanisms of growth such as spirals without as well as with consistent crystallographic orientation. But the forms and habits of crystals also exhibit the statistical consistency expressed in the surveys referred to below. This statistical consistency, this repeated preference

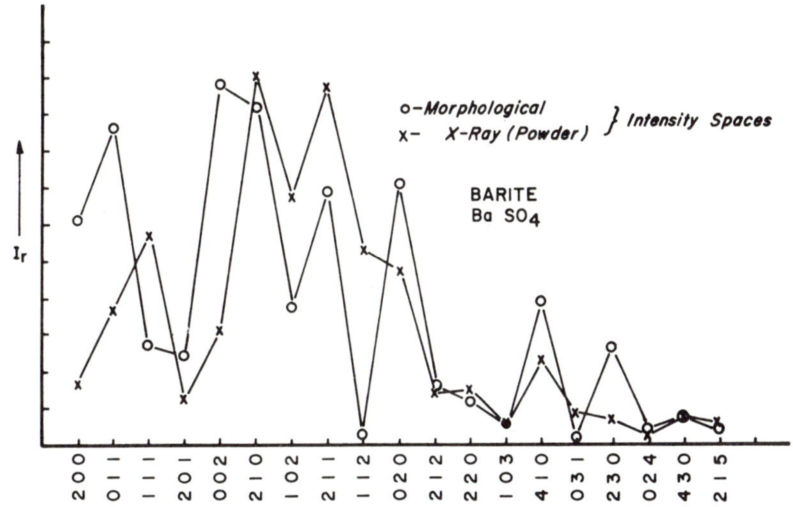

FIG. 5. Morphological and diffraction intensity spaces. The non-zero reciprocal lattice weights (P') for barite observed by Braun (1932) and the N.B.S. standard X-ray powder diffraction intensities compared. Magnitudes are plotted in order of decreasing interplanar spacings (d).

of some forms over others throughout the very great diversity of habit and environment characteristic of the provenance of a mineral species, is evidence of a crystallographically oriented distribution of surface energies.

If the forms and habits which admittedly vary widely within a mineral species, did not exhibit a statistical consistency but were as random as they appear to the nonmineralogist, then there would be no significant relationship between crystal form and crystal structure. Growth would be solely a function of microscopic mechanisms. If, on the contrary, significant patterns in the occurrence of crystallographic forms of a mineral species emerge through the study of crystals grown in a wide range of environments, then a mechanism *in species* is also in operation, and the environment is not the sole determinant of the crystal form. To decide the question we may seek such evidences in the laboratory, but in fact it already exists in the geological field and its examination falls within the science of mineralogy.

In the long geologic past and throughout the earth's crust, it may be assumed that a very wide range of experimental conditions, physical and chemical, have permitted the growth of crystals of all mineral species. In effect, the earth's crust is an enormous laboratory in which an experiment on a scale vast enough to provide a statistical basis for conclusions has already been carried out. The geological field is the result of this effort through geological time, which may be called the experiment of the world. As our record of the world-experiment becomes more complete it should form an increasingly close approximation to statistical equilibrium, *cf.* definition below.

The Classical Conditions for Equilibrium. J. W. Gibbs in 1878 (1961, p. 322) and P. Curie (1885), p. 147) proposed that the equilibrium shape of the crystal would be such as to minimize the sum of the surface energies. If σ_i is the specific energy of the ith face with area F_i, then the condition for equilibrium is that the sum $\Sigma_i F_i \sigma_i$ shall be a minimum for the volume of the crystal. Sheftal and Gavrilova (1966) modified this to consider the equilibrium shape with respect to the bulk free energy. Wulff (1901), taking the velocities of growth represented by h_i, the face-normals from a central point, as directly proportional to the specific surface energies σ_i, proposed that for the equilibrium crystal shape:

$$h_i/\sigma_i = h_j/\sigma_j = h_k/\sigma_k = \cdots = C. \qquad (1)$$

The Wulff theorem was derived from first principles by von Laue in 1943. It should be emphasized that the shape referred to is a relationship between specific surface energies and the relative magnitudes of surface planes, and not necessarily the shape of a real crystal.

The Wulff theorem is applicable to a single crystal. Any rule or "law" of mineralogy (as contrasted with physical law) applies to an ensemble of crystals. If a very large quantity of a particular compound is crystallized as one huge crystal, the equilibrium morphology will be defined by Wulff's theorem. If the same quantity of material is crystallized as a very large number of individual crystals, experience, as well as statistical mechanics dictates that the probability of a variety of combinations of forms is overwhelmingly greater than the probability that all the crystals will have identical forms. The same high energy form which is prohibited by thermodynamic considerations, is virtually assured of appearance by probability calculations. The diversity which is *a priori* implicit in the second law is described *a posteriori* by the statistical surveys of crystal morphology.

Valid objections have been raised to the concept of growth as an equilibrium process, beginning with queries by Gibbs and Curie in the original papers in which the equilibrium shape was defined.

Experimental Critique. To try to determine the extent to which the forms of single crystals adapt during growth to achieve equilibrium, we undertook a number of elementary experiments. These experiments demonstrated qualitatively that growth rates and therefore shapes are not obviously affected by the force of gravity, and therefore that thermodynamic equilibrium was not achieved for individual crystals.

Small seeds of Rochelle salts single crystals cut in the shape of cubes approximately 0.5 cm on a side with cube axes parallel to the orthorhombic crystallographic axes, were allowed to grow in a centrifuge under accelerations corresponding to as much as 1000 times normal gravity. Crystals were grown in various orientations with respect to the field so that the *a*, *b*, and *c* axes were in turn vertical. In several trials, three seeds were mounted along the vertical axis of the same centrifuge tube so that adjacent seeds were subject to very different gravitational force. In no instance was a significant difference in growth rate observed which could be related to the gravitational field. A potential energy difference between the top and bottom cube surfaces of a seed growing at an acceleration of 500 g (g is the normal force of gravity) is 500 $\Delta\rho gh$, (h is the height of the cube, $\Delta\rho$ is the difference in density between the crystal and the medium).[1] Reduction of the potential energy of this order of magnitude would be achieved if the crystal tended toward a plate-like habit. Nothing like this could be observed in our experimentation. The forces involved in crystal growth are evidently of a much greater order of magnitude. Thermodynamic equilibrium could not be demonstrated.

Wulff (1901) measured relative differences in growth rates from face to face from the total increments in the Wulff vectors in each experiment. Our own experiments measured lengths along the directions of the crystallographic axes quasi-continuously by time-lapse photography. In the curves of length vs time, growth began rapidly and somewhat erratically, slowed into constant rates (linear) and later curved as growth rates began to slow still further. But it is not clear that these results may not have been related to the particular conditions of the experimentation such as size of chamber, etc. We can only conclude that relative growth rates and therefore crystal shapes, do change during growth, *cf.* Figure 6, and that a single privileged shape was not determined.

Statistical Equilibrium.

Definition. In his memoir of 1875, Gibbs (1961, p. 56) defines the condition of thermodynamic equilibrium of a conservative system as " . . .necessary and sufficient that in all possible variations of the state of the system, the variation of its entropy shall either vanish or be negative." The condition of statistical equilibrium stated by Gibbs in 1902 (1960, p. 8) for a conservative system is that the probability distribution shall be constant in time—a condition satisfied when the distribution is that of maximum probability, or for the velocities of classical particles, the Maxwell-Boltzmann distribution, Figure 7, curve a, (*cf.* Lindsay and

[1] This corresponds to a potential difference between the top and bottom of a 1 cm-high crystal in a 500 g field of $\simeq$.01 *cal* for each gram/ml of $\Delta\rho$; *cf.* below and Appendix for relative specific surface energies.

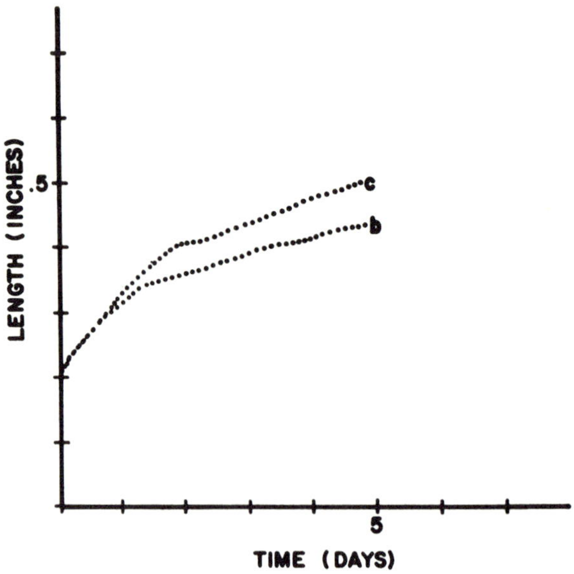

FIG. 6. Growth of Rochelle Salts crystal along *b* and *c* axes. A single crystal of Rochelle Salts growing freely at 30°C from a seed cut to a cube with axes parallel to the crystallographic axes, observed photographically at fifteen minute intervals; length in inches vs. time in days.

Margenau, p. 224). The probability of a mechanical distribution and the entropy of a thermodynamic system are related.

$$S = k \ln P; \quad \text{(Whittaker, 1960, p. 82)} \qquad (2)$$

In curve a, Figure 7, the vertical axis represents the number of particles with velocities between the abscissa values v and $v+\Delta v$. According to classical statistical mechanics, it is the most probable distribution and the distribution toward which any system tends with increasing time and number of samples. It is therefore the form toward which the distribution of energy among the external forms of a crystal species must tend.

Niggli's Method of Measuring Persistence.

Niggli (1941) defined the persistence- or P-value of a crystal form $\{hkl\}$, as the number of combinations of forms in which it appeared in the specimens of a given mineral species.

Braun (1932, p. 208) reported one specimen of barite with only forms {101}, {210}; another with these two forms plus {001}, {010}, and {102}; another with {101}, {210}, {001}, {010}, and {011}. Each of the three specimens is enclosed by a distinct combination of forms. The first two forms are counted three times; {001}, and {010} are counted twice; {102} and {011} are counted once each. If another specimen is found with only the first two forms, {101}, and {210}, this will increase the value for the frequency of occurrence of the two forms, but it does not count for the *P*-value. In this way the statistics are less likely to be effected by bias in the selection of specimens, by the quantities of specimens collected, cr the size of the deposit, etc.

The persistence-value so defined does not take into account the relative development of the faces, counting a single minute facet and a fully

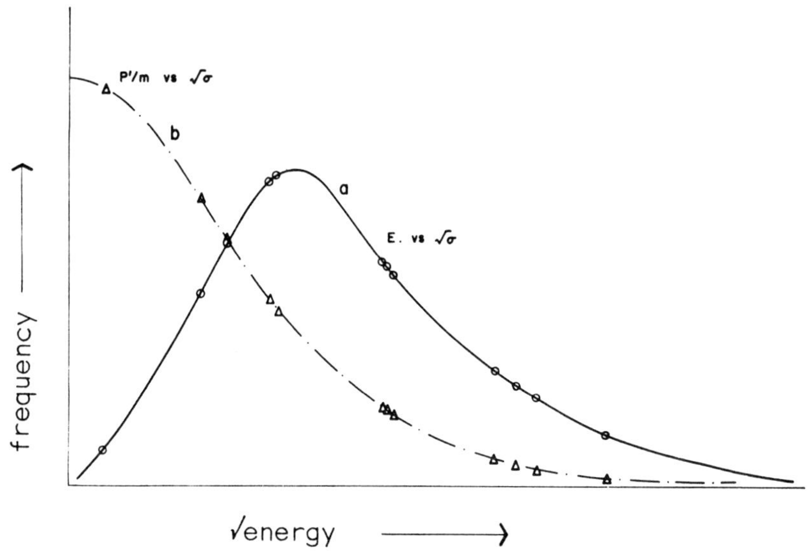

FIG. 7. The Maxwell-Boltzmann frequency distribution. Curves for the proportion of particles vs. √kinetic energy. The barite *P* /*m* vs. $\sqrt{\sigma}$ from equation (8) is shown on curve b and the total energy for each form E_{hkl} from equation (4) vs. $\sqrt{\sigma}$ is shown on curve a. Points are coincident with the curve because the values of σ and therefore $\sqrt{\sigma}$ are obtained from the equations.

developed prism equally as a single occurrence. To take this partly into account, Niggli used an amended type of persistence-value which he designated *P'*, and which is the value used in this research. The *P'*-value counts a form once for each time that it appears in a distinct combination of forms in any habit (habits distinguished qualitatively). Since a markedly different development of forms means a distinct habit, forms persist through changes of habit in proportion approximately relative to their size.

Braun (1932, p. 200) reports the same combination of forms {001}, {010}, {011}, {010}, {210} in the habit which he designates Rechteckig Tafliger-Typ I as well as in his habit,

Erz-Typ IIIa. Each of these forms is counted twice for the P'-value because the combinations though the same, are in different habits. They would be counted only once for the P-value. In order to visualize the effect of the unequal development of forms on the P'-values, consider a graph in which the proportionate surface area of each form is plotted along a different Cartesian coordinate axis. If a crystal combination consisted of a pinacoid (x) and a hexagonal prism (y), a plate or tabular habit would be represented by a point on the x–y plane with the value for the surface of the pinacoid (x) high and the value for the surface of the prism (y) low. A needle-like habit would be represented by a point far out along the prism (y) axis but relatively close to the origin of the pinacoid (x) axis. A stubby prismatic habit would be represented by a point somewhere along a line bisecting the two axes. If a third face is to be considered, a third axis would be added. For the complete system *in species* of n possible forms, each habit would be represented by a small region in an n-dimensional graph. Minor modifying facets would not displace the point representing the crystal from the small region of the habit. A major development however would mean a new habit and the form would be counted again. Restricting attention to a limited number of defined habits is essentially using a grab sample procedure for the n-dimensional graph. The graphs of snow crystals in environmental coordinates (Nakaya, 1954, p. 244) resemble projections on a plane through the n-dimensional space of such a graph (Fig. 8).

In any event, when statistical surveys of both sizes of faces and their persistence are reported, there is little difference (*cf.* Gebhardt, 1933).

Results. Figure 9, after Niggli (1941) represents the persistence values of the forms of ferberite arranged in order of decreasing magnitude. In 12 other examples given by Niggli, curves of the forms of various mineral species arranged in the order of descending values of P' also approximate the curve of Figure 9. Obenauer (1932) found the same curve for his surveys of 101 and 131 galena crystals. Nine individual frequency curves given by Braun (1932) for 9 distinct habits of barite show close adherence to the same curve. Also the order of the frequency of the particular forms in the 9 curves is amost identical. It would make little difference if any of the individual 9 curves or if the initial P-values were substituted for the P'-values used in the Fourier transform.

It is characteristic of all of these curves that there are few forms of high persistence, more forms of intermediate persistence and indefinitely many forms of low persistence. The number of forms in a given range of persistence increases as the persistence decreases. The same relationship characterizes an equilibrium distribution of particles among energy states. But if random numbers are assigned to forms which are then arranged in order of decreasing magnitude, there should be as many forms within a range of high magnitude as within a range of low magnitude. As an example, in selecting n random numbers between 1 and 10^6, the ratio of n_a picked between 9×10^5 and 10^6, to the number n_b picked between 1×10^5 and 2×10^5, should approach unity as the sample n gets larger. I conclude that the distribution of Figure 9 is not random.

If Gibbs' assumption of a particular and morphologically significant

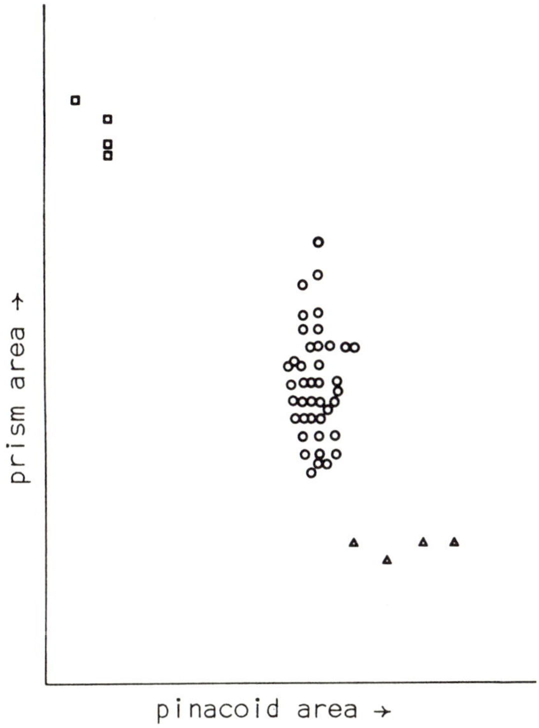

FIG. 8. Crystal habits represented on a prism vs. pinacoid surface graph. The effect of markedly differential development of forms is represented by this projection on a prism surface vs. pinacoid surface for snow. The points for three different snow crystal habits are taken from the temperature water vs. air graph of Nakaya (1954). Needles (prisms) are represented by □; plates (pinacoids) by Δ; and dendrites with more equal development of side and upper and lower bounding faces (prisms and pinacoids) by ○. A major change in the proportionate development of a form carries it into a new habit and changes its persistence-value.

value for the specific surface energy (our $\bar{\sigma}$) of a form of a mineral species was wrong, there would be no significant differences in values of P' for different forms, since on the average one would be as likely to occur as another. If there were no structural control over the occurrence of the forms of crystals, curves such as Figure 9 would not be repeated closely for 9 different habits of barite, or for the three minerals of the barite group examined.

Derivation of Relative Surface Energies from Persistence Values

Theory. Following the precedent of Gibbs (1961, pp. 224, 316) I have considered that there is a particular value of the specific surface energy

σ_{hkl} associated with each form $\{hkl\}$ of a mineral species. It is equal to the energy required to increase the surface $\{hkl\}$ by one unit of area, the volume being held constant. Environmental effects, variations in the pressure, temperature, and chemistry of the system in which the crystal is growing should cause the observed surface energies to cluster about mean values, much as any suite of measurements of a physical quantity scatters about a mean. Any consistent sampling of the world-experiment represents such a suite of measurements. The P_{hkl}' values correspond to the probability of occurrence of a form $\{hkl\}$ through the ranges of pressure, temperature, chemistry, and time of growth of crystals of the mineral species. They are either functions of the crystal structure which is invariant for the species, or else they are random numbers. The effects of environment may be separated into two categories: those effecting all forms equally therefore not effecting relative persistence (P'), and those selectively effecting particular forms because of interaction with internal factors such as the chemistry and structure of the crystal. P' values have been regarded as functions of internal factors and independent of external factors by Niggli (1920) and his followers, by Donnay and Harker (1937), and recently by Hartmann and Perdok (1955), who have used the same P'-values used here. Results from laboratory measurements suggest that synthetic crystals may obey the same laws of distribution as natural crystals. The relative growth rates of synthetic quartz were reported to correlate sensibly with the relative preponderance of the forms in nature (Laudise, 1958, p. 460).

Let the abscissa of Figure 7 represent the n possible forms of a mineral species, plotted in the order of increasing relative specific surface energies, σ_{hkl}. Then if the total surface energy is distributed by chance among the n possible forms of the crystal and if the sample space is large enough, the curve of the total energy associated with each form vs. $\sqrt{\sigma hkl}$ should

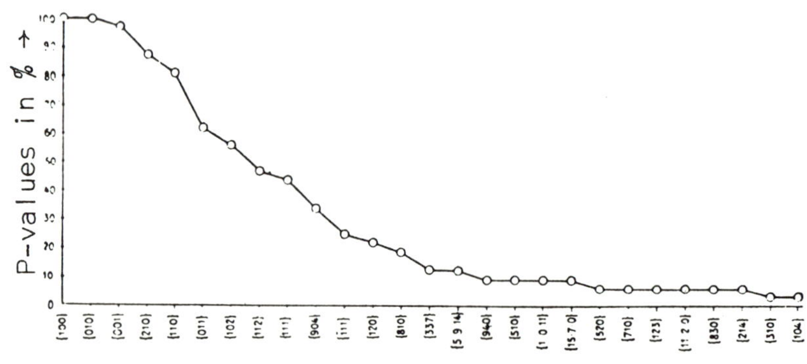

FIG. 9. Niggli's (1941) persistence-value curves for ferberite.

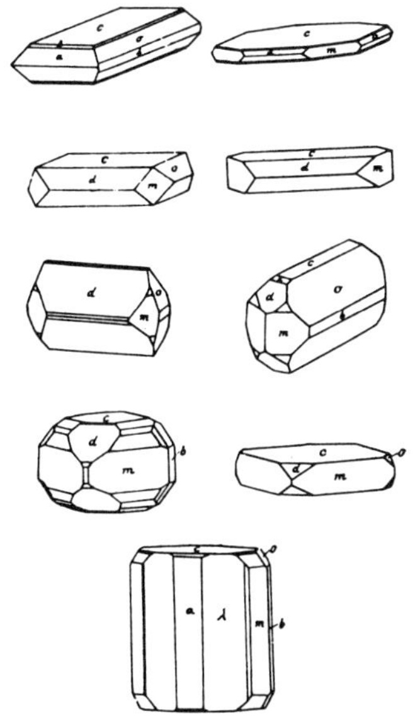

FIG. 10. The nine habits of barite crystals distinguished by Braun (1932). Fourier coefficients derived from the persistence values of these forms (Table 1) were used for the maps of Figures 1–3a. The nine habits illustrate the source morphology for the transforms to the structure.

be the familiar skewed curve of the Maxwell-Boltzmann distribution.[1] Let F_i be the surface area of the ith specimen of the form $\{hkl\}$. Then the total surface energy E_{hkl} associated with the P' occurrences of the form is given by;

$$E_{hkl} = \sum_{1}^{P'_{hkl}} F_i \sigma_{hkl} \tag{3}$$

Replacing $F_i\sigma_{hkl}$ by an average value over P'_{hkl} for the surface energy, $\bar{\sigma}_{hkl}$ (the average energy associated with the form $\{hkl\}$ per crystal of the sample population) then;

$$E_{hkl} = P'_{hkl}\bar{\sigma} \tag{4}$$

and;

$$P'_{hkl} = E_{hkl}/\bar{\sigma} \tag{5}$$

[1] For comparison with the kinetic-molecular curves of frequency vs. velocity, the abscissa should have proportional dimensions. Therefore total energy and persistence are plotted against $\sqrt{\bar{\sigma}}$ rather than $\bar{\sigma}$

Divide each value of E_{hkl} of curve a, Figure 7, by the corresponding value of σ to obtain P' values for curve b. Curve b is the ideal frequency distribution curve corresponding to curve a. It is the curve to expect as the distribution of total surface energy among the forms of a mineral species approaches equilibrium. The empirical values of P'/m for barite vs. the values for $\sqrt{\bar{\sigma}}$ obtained from equation 10 (below) have been marked on curve b. Multiplying $\bar{\sigma}$ by P'/m gives E_{hkl} (Equation 4) plotted on curve 7a as a function of $\sqrt{\bar{\sigma}}$. The curves are ideal. The observed P'-values are an exact fit because the $\bar{\sigma}$ are obtained from Equation 10 below which expresses the equilibrium relationship. That is, the assumption was made that the observed statistical persistence (P') could be equated with the theoretical probability of occurrence. On this assumption of an equilibrium distribution, the relative $\bar{\sigma}$ are easily determined.

Calculation of $\bar{\sigma}$ from Persistence Values. Following the precedent for analysis of the intensities of X-rays diffracted by a powder, the observed values for P' were divided by a multiplicity factor (m) for the number of faces of a particular form generated by the operations of symmetry. The symmetry of minerals of the barite group is described by the point group *mmm*; the general form is the orthorhombic dipyramid of 8 faces. There are 8 possibilities for the observation of a pyramid face as compared with 2 for the observation of a pinacoid or 4 for a dome or prism. The multiplicity of the probability of observation is a sampling factor having nothing to do with the relative surface energies and the multiplicity correction is therefore made as an adjustment to the P'-value before using it to derive an observed potential energy.

According to classical statistical mechanics, the proportionate numbers of particles n_1/n_2 in two energy states σ_1, σ_2 of a system in statistical equilibrium, is given by;

$$n_1/n_2 = e^{-\Delta\bar{\sigma}/kT} \tag{6}$$

$\Delta\sigma$ is the difference in energy between the two states σ_1 and σ_2, k is the Boltzmann constant, T is the absolute temperature or in this case, average temperature. If n_2 is the total number of distinct combinations of forms in distinct habits, and n_1 is the number of times the form $\{hkl\}$ was counted;

$$n_1/n_2 = P'_{hkl}. \tag{7}$$

$$\sigma_{hkl} = -\mathrm{kT}\ln P'_{hkl}. \tag{8}$$

Equation (8) is (6) rewritten in logarithmic form with substitution of equation (7); σ_{hkl} is now the potential energy of the form $\{hkl\}$ with respect to the potential energy of a form occurring in all specimens counted ($P' = 100$).

As an example if the average temperature of growth of barite crystals was 202°C or 475°K, the difference in σ between a prism like {210} and the pinacoid {001} would be 0.718 kcal/mol (Appendix). The value for room temperatue would be about 0.5 kcal/mol. This means that a single *gram-molecule* of barite in solution would effect a reduction of potential energy of about 0.7 kcal by adhering to the surfaces of prisms {210} rather than pinacoids {001}, at a temperature of 202°C. The same process for a crystal growing at room temperature, would effect a lowering of potential of half as much. Therefore the diversity of forms on a crystal should be greater for crystals grown at low temperature relative to those grown at high temperature since at low temperature there is less driving energy for one form to be chosen over another. In addition there is an entropy of mixing of forms which favors diversity in the forms of the crystal.

Niggli (1941) and Braun (1932) counted the numbers of forms per combination and the numbers of combinations with every number of forms, (z-value). In a survey of more than 62 forms in 185 combinations for barite, Braun found that z, the number of combinations plotted against the number of forms per combination, is a curve roughly approximating the Maxwell-Boltzmann distribution curve. Twenty-four combinations were observed with 6 forms; only 2 with 2 forms, 1 each with 18 and 21 forms. In our experiments with the growth of Rochelle salts at temperatures from 0°C to 35°C we obtained a distinctly richer morphology from the crystals grown at 0°C.

In order to convert observed surface energies to morphological structure factors, it is necessary to take into account the interplanar spacings, d_{hkl}.

For ordinary copper with atoms in the face-centered cubic arrangement, structure factors are 4 for {111}, {200}, {220}, {311}, etc. and 0 for planes of intermediate spacings. The effective surface energies for the planes, conceived as the energy required to extend a surface by one unit, cannot be the same for planes of different interplanar spacings. The Bravais and Donnay-Harker rules provided that morphological importance would be proportional to interplanar spacing. Surface energy would therefore be inversely proportional to interplanar spacing.

To obtain σ in appropriate units (energy per unit of area) d_{hkl} should be squared;

$$\sigma_{\mathrm{hkl}} = -C/d^2_{hkl} \qquad (9)$$

Combining (9) with (8), including the correction for multiplicity (m_i), with C' as a constant of proportionality, the expression for the observed morphological structure factor for the ith form of a species is:

$$\sigma_{\mathrm{i}} = -C'(\ln P_i'/m_i)/d_i^2. \qquad (10)$$

It is these values of σ_{hkl} which were used as the coefficients of the Fourier transforms with phases supplied a priori for the electron density transforms. For the Patterson (Harker) transforms, the σ_{hkl} were squared and no a priori information was required.[1] It should be emphasized that equation 9 is one possible modification of equation 8, and not necessarily the only way in which to transform surface energies to structure factors.

Conclusion

The contradiction between thermodynamic equilibrium which specifies a sharply limited number of simple forms for the polyhedral envelope of a single crystal, and statistical equilibrium which may encompass forms of unusually high specific surface energy, is only apparent. The persistence values of forms are a measure of their specific surface energies. The proportionate energies are in qualitative agreement with general observations of the growth of crystals under different conditions. They are in quantitative agreement with potential energies calculated as structure factors, as is demonstrated by the calculated residuals. Their validity and significance for further studies is warranted by their use in the two kinds of Fourier transforms from morphological space, first to the electron density and second, to the vector space. The transforms to the structure suggest the inverse,—the calculation of the morphological distribution from the known structure.

The factors of error are high when compared to those for diffraction methods. Does this result from deficiencies in the procedures for sampling, or does it represent the degree of departure from the state of statistical equilibrium? Better methods of proceeding from the morphological intensity space to the structure factor space are now under consideration.

The significance of the transform is not in the possibility of a return to morphological methods for the determination of crystal structures but in the understanding that emerges of the relationship between form and structure and in the further elucidation of what Friedel (1905) termed "the reticular hypothesis." The analytic techniques of diffraction which developed after 1912 were greatly indebted to the mineralogical studies which preceded them. The application of these techniques to the

[1] The σ_i used for the maps were calculated with the factor $1/d_i$ rather than $1/d_i^2$. The residual calculated for anglesite with equation (10) was 0.187 as compared to a residual of 0.205 with the earlier factor. McLachlan (1952) expressed "the dominance of the crystal faces" by a laminarity factor, L_{hkl} proportional to d_{hkl} times a function of the structure factor. If each atom in the calculation of the laminarity is weighted with the atomic number, then these residuals are the agreement of the set of σ with $\sqrt{Ld}$ and $\sqrt{L/d}$ respectively.

problems of the experiment of the world is an operation of symmetry. In effect, the geologic distribution of a form of a mineral species is determined by a specific energy which is a function of the structure factor. The rule at which we finally arrive is this: *The morphological probability distribution is a function of the structure, and the structure is a Fourier transform of the morphological probability distribution.*

ACKNOWLEDGEMENTS

Ruben Silva, Allen Hayes, and Robert Couture assisted with this research. The Graduate School and the Computation Center of the University of New Hampshire, and the National Science Foundation provided material support. A modified Busing-Levy ORFLS (1962) program was used. Professor G. Busch and the Institute for Solid State Physics of the E.T.H. of Zürich provided hospitality in the initial stages of this work. Professor G. Schiavinato and the Istituto di Mineralogia of the Universita degli Studi di Milano provided me with the facilities for completing this paper.

This research was based on the early studies of Professors J.D.H. and G. Donnay, David Harker, the work on transforms of Professor M.J. Buerger and an essay by A. Lande. I am particularly indebted to these individuals and institutions for inspiration and support.

REFERENCES

BARLOW, W. (1883) Probable nature of the internal symmetry of crystals. *Nature*, **29**, 186–188, 205–207.

———, AND W. J. P. POPE (1907) The relation between the crystalline form and the chemical constitution of simple inorganic substances. *Trans. Chem. Soc.* (London), **91**, 1150–1214.

BOLTZMANN, L. (1964) *Lectures on Gas Theory*. Univ. Calif. Press, Berkeley.

BRAUN, F. (1932) Morphologische, genetische und paragenetische Trachtstudien an Baryt *Neues Jahrb. Mineral.* **65A,** 173–222.

BRAVAIS, M. A. (1866) *Etudes Cristallographiques*. Gauthier-Villars, Paris.

BUCKLEY, H. E. (1951) *Crystal Growth*. John Wiley and Sons, New York.

BUERGER, M. J. (1947) The relative importance of the several faces of a crystal. *Amer. Mineral.* **32,** 593–606.

——— (1959) *Vector Space*. John Wiley and Sons, New York.

BUSING, W. R., K. O. MARTIN, AND H. A. LEVY (1962) ORFLS, a fortran crystallographic least-squares program. *U. S. Clearinghouse Fed. Sci. Tech. Info. Doc.* **ORNL-TM-305.**

CABRERA, N., AND D. A. VERMILYEA (1958) The growth of crystals from solution. *In* R. H. DOREMUS, B. W. ROBERTS, AND D. TURNBULL, (eds.) *Growth and Perfection of Crystals*. John Wiley and Sons, New York, 393–408.

CURIE, P. (1885) Sur la formation des cristaux et sur les constantes capillaires de leurs differentes faces. *Bull. Soc. Fr. Mineral.* **8,** 145–150.

DONNAY, J. D. H., AND D. HARKER (1937) A new law of crystal morphology extending the law of Bravais. *Amer. Mineral.* **22,** 446–467.

EWALD, P. P., (1921) Das "reziproke Gitter" in der Strukturtheorie. *Z. Kristallogr.* **56,** 129–156.

Friedel, G. (1905) Sur les bases expérimentales de l'hypothèse réticulaire. *Bull. Soc. Fr. Mineral.* **18,** 95–149.

Gebhardt, T. (1933) Über Zinkblende. *Neues Jahrb. Mineral.* **42A,** 1–52.

Gibbs, J. W. (1960) *Elementary Principles in Statistical Mechanics.* Dover Press, New York.

——— (1961) *The Scientific Papers.* Dover Press, New York.

Goldschmidt, V. (1923) *Atlas der Kristallformen.* Winter, Heidelberg.

Hartman, P. and W. G. Perdok (1955) On the relations between structure and morphology of crystals. *Acta Crystallogr.* **8,** 49–52, 521–524, 525–529.

James, R. W. (1954) *The Optical Principles of the Diffraction of X-rays.* Bell, London.

Lande, A. (1949) The physical significance of the reciprocal lattice of crystals. *Amer. Sci.* **37** (**3**), 414.

Laudise, R. A. (1958) Hydrothermal crystallization. *In* R. H. Doremus, B. W. Roberts, and D. Turnbull (Ed.) *Growth and Perfection of Crystals.* John Wiley and Sons, New York, p. 458–464.

Laue, M. von (1943) Der Wulffsche Satz für die Gleichgewichtsform von Kristallen. *Z. Kristallogr.* **105,** 124–133.

Lindsay, R. B., and H. Margenau (1957) *Foundations of Physics.* Dover Press, New York.

McLachlan, D., Jr., (1968) *Statistical Mechanical Analogies.* Prentice-Hall, Englewood Cliffs.

———, C. J. Christiansen, and A. Carlson (1952) Some factors in the growth of crystals. *Bull. Univ. Utah Exp. Sta.* **57,** 1–12.

Nakaya, U. (1954) *Snow Crystals.* Harvard Univ. Press, Cambridge.

Niggli, P. (1920) Beziehung zwischen Wachstumformen und Struktur der Kristalle. *Z. Anorg. Chem.* **110,** 55–81.

——— (1941) *Lehrbuch der Mineralogie und Kristallchemie, I* Borntraeger, Berlin.

Obenauer, K. (1932) Zur Tracht und Paragenese des Bleiglanzes. *Neues Jahrb. Mineral.* **67A,** 87–117.

Schneer, C. J. (1959) Proposed rule of crystal habit. [Abstr.] *Bull. Geol. Soc. Amer.* **70,** 1670.

——— (1968) Crystal form and crystal structure. *Helvetica Phys. Acta* **41,** 1151–1155.

——— (1970) *Mind and Matter.* Grove Press, New York.

———, and R. Silva (1966A) A method of determination of relative surface energies for crystal growth. [Abstr.] *Geol. Soc. Amer. Spec. Pap.* **87,** 149.

——— (1966B) Vectorial growth rates, statistical morphology, and the internal structure of minerals. [Abstr.]*Symposium I Int. Mineral. Ass., Cambridge.* Unpaged.

Sheftal, N. N., and I. V. Gavrilova (1966) Equilibrium shape of a crystal in relation to the bulk free energy. *In* N. N. Sheftal and A. V. Shubnikov (eds.) *Growth of Crystals.* **4,** Consultants Bureau, New York.

Swanson, H. E., R. K. Fuyat, and G. M. Ugrinic (1954) Standard X-ray diffraction powder patterns. *Nat. Bur. Stand.* [*U. S.*] *Circ.* **539,** *Vol.* 30.

Wells, A. F. (1946) Crystal habit and internal structure. *Phil. Mag.* **37,** 184–199, 217–235, 605–630.

Whittaker, E. (1960) *Aether and Electricity.* Harper, New York.

Wyckoff, R. W. G. (1963) *Crystal Structures. Vol.* **3**. Interscience, New York.

Wulff, G. (1901) Zur Frage der Geschwindigheit des Wachstums und der Auflösung der Kristallflächen. *Z. Kristallogr.* **34,** 449–530.

Manuscript received, March 18, 1969; accepted for publication, March 23, 1970.

Appendix: Calculation of Specific Surface Energies

Niggli (1941) gives $P'_{001}=98$ and $P'_{110}=92$ for barite. Using the structural cell (Wyckoff, 1963), $\{110\}$ becomes $\{210\}$. The multiplicity of $\{210\}$ is twice that of $\{001\}$ and therefore P'_{210} is halved and becomes 46. The Boltzmann factor k per mole becomes R (the gas constant $\simeq 2$ *cal/mol-deg*). A temperature of 202°*C*. is 475°*K*. From (8), $|\sigma_{001}-\sigma_{210}|=RT(\ln 98-46)\simeq 718$ *cal/mol.*

31

Reprinted from *Mineral. Soc. Japan, Spec. Paper 1*, 222–225 (1971)

Morphological Distribution Curves For Isostructural Series

by Cecil J. Schneer

University of New Hampshire, Durham, New Hampshire 03824, U.S.A.

Abstract

Statistical surveys of the persistence of the forms of crystals of a mineral species based primarily on Goldschmidt's *Atlas der Kristallformen* (1918), show characteristic patterns of correlation with lattice type (Bravais, 1866), space-group (Donnay-Harker, 1937) and chemical composition (Schneer, 1968). The logarithm of the persistence of the seven most frequent forms (P′ of Niggli, 1941) of 208 crystals of native copper, silver, and gold, is directly proportional to the interplanar spacing d of the forms, a result empirically confirming the Bravais rule. The plot of the ratios of the logarithms of P'_{111} to P'_{200} against $|Z_A - Z_X|/|Z_A + Z_X|$, (Z is atomic number) for AX compounds with the halite structure indicates that morphological persistence is dependent on the electron density distribution and not simply the reticular density of the rule of Bravais. This is confirmed strikingly by the corresponding plot for the persistences of the same two forms for 288 crystals of the aragonite group. The same persistences used as Fourier coefficients for Patterson-Harker 0–level sections for sphalerite and fluorite show distinctly the differences between the S–S, Zn–Zn vector density of sphalerite on the one hand, and the Ca–Ca, F–F, *vs.* the F–F vector densities of fluorite on the other. A comparison of the morphologically derived ¼ level Patterson-Harker sections of sphalerite and galena, show the Zn–S vector termination in the center of the tetrahedron for sphalerite, and nothing for galena.

These transforms and graphs are empirical support of the fundamental assumption of the functional dependence of crystal morphology on crystal structure. They support the conclusion that the ensemble of the forms of crystals of a mineral species is a reversible transform of the electron density.

* * *

In order to provide a quantitative measure of proposed transforms from morphology to structure (the Bravais and Donnay-Harker rules, Fourier synthesis), we have undertaken a number of statistical surveys of mineral morphologies. The rule of Bravais is usually stated in qualitative terms relating the morphological importance of a face to the density of lattice points. The density in turn varies with the interplanar spacing. The rule as modified by Donnay and Harker (1937) relates the morphological importance of a form (hkl) to the interplanar spacing as modified by the geometry of equipoints introduced by space group operations. This is equivalent to subjecting the list of form indices to the space group extinction rules.

Considering the persistence P′ of a form as the quantitative measure of its morphological importance, the Bravais rule is examined by plotting P′ *vs.* d for the three native metals copper, silver, gold. Persistence, P′ was defined by Niggli as the number of times a given form appeared on lists of the observed combinations of forms for crystals of a given mineral species compiled for each distinct habit (Niggli, 1941; Schneer, 1970). Persistence values were derived from the compilations of Goldschmidt's *Atlas der Kristallformen* (1918) for the forms of 86 copper, 36 silver, and 86 gold crystal types. Statistical mechanical reasoning (Schneer, 1970) leads to the conclusion that the negative logarithm of the persistence should be proportional to the interplanar spacing. Figure 1 shows the negative logarithm of the persistence plotted *vs. d*. The native metal minerals are chosen as simple examples of mineral structures unaffected by complications of space group equipoints or the chemistry of the internal distribution of matter. The Bravais rule is directly applicable and in the form,

$$-\ln P_{hkl'} = kd_{hkl} \qquad (1)$$

becomes the Bravais Law.

This relationship (1) stems directly from the curve of Figure (1) and is therefore empirical as well as theoretical.

Figure 2 illustrates the comparative morphology of isostructures of halite (with sphalerite for comparison). The native metals are considered as halite structure AX compounds with the electron count, or other measure of the properties of the element X taken as 0. It is at once apparent that the chemistry as well as the structure effects the relative persistences of different forms as for example the cube and the octahedron. These are effected in the same way as the intensities in the X-ray powder diffractograms (a graph comparing the diffracted intensities with the persistences for barite is included in Schneer, 1970). For the octahedral direction of the halite structure, planes of A alternate with planes of X interfering destructively and weakening the diffracted beam. For the first order cube direction, (100), the destructive interference is complete and the diffracted beam is cancelled. For the second order cube, (200), the interference is constructive and the diffracted beam intensity is governed by

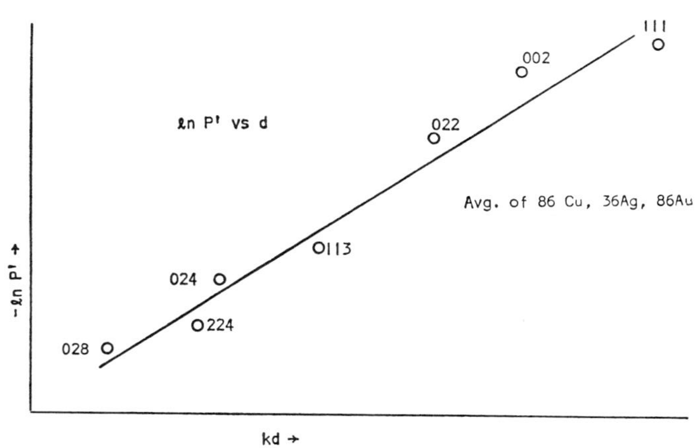

Fig. 1

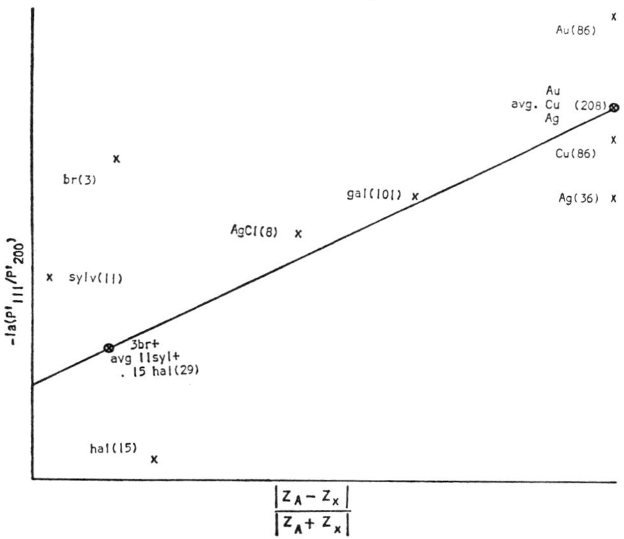

Fig. 3

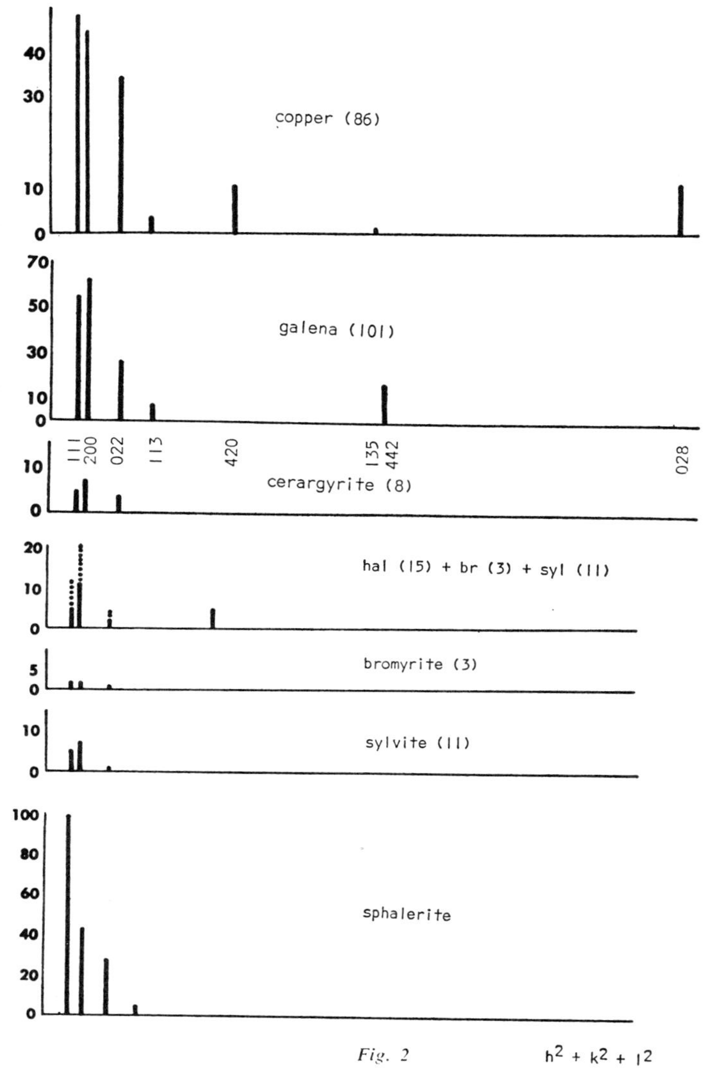

Fig. 2

the sum of the electron counts of the A and X atoms in the cell.

Theoretical considerations in this instance also, led to the anticipation of a relationship similar to that of X-ray diffraction for vectorial growth rates and morphological persistences. Comparisons of specific surface energies derived from P'-values with a *priori* calculated structure factors yielded residuals of less than 0.2 to 0.4 for anglesite, barite, and celestite (Schneer, 1968, 1970). Figure 3 shows the logarithms of the persistences P' for the forms of the halite isostructures as a function of the chemical (electron count) distribution within the cell. The ratio of the negative logarithms of the persistences for the octahedron to those of the cube are shown plotted against the ratio of the differences of the atomic numbers ($Z_A - Z_X$) to the sums ($Z_A + Z_X$). The statistics were again assembled from the Goldschmidt *Atlas,* except for sphalerite taken from Gebhardt (1933). Figure 3 measures the degree of the functional dependence of the morphology upon the electron density. The same relationship observed for the halite isostructures is shown for four minerals of the aragonite group, 50 strontianite, 108 aragonite, 30 witherite, and 100 cerussite crystals, (Figure 4). Similar relationships have been found for the corundum group. The graphs are an empirical confirmation of the Donnay-Harker rule (1937) relating the morphological importance of forms to the structure (as opposed to the lattice). They also, independently and empirically confirm the functional dependence of the specific surface energies of the forms on the electron density (the chemical distribution within the cell).

Previously published Fourier transforms between the morphology with specific surface energies derived from persistences used as the weights of reciprocal lattice points, yielded crude Fourier maps (Schneer and Silva, 1966). To meet the objection that these required a *priori* input of detailed symmetry as well as the signs of the coefficients, Patterson-Harker projections were

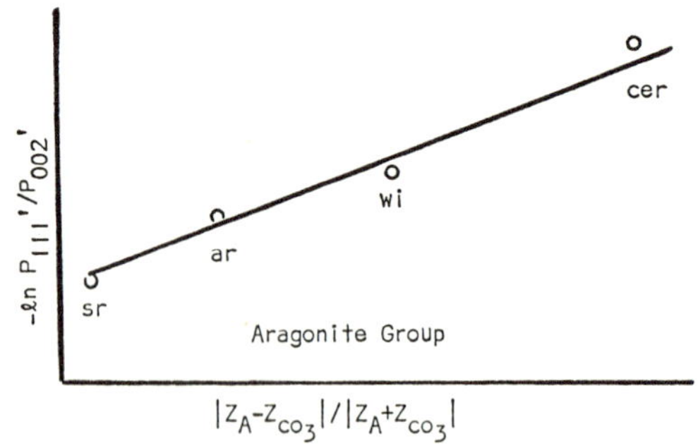

Fig. 4

prepared for barite (Schneer, 1968, 1970). Figure 5 compares the Patterson using specific surface energies $|\sigma|^2 = |kT\ln(P'/d)|^2$ for Fourier coefficients, for sphalerite with that for fluorite (0 level). The terminus of the Zn–Zn and S–S vector is clear as is the distinction between the termini of the F–F and the F–F with Ca–Ca vectors of the fluorite. Figures 6 and 7 compare the ¼ level morphologically derived Pattersons for sphalerite and galena. The Zn-S vector termination at the center of the sphalerite tetrahedron is distinct. The same position on the galena map is clear. Preliminary and tentative optical transforms (Bragg, 1965, pp. 229–231) from galena and halite morphologies show directly the differences in the density of the A and X atoms—a difference very much greater for galena than for halite.

These results are empirical corroboration of the functional dependence of mineral morphology upon lattice type (Bravais), space-group symmetry (Donnay and Harker), and chemistry (Schneer, 1968). They confirm the significance of the statistical persistence of Niggli which leads to the electron and vector densities in the cell. They open the way to the further investigation of the transforms between the external poly-

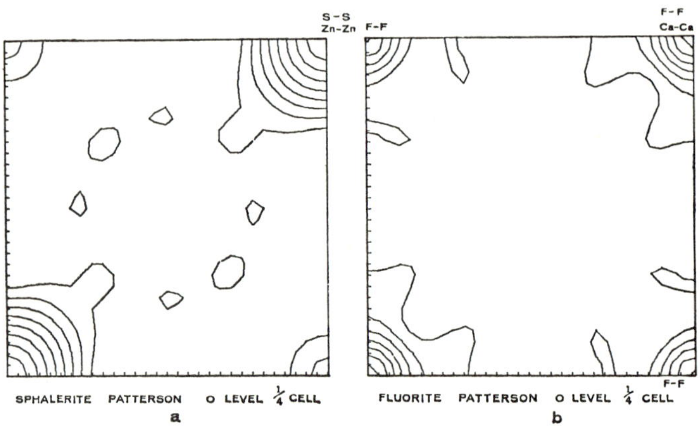

Fig. 5

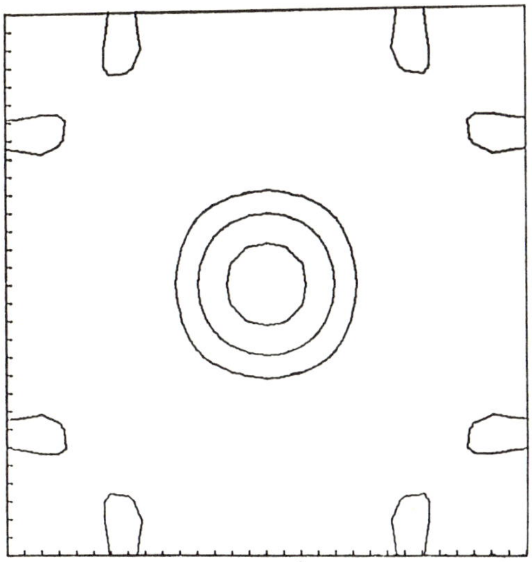

Fig. 6

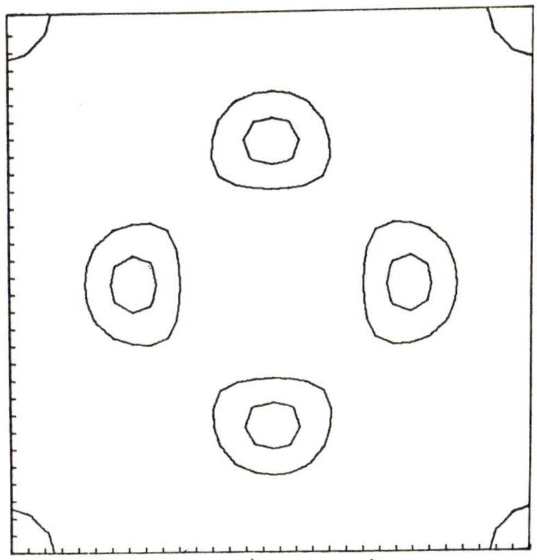

Fig. 7

hedral envelop of the crystal and the periodic distribution of matter within.

* * *

REFERENCES

BRAGG, L. (1965) *The Crystalline State*, Ithaca; Cornell University Press.

DONNAY, J.D.H., AND D. HARKER, (1937) A new law of crystal morphology extending the law of Bravais. *Amer. Mineral* **22**, 446-467.

GEBHARDT, T. (1933) Uber zinkblende. *Neues Jb. f. Min. LXVII, A*, 1-52.

GOLDSCHMIDT, V. (1918) *Atlas der kristallformen.* Heidelberg: Winter.

NIGGLI, P. (1941) *Lehrbuch der mineralogie und kristallchemie, I*, Berlin: Borntraeger.

SCHNEER, C.J. (1968) Crystal form and crystal structure. *Helvetica Phys. Acta* **41**, 1151-1155.

——————, (1970) Morphological Basis for the Reticular Hypothesis. *Amer. Mineral* **55**, 9-10, in press.

——————, AND R. SILVA, (1966) Abstract, Vectorial growth rates, statistical morphology, and the internal structure of minerals. *Symposium I, Bonding forces and crystal growth in minerals*, Cambridge; International Mineralogical Association.

32

Reprinted from *Bull. Univ. Utah Eng. Exptl. Sta.*, **57**, 1–12 (1974)

SOME FACTORS IN THE GROWTH OF CRYSTALS: Part I. Extension of the Donnay–Harker Law

Dan McLachlan, Jr.

Perhaps the most important observation relating the habits of crystals with their suspected structures was made by Bravais.[1] The ultimate publication of many examples by G. Friedel was so convincing that the observation became known as the law of Bravais. The observations or the law may be briefly stated as follows: the most prominent external faces of a crystal (i.e., those faces which have the greatest area) are those faces which are parallel to the internal planes (or families of planes) having the highest reticular density. By "reticular density," we mean the number of nodes per square centimeter on a plane of nodes within the lattice. It may be shown that the reticular density of each plane of a family of planes within a crystal is directly proportional to the interplanar spacing of that family of planes. Knowing the unit cell dimensions and angles between the edges of the faces, the interplanar spacings may be readily computed. For example, the interplanar spacing for the monoclinic crystal may be computed from the equation

$$d_{(hkl)} = \sqrt{\left\{\left[\left(\frac{h}{a}\right)^2 + \left(\frac{l}{c}\right)^2 - \frac{2hl}{ac}\cos\beta\right]\Big/\sin^2\beta\right\} + \left(\frac{K}{b}\right)^2} \tag{1}$$

where h, k, and l are the Miller indices, a, b, and c are the unit cell dimensions, and β is the obtuse angle between the a and c edges. Although the examples furnished by Bravais and Friedel were chosen largely from the cubic and orthorhombic systems, the law is useful in any system. The law predicts that the most prominent faces are those having small Miller indices, h, k, and l.

A very remarkable achievement of Bravais was his recognition that not only does the law apply to any one of the seven shape determination lattices (i.e., triclinic, monoclinic, orthorhombic, tetragonal, cubic, hexagonal and rhombohedral, but it also applies to the entire 14 Bravais lattices that are derived from the seven primitive lattices by assuming judiciously the possibility of face and body centers. Bravais was able to show that the existence of body and face centers in a lattice halves certain of the spacings, thereby reducing the reticular density and predicting, for those crystals, a different order of predominance than is found in crystals having the primitive lattice.

There occurred subsequently frequent discoveries of crystals that did not adhere to the law of Bravais as used in his works. It must be remembered, however, that the law is based on the general concepts of structure, and that structure as we know it today (from X-ray and electron diffraction data) was not available to Bravais. The face and body centers that cause the halving of interplanar spacings also cause extinctions in X-ray diffraction patterns; but face and body centers are not the only cause of extinctions in diffraction patterns; the presence of glide mirrors and screw axes also cause extinctions in these patterns. Donnay and Harker,[2] recognizing that extinctions in diffraction patterns are always criteria for exact halving in interplanar spacings, extended the law of Bravais so successfully to account for many of the crystals which deviated from the law of Bravais that this extension is known as the Donnay–Harker law. The Donnay–Harker law therefore is not restricted to the 14 Bravais lattices, but covers 97 morphological aspects of the 230 space groups. As tabulated by these authors, the 97 morphological aspects are based on the symmetry operations associated with a glide, either planar or axial. It was hoped that the morphology or habit of a crystal could be used not only as evidence for the existence of the simple mirror planes and rotation axes, but also as a clue for the presence of gliding of the planes and axes. This hope has been fulfilled.

If any objection is to be raised against the laws of Bravais and of Donnay and Harker it is not that they do not hold remarkably well over a wide choice of the world's crystals, especially minerals, but that any attempt to derive these laws atomistically in a rigorous manner requires an assumption that the atoms are in special positions.[3] Let us see if we can derive an equation based on generalized structures that predicts the prominence of the face of a natural crystal. First, let us acknowledge that the "reticular density" is very difficult to define when we consider atoms that are not located exactly on planes. Instead of reticular density, let us substitute the concept of "laminarity," which will be shown to be the product of the interplanar spacings d_{hkl} and the congestion C_{hkl}. In expressing mathematically the congestion C_{hkl}, we must search for a function that distinguishes between the conditions shown in Figure 1a, wherein all the atoms are located on the planes, and the condition shown in Figure 1b, wherein the atoms are scattered between the planes. In other words, the selected function must have a higher value when the perpendicular distances, p_j, from the atoms to the planes are all zero, as in Figure 1a, than when the perpendicular distances p_j have numerous nonzero values, as in Figure 1b.

A function that follows this trend is the cosine function $\Sigma_j \cos 2\pi p_j/d_{hkl}$, because for Figure 1a the function becomes $\cos 0/d_{khl} = 1$ (since the cosine of zero equals unity) and any nonzero values of p_i must cause a decrease in the value of the cosines. In Figure 1b, the function is

$$\sum_j \cos \frac{2\pi p_j}{d} = \cos \frac{2\pi p_1}{d} + \cos \frac{2\pi \bar{p}_1}{d} = 2 \cos \frac{2\pi p_1}{d}$$

However, the function distinguished unfavorably between configurations of equal congestion, as illustrated in Figures 2a and 2b. Whereas the func-

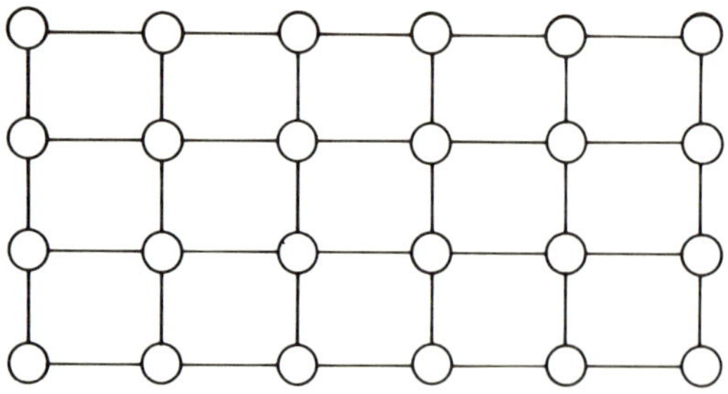

Figure 1a

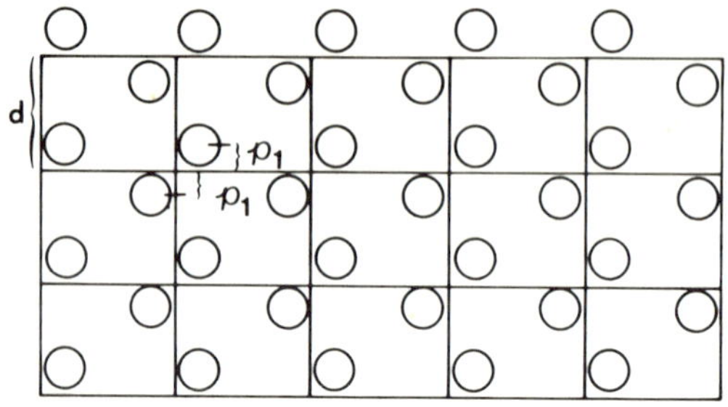

Figure 1b

tion sought after should have equal values for the configurations in Figures 2a and 2b, the cosine function gives for Figure 2a a value of

$$\sum_j \cos\frac{2\pi p_j}{d} = \cos\frac{2\pi 0}{d} + \cos\frac{2\pi p_1}{d} = 1 + \cos\frac{2\pi p_1}{d}$$

and for Figure 2b a value of

$$\sum \cos\frac{2\pi p_j}{d} = -\left(1 + \cos\frac{2\pi p_j}{d}\right)$$

What is really desired is a function that is unaltered by the phase of the congestion planes with respect to the planes from which the measurements p_j are made. Such a function is

$$C'_{hkl} = \left(\sum_j \cos\frac{2\pi p_j}{d_{hkl}}\right)^2 + \left(\sum_j \sin\frac{2\pi p_i}{d_{hkl}}\right)^2 \tag{2}$$

For Figure 2a, equation (2) becomes

$$C' = \left(\cos\frac{2\pi o}{d} + \cos\frac{2\pi p_j}{d}\right)^2 + \left(\sin\frac{2\pi o}{d} + \sin\frac{2\pi p_j}{d}\right)^2$$

$$= 2\left(1 + \cos\frac{2\pi p_j}{d}\right)$$

and for Figure 2b, equation (2) becomes

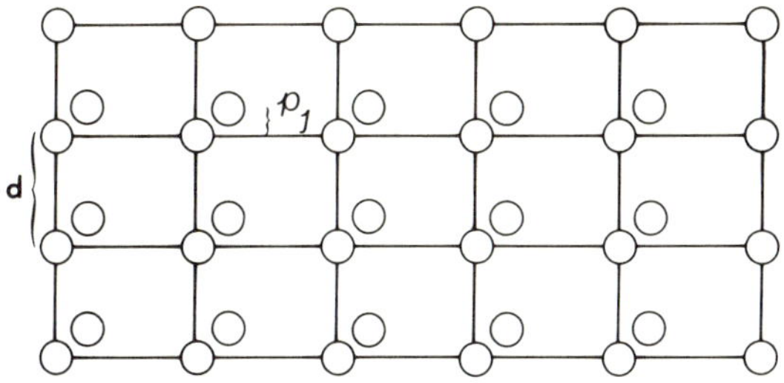

Figure 2a

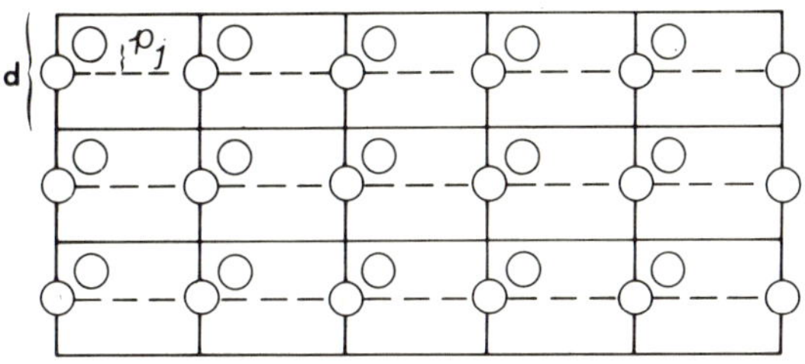

Figure 2b

$$C' = \cos\left[\frac{2\pi}{2} + \cos 2\pi\left(\frac{P_j}{d} + \frac{1}{2}\right)\right]^2 + \left[\sin\frac{2\pi}{2} + \sin 2\pi\left(\frac{P_j}{d} + \frac{1}{2}\right)\right]^2$$

$$= 2\left(1 + \cos\frac{2\pi p_j}{d}\right)$$

This equation gives the value C_{hkl} for any congestion of atoms on equally spaced planes no matter how the atomic planes are spaced with respect to the reference planes. In other words, the congestion function is not phase sensitive. For example, if the N atoms are all at the same distance $p_j = s$, equation (20) becomes

$$C'_{hkl} = \left(N\cos\frac{2\pi s}{d_{hkl}}\right)^2 + \left(N\sin\frac{2\pi s}{d_{hkl}}\right)^2$$

$$= N^2\left[\left(\cos^2\frac{2\pi s}{d_{hkl}}\right) + \left(\sin^2\frac{2\pi s}{d_{hkl}}\right)\right] = N^2$$

Even equation (2) is not totally satisfactory, because it treats all atoms alike regardless of the atomic number, binding forces, potential fields, valence, or any other atomic factor that may influence the growth of the crystal. We may improve the congestion factor as expressed in equation (2) by assigning to each atom a weighting factor G_j that expresses the effect of the atom, j, upon the growth of the crystal in which it exists. This is done arbitrarily without designating the factors of which G_j may be a function:

$$C_{hkl} = \left(\sum_j G_j \cos \frac{2\pi p_j}{d_{hkl}}\right)^2 + \left(\sum_j G_j \cos \frac{2\pi p_j}{d_{hkl}}\right)^2 \tag{3}$$

The laminarity L_{hkl} with which we wish to express the dominance of the crystal faces must distinguish (as do the laws of Bravais and Donnay–Harker) between the two configurations shown in Figures 3a and 3b,

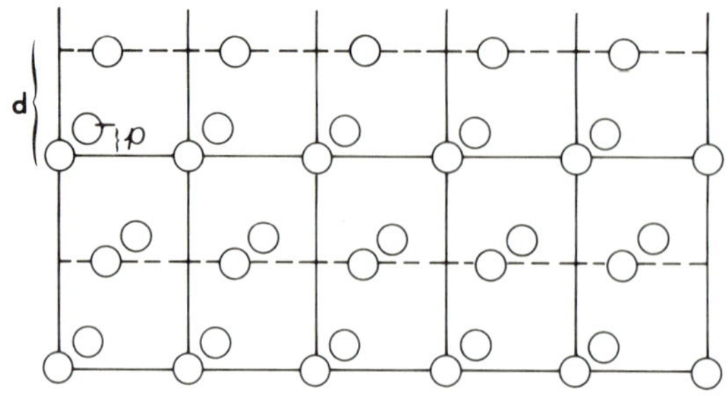

Figure 3a

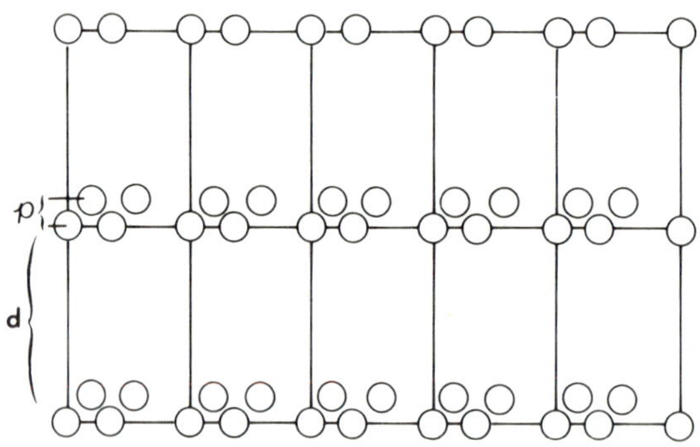

Figure 3b

wherein an attempt has been made to show two of the projections of two atomic configurations having the same bulk density but having interplanar spacings proportional to the reticular density. Equation (3) does not do this. The laminarity of a crystal should be proportional to the product of the interplanar spacing and the planar congestion, or

$$L_{hkl} = KC_{hkl}d_{hkl}$$

$$= Kd_{hkl}\left[\left(\sum_j G_j \cos \frac{2\pi p_j}{d_{hkl}}\right)^2 + \left(\sum_j G_j \sin \frac{2\pi p_j}{d_{hkl}}\right)\right]^2 \tag{4}$$

where K is an arbitrary constant.

This equation, although expressing all that we are at present capable of saying in regard to crystal habit, may be presented in another form, which involves what we may know about the structure of a crystal. It may be

shown that p_j/d_{hkl} (i.e., the ratio of the perpendicular distance from an atom to a plane and the interplanar spacings of the family of planes) may be expressed in terms of the Miller indices h, k, and l and the atomic positions, X_j, Y_j, Z_j by the equation[7]

$$\frac{p_j}{d_{hkl}} = hx_j + ky_j + lz_j, \tag{5}$$

where x_j, y_j, and z_j are fractions X/a, Y/b, and Z/c. And d_{hkl} may be expressed in terms of the Bragg angle and the wavelength λ of the X-rays from which d_{hkl} is experimentally measured.

$$d_{hkl} = \frac{\lambda}{2 \sin \theta_{hkl}} \tag{6}$$

Substituting equations (5) and (6) into equation (4), we obtain

$$L_{hkl} = \frac{K\lambda}{2 \sin \theta_{hkl}} \left\{ \left[\sum_j G_j \cos 2\pi(hx_j + ky_j + lz_j) \right]^2 + \left[\sum_j G_j \sin 2(hx_j + ky_j + lz_j) \right]^2 \right\} \tag{7}$$

To demonstrate that equation (7) obeys the laws of Bravais and Donnay– Harker, let us present three examples.

Example 1

Given a primitive lattice in which all the atoms are at special positions $x_j y_j z_j = 0$. Then equation (7) becomes

$$L_{hkl} = \frac{K\lambda}{2 \sin \theta_{hkl}} \{[G \cos 2\pi(0 + 0 + 0)]^2 + [G \sin 2\pi(0 + 0 + 0)]^2\} \tag{8}$$

and the prominent faces are those having the greatest d_{hkl} values, as proposed by Bravais.

Example 2

Given a body-centered lattice in which there are two atoms, one at 000 and one at ½½½. Equation (7) becomes

$$\begin{aligned} L_{hkl} &= \frac{K\lambda}{2 \sin \theta_{hkl}} \left\{ G^2 \left[\cos 2\pi(0 + 0 + 0) + \cos 2\pi\left(\frac{h}{2} + \frac{k}{2} + \frac{l}{2}\right) \right]^2 \right. \\ &\quad \left. + G^2 \left[\sin 2\pi(0 + 0 + 0) + \sin 2\pi\left(\frac{h}{2} + \frac{k}{2} + \frac{l}{2}\right) \right]^2 \right\} \\ &= \frac{K\lambda G^2}{2 \sin \theta_{hkl}} \left\{ \left[1 + \cos 2\pi\left(\frac{h}{2} + \frac{k}{2} + \frac{l}{2}\right) \right] \right. \\ &\quad \left. + \sin^2 2\pi\left(\frac{h}{2} + \frac{k}{2} + \frac{l}{2}\right) \right\} \end{aligned} \tag{9}$$

Extensions of the Donnay-Harker Law

There are two types of families of planes to be especially considered: (a) those for which $h + k + l$ is odd, and (b) those for which $h + k + l$ is even. For the odd set,

$$L_{hkl} = \frac{K\lambda G^2}{2 \sin \theta_{hkl}}\{(1 - 1)^2 + (0)^2\} = 0 \tag{10}$$

and, for the even set,

$$L_{hkl} = \frac{K\lambda G^2}{2 \sin \theta_{hkl}}\{(1 + 1)^2 + (0)^2\} = \frac{4K\lambda G^2}{2 \sin \theta_{hkl}} = 4KG^2 d_{hkl} \tag{11}$$

Therefore, for a simple body-centered lattice, only the planes for which $(h + k + l)$ is even need to be considered, as shown by Bravais.

Example 3

Given a primitive lattice containing a twofold screw axis along z and the atoms at special positions $x_j = y_j = z_j = 0$. Then there are atoms 000 and 00½.

$$L_{hkl} = \frac{K\lambda}{2 \sin \theta_{hkl}}\left\{G^2\left[\cos 2\pi(0 + 0 + 0) + \cos 2\pi\left(0 + 0 + \frac{l}{2}\right)\right]^2 + G^2\left[\sin 2\pi(0 + 0 + 0) + \sin 2\pi\left(0 + 0 + \frac{l}{2}\right)\right]^2\right\}$$

$$= \frac{K\lambda G^2}{2 \sin \theta_{hkl}}\left\{\left(1 + \cos 2\pi\frac{l}{2}\right)^2 + \sin^2 2\pi\frac{l}{2}\right\} \tag{12}$$

The two types of planes to be considered are (a) those for which l is even and (b) those for which l is odd. For the odd set,

$$L_{hkl} = \frac{K\lambda G^2}{2 \sin \theta_{hkl}}[(1 - 1)^2 + (0)^2] = 0 \tag{13}$$

and, for the even set

$$L_{hkl} = \frac{K\lambda G^2}{2 \sin \theta_{hkl}}[(1 + 1)^2 + (0)^2] = \frac{4K\lambda G^2}{2 \sin \theta_{hkl}} = 4KG^2 d_{hkl} \tag{14}$$

Equation (7), which we wish to designate as the mathematical expression of our extension to the Donnay–Harker law, deserves some comment, particularly with regard to G_j. Suppose that by some unknown chance $G_j = f_j$, where f_j is the scattering factor of the atom for X-rays; then L_{hkl} would become

$$L_{hkl} = \frac{K\lambda}{2 \sin \theta_{hkl}} F_{hkl}^2 \approx \frac{F_{hkl}^2}{2 \sin \theta_{hkl}} \tag{14a}$$

In this form, equation (14) reminds one that there was a statement made by Wells[4] in 1946 that the intensity of diffraction of X-rays from any family of planes should be a measure of the expected prominence of growth planes. He also suggested dividing the intensities by the sine θ. However, our derivation shows that intensities are a measure of something that is related to laminarity as we define it, but we can find no proof that G_j is any more related to the X-ray scattering factor than to the electron-scattering factor or neutron-scattering factor. It is the author's suspicion that G_j has the following possible meanings:

a. For growth planes—bonding energy.

b. For glide planes in metals—activation energy.[5,6]

c. For cleavage planes—bonding force.

A complete analysis of the meaning of G_j should be a very interesting investigation.

ADDENDUM BY THE AUTHOR (1974)

Since the first printing of the foregoing paper in 1952, much thought has been given to the meaning of G in equation (4). To summarize these thoughts, some discussion will herein be devoted to (a) form factors, (b) spike functions, and (c) morphology and structure.

(a) Form Factors

Using the principles of Fourier series,[8] we can represent any repeating pattern in one dimension by

$$\rho(p) = \frac{1}{d}\sum_{1}^{\infty} G(n)e^{-2\pi inp/d} \tag{15}$$

where d is the repeat distance, p is the distance from the origin, and the n's are integers. Figures 4a and 4b are examples of such repeating patterns. The values of $G(n)$ in equation (15) are deduced from the equation

$$G(n) = \int_0^d \rho(p)e^{2\pi inp/d}\,dp\,. \tag{16}$$

These Fourier coefficients, $G(n)$, are of greatest interest for the present discussion. Considering the so-called square pulse shown in Figure 4a, we have $\rho(p) = A$ from $-(\Delta p)/2$ to $+(\Delta p)/2$ and zero elsewhere. Therefore, equation (16) becomes

$$\begin{aligned} G(n) &= \int_{-\Delta p/2}^{\Delta p/2} Ae^{2\pi inp/d}\,dp \\ &= \left\{\frac{A2\pi in}{d}e^{2\pi inp/d}\right\}_{-\Delta p/2}^{\Delta p/2} \\ &= \frac{A\,\Delta p\,\sin(2\pi n\,\Delta p/2d)}{(2\pi n\,\Delta p/2d)} \end{aligned} \tag{17a}$$

A series of square pulses each at position p_i in the unit cells, rather than at $p = 0$, has a $G(n)$ term as

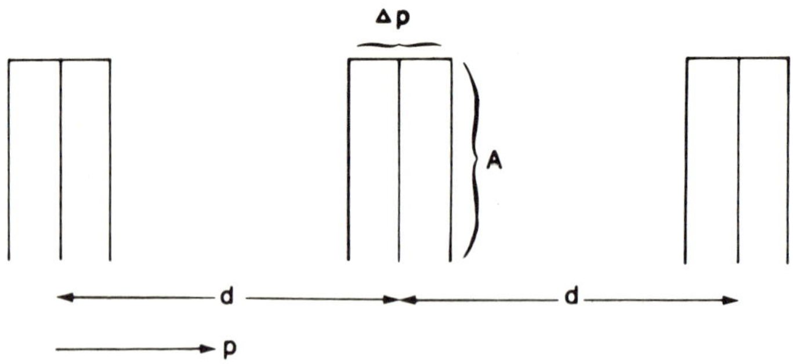

Figure 4a

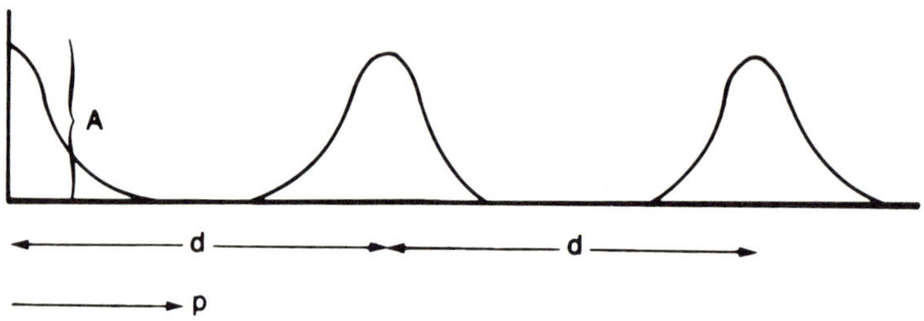

Figure 4b

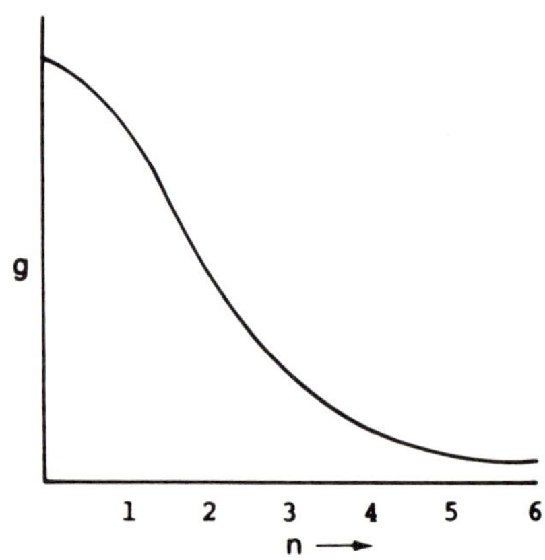

Figure 4c

$$G(n) = \int_{p_1-(\Delta p/2)}^{p_1+(\Delta p/2)} Ae^{2\pi inp/d}\,dp$$

$$= \frac{A\,\Delta p\ \sin(2\pi n\,\Delta p/2d)}{(2\pi n\,\Delta p_i/2d} e^{2\pi ip_i/d} \tag{17b}$$

If each unit cell has m identical pulses, each at $p_1, p_2, p_3, \ldots, p_j, \ldots, p_{m-1}$, the Fourier coefficients would be

$$G(n) = \frac{A\,\Delta p\ \sin(2\pi n\,\Delta p/2d)}{2\pi n\,\Delta p/2d} \sum_{j=1}^{m-1} e^{2\pi inp_j/d} \equiv g \sum e^{-2\pi inp_i/d} \tag{17c}$$

Note that the terms in brackets in equations (17a), (17b), and (17c) are identical and are independent of the positions of the pulses. We will call this term the "form factor" of a square pulse. It depends on the shape of the pulse only. As another example, if the pulses were Gaussian (rather than square), as shown in Figure 4b, i.e., of the form

$$Ae^{-\mu^2 p^2} \tag{18}$$

then we could show by the same procedure of integration as we used above [see equation (16)], that for single pulses at the origins of the cells,

$$\begin{aligned} G(n) &= \int_{-\infty}^{\infty} Ae^{-\mu^2 p^2} e^{2\pi i n p/d}\, dp \\ &= [A\sqrt{\pi} e^{-(\pi/d\mu)^2 n^2}] \end{aligned} \tag{19a}$$

If the pulses were all at positions p_1,

$$G(n) = [A\sqrt{\pi} e^{-(\pi/d\mu)^2 n^2}] e^{2\pi i n p_1/d} \tag{19b}$$

And for m identical Gaussian pulses at positions p_j,

$$G(n) = [A\sqrt{\pi} e^{-(\pi/d\mu)^2 n^2}] \sum_{j=1}^{m-1} e^{2\pi i p_j/d} \tag{19c}$$

Note that the terms in brackets in equations (19a), (19b), and (19c) are identical just as were those in brackets in equations (17a), (17b), and (17c). So we can call these latter terms the "form factor" for a Gaussian pulse. We can designate the general symbol for the form factor g_j, so that for a repeating pattern containing many pulses of differing shapes we get*

$$G(n) = \sum_{j=1}^{m-1} g_j e^{2\pi i p_j/d} \tag{20}$$

Now we wish to say something about form factors for atoms. Figure 5a shows the function $u(r)$ plotted against r for the particular case of the atom neon,[10] where $u(r)$ is defined as the probability of finding an electron at any distance r from the center of the atom. Notice that it is almost Gaussian. In X-ray studies a true form factor is not sought after in the same sense as was discussed above, but rather an atomic scattering factor f derived from $u(r)$, through the equation[11]

$$f = \int_0^{\infty} 4\pi r^2 u(r) \frac{\sin kr}{kr}\, dr, \qquad \text{where } k = \frac{4\pi}{\lambda} \sin\theta \tag{21}$$

Such a curve is shown for neon in Figure 5b. The reason for the use of scattering factors instead of form factors is that structure analysts want to use the f values to predict the F values, i.e., crystal structure factors repre-

* Note the resemblance between equation (20) and the familiar X-ray[(19)] analogue,

$$F(h) = \sum_j f_j e^{2\pi i h x_j/d}$$

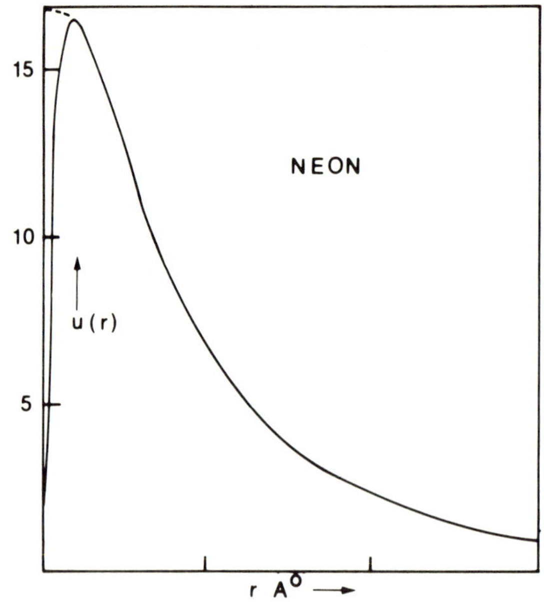

Figure 5a

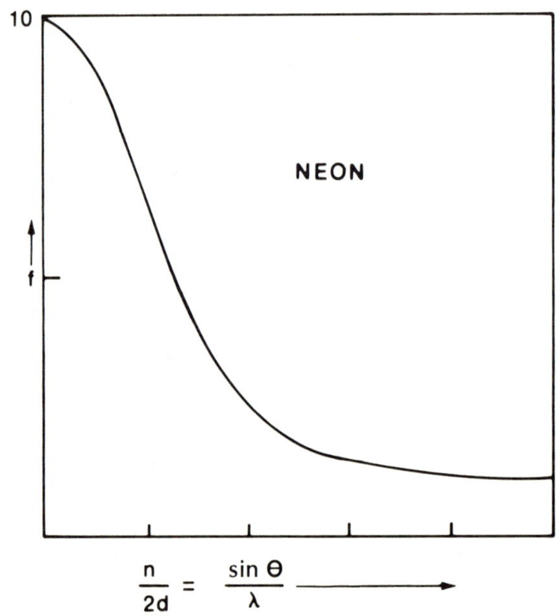

Figure 5b

senting the amplitudes of X-rays scattered from an entire crystal. Since the $u(r)$ function of Figure 5a is of the Gaussian type, it is not surprising that the f function in Figure 5b is also similar to a Gaussian (compare Figures 4c and 5b).

(b) Spike Functions and Point Atoms

Returning to Figure 4a, we see that as the width Δp of the square pulse approaches zero, the pattern approaches a spike pattern, i.e., a vertical line at each pulse position. The terms in brackets in equations (17a), (17b), and (17c) reduce to unity as $\Delta p \rightarrow 0$ ($\sin\theta \rightarrow \theta$ as $\theta \rightarrow 0$)

$$\begin{aligned} g_j &= \frac{A\Delta p \sin(2\pi n \Delta p/2d)}{(2\pi n\Delta p/2d)} \\ &\rightarrow \frac{A\Delta p \sin(2\pi n 0/2d)}{(2\pi n 0/2d)} \rightarrow 1 \\ &= A\,\Delta p \qquad \text{for all values of } n \end{aligned} \tag{22}$$

Also in equations (19a), (19b), and (19c) the terms in brackets reduce to a constant, independent of n as μ approaches infinity.

$$\begin{aligned} g_j &= A\sqrt{\frac{\pi}{\mu}}e^{-(\pi/d\mu)^2} \rightarrow A\sqrt{\frac{\pi}{\mu}}e^{-(\pi/d_\infty)^2 n^2} \\ &= A\sqrt{\frac{\pi}{\mu}} \end{aligned} \tag{23}$$

This again represents a spike function, since the higher the value of μ in equation (18) the narrower and sharper the Gaussian curve in Figure 4b. In other words, the form factors for spike functions are constants independent of n.

Approaching spike functions from a different viewpoint, equations (17c) and (19c) can be divided by their respective g values to produce a new equation,

$$\hat{G}(n) \equiv \frac{G(n)}{g} = \sum_{j=1}^{m-1} e^{2\pi i n p_j/d} \tag{24}$$

Equation (24) is highly reminiscent of the procedure proposed by A. L. Patterson[12] for projecting a structure of "point atoms," i.e., by dividing all $F(hkl)$ values by f which, when introduced into the Fourier syntheses, tends to give points at which the atomic centers are located rather than the electron density, which is diffused. This is also closely related to the "unitary" structure factors of Harker and Kasper[13] and referred to by Karle and Hauptman and others.

Now we come to the question of the meaning of G in equations (3) and (4). Since Bravais referred to "nodes" in thinking about atoms at special positions and Donnay and Harker referred to atoms, it is possible that reticular density really means atomic density rather than electron density. So it seems more appropriate to think about

$$\hat{G}_{(hkl)} \quad \text{or} \quad \hat{F}_{(hkl)} = \frac{F_{(hkl)}}{f} \tag{25}$$

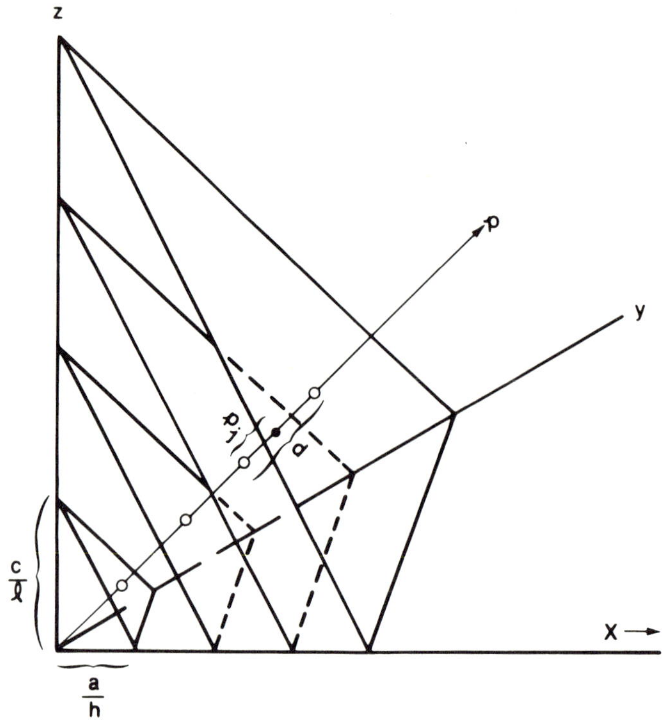

Figure 6

(c) Morphology and Structure

When comparing equation (3) with the accepted equations for the crystal structure factors, *F*, and after showing that there is a close theoretical connection between *G* and *F*, one might wonder if the areas of the faces of a crystal might be a measure of the *F* values. Schneer[14] actually has made some electron density maps using the prominence of faces in the Fourier series instead of using the measured X-ray data. The results were surprisingly good although the details were poorly resolved. The poor resolution can be explained without detracting from Schneer's novel idea.

Referring to Figure 6, the line $\overline{op}$ is perpendicular to a family of planes whose indices are h, k, l. All planes in a family are parallel to one another and to the crystal face {hkl} to which they pertain. Examples of planes of the same family are 100, 200, 300, etc.; 120, 240, 360, etc., and 312, 624, 936, etc. In general, a family of parallel planes comprise all planes whose indexes are nh', nk', nl', where the n's are integers and h', k', and l' are the indexes of the plane of greatest spacing. The equation

$$\rho(p_{h'k'l'}) = \frac{1}{d}\sum_{n} F_{nh'nk'nl'} e^{-2\pi i[(nh'/a)+(nk'/b)+(nl'/c)]} \tag{26}$$

represents the electron density projected on the line $\overline{op}$, just as the equation

$$\rho(p_{100}) = \frac{1}{a}\sum_h F_{hoo}e^{-2\pi i(hx/a)} \qquad (27)$$

represents the electron density projected in the x axis.

When using the prominence of faces as a measure of the F values, one is handicapped by the fact that the face has only index $\{h'k'l'\}$ corresponding to $n = 1$ in equation (26) and in the Fourier summations; all other possible F values must be left out. Thus the higher indexes (which contribute most to the fine detail of the projection) are left out, thus accounting for the poor resolution in Schneer's[14] projection.

REFERENCES

1. For a description of the work of Bravais, see F. C. Phillips, *An Introduction to Crystallography*, Longmans, Green & Co. 1946, p. 278.
2. J. D. H. Donnay and D. Harker *Am. Mineralogist*, **22**, 446 (1937).
3. R. W. G. Wyckoff, *Analytical Expression of the Results of the Theory of Space Groups*, published by Carnegie Institute of Washington, 1930.
4. A. F. Wells, *Phil. Mag*, **37**, 217 (1946).
5. See Eyring's "Theories of Viscosity," in Gladstone, Landler, and Eyring, *The Theory of Rate Processes*, McGraw-Hill Book Company, New York, 1941, and subsequent works of Eyring.
6. N. F. Mott, "Slip at Grain Boundaries and Grain Growth in Metals," *Phys. Soc. Proc.*, **60**, 391 (1948).
7. W. P. Davey, *Study of Crystal Structure and its Applications*, McGraw-Hill Book Company, New York, 1934, p. 38.
8. Fredrick S. Woods, *Advanced Calculus*, Ginn and Company, Boston, 1926, p. 295.
9. Martin J. Buerger, *Crystal Structure Analysis*, John Wiley & Sons, Inc., New York, 1960.
10. Arthur H. Compton and Samuel K. Allison, *X-Rays in Theory and Experiment*, pages 143–155, D. Van Nostrand Company, Inc., New York, 1948, pp. 143–155.
11. B. E. Warren, *X-Ray Diffraction*, Addison-Wesley Publishing Company, Reading, Mass., 1969, p. 9.
12. A. L. Patterson, *Z. Krist.*, A **90**, 517–542 (1935).
13. David Harker and J. S. Kasper, *J. Chem. Phys.*, **15**, 882 (1947).
14. Cecil J. Schneer, *Am. Mineralogist*, **55**, 1466 (1970).

33

Reprinted from *Bull. Univ. Utah Eng. Exptl. Sta.*, **57**, 24–38 (1952)

SOME FACTORS IN THE GROWTH OF CRYSTALS: Part III. Dendritic Growth

Dan McLachlan, Jr., and Allan Carlson

Crystals may grow in such a great variety of shapes under differing conditions that it would appear that the laws governing all of them are either very complicated or nonexistent. However, there are two extreme types of growth which lend themselves to analysis: (a) the type which we will call "normal" growth, and (b) "dendritic" growth. By normal growth we mean that type of growth which results in crystal forms having a few well-defined faces and all external angles convex. This is the type of growth which results in the so-called characteristic habits of crystals. Normal growth has been studied most extensively, and it has been observed to be a slow process. Dendritic growth, on the other hand, has invariably been observed to be very rapid and is characterized by the tree-like patterns which result. The shapes of dendritic snow crystals, $KClO_4$,[1] antimony, and bismuth are illustrations of this type of growth. Vogel[2] in 1921 proposed a theory of dendritic growth based on the idea that the escape of heat from the corners of a crystal is faster than from the edges, which in turn permit heat escape faster than do the faces. According to Vogel's theory, dendrites form when the conditions for growth are such that the limiting factor in the growth rate is the rate at which heat of crystallization can escape from the various portions of the crystal. This theory leads to the prediction that under these conditions the growth is a "runaway" process in which sharp points become sharper, long spines become longer, and acute angles become more acute.

We would like to endorse Vogel's theory and refine it by introducing more modern field theory. We would also emphasize that the escape of heat is not the only factor, although the other factors which we consider involve the same mechanism as Vogel describes. The factors are:

1. The diffusion of heat.
2. The diffusion of solute or solvent.
3. The diffusion of electric charge.

Vogel's reasoning has been described by Buckley[3] and is illustrated in Figure 1a and 1b. In Figure 1a is shown a cube of material representing a crystal in the process of growth. If this cube is built up of many equal-sized blocks, there are three kinds of such blocks on the surface of the cube: the corner blocks, of which one is marked *A*; the edge blocks, such as *B*; and face blocks *C* (all shown in lower left corner of figure). About the large cube there may be imagined a layer of thickness *t*, through which the heat of crystallization must diffuse whenever a block of material deposits on the cube. Because different volumes of the diffusion layer are shared by the different kinds of blocks, it is expected that the corner blocks may lose their heat faster than edge blocks, and edge blocks faster than face blocks. As shown in the figure, the corner block has the cubical volume *A* into which heat may escape, the edge blocks have the tabular volumes B_1, B_2 and B_3, while the face blocks have the columnar volumes C_1, C_2, and C_3. Figure 1b shows a two-dimensional version of Vogel's principle. It is easy to visualize that as the corners and edges grow faster they become more acute, and thus the differential diffusion effects become enhanced, so that the growth results in spines or fins.

We can readily see that the diffusion of heat enhances maximum growths at points on the crystal where the bonding energies (Kossel[4] and

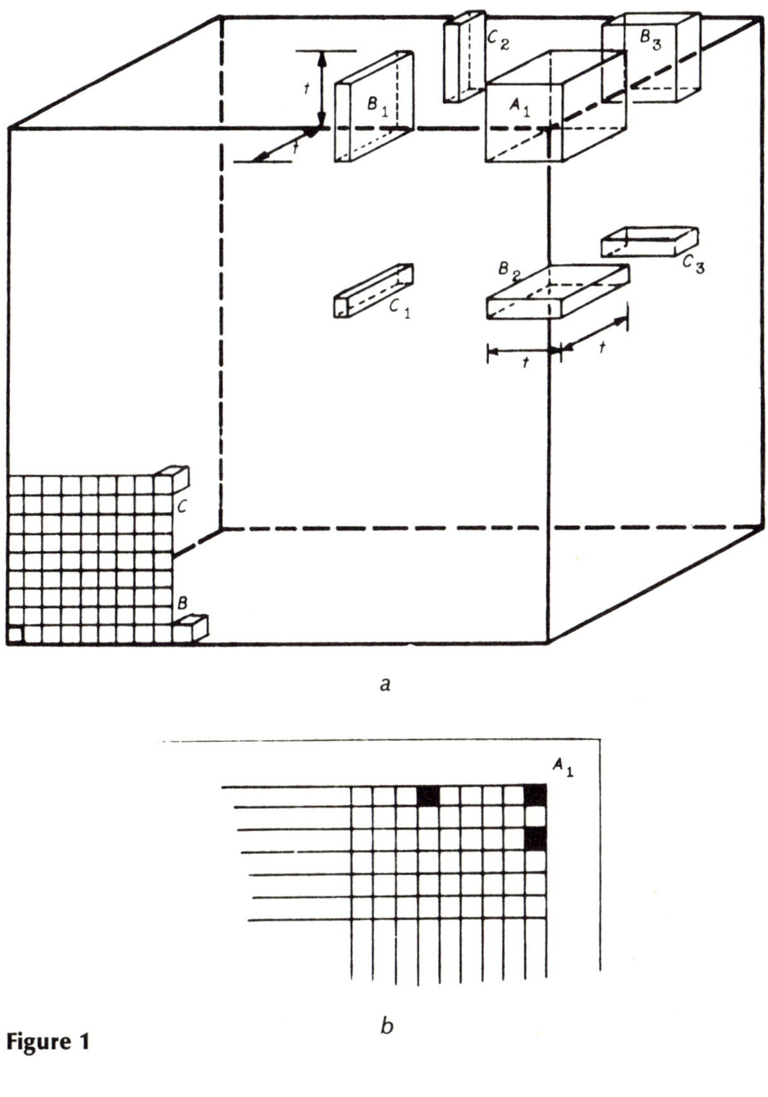

Figure 1

Stranski[5]) enhance the growths to the smallest extent. This is the reason that we emphasize that extremely different environments are required for normal growth and dendritic growth. We believe that normal growth occurs when circumstances are such that the rate of growth is a function of the crystal structure alone and that dendritic growth occurs when the rates of diffusion are the controlling factors. In growth from the vapor phase, the diffusion of heat is of importance; in growth from solution, the diffusion of solute or solvent may be the controlling factor; in electrolytic deposition of metals such as tin, silver, or zinc, the diffusion of ions or electrons in the electric field is the controlling factor, in fact so much so that it is only with extreme precaution that dendritic growth can be avoided.

Although we have no objections to Vogel's qualitative picture, we should like to present a more accurate method of treating the diffusion mechanisms giving rise to dendrites. It is very fortunate that there are three fields of science in which the same type of mathematics is used: hydrodynamics, electrostatic field theory, and heat flow. Although the mathematical methods have been worked out completely only as far as the two-dimensional problems, they are very effective when applicable.

In problems in electricity, hydrodynamics, and heat flow, one finds oneself confronted with the solution of an equation of continuity known as Laplace's equation, which in two dimensions is

$$\frac{\partial^2 V}{\partial x^2} + \frac{\partial^2 V}{\partial y^2} = 0$$

where x and y are coordinates of the space being investigated and V is the potential. When solving the equation one always finds that one has two kinds of curves in the graphical solution: (a) equipotential curves and (b) streamlines.

The equipotential curves are lines of constant voltage in electrostatics, constant temperature in heat flow, or constant pressure in fluid flow. The streamlines are the directions of the lines of force in electrostatics, the direction of the escape of heat, or the direction of the flow of fluid. It is found that the equipotential lines are always perpendicular to the streamlines in any system considered.

A very useful way of solving the Laplace equation for rather complicated boundary conditions is to transform the solution for the simplest known system (namely, parallel plates) to the more difficult system. This process is known as conformal mapping.[6] To illustrate the use of conformal mapping let us assume that we have a mass of solid material (shown in the cross-hatched area of Figure 2) having an irregular surface BB'. At some distance from this surface there is a planar surface AA', which is at constant temperature (voltage of concentration of solute). We wish to know the shapes of the curves of, say, constant voltage such as CC', and also the directions of flow of current as indicated by generally vertical lines. We wish to keep the generality of the problem by remembering that the solutions of the problem are the same whether AA' represents a

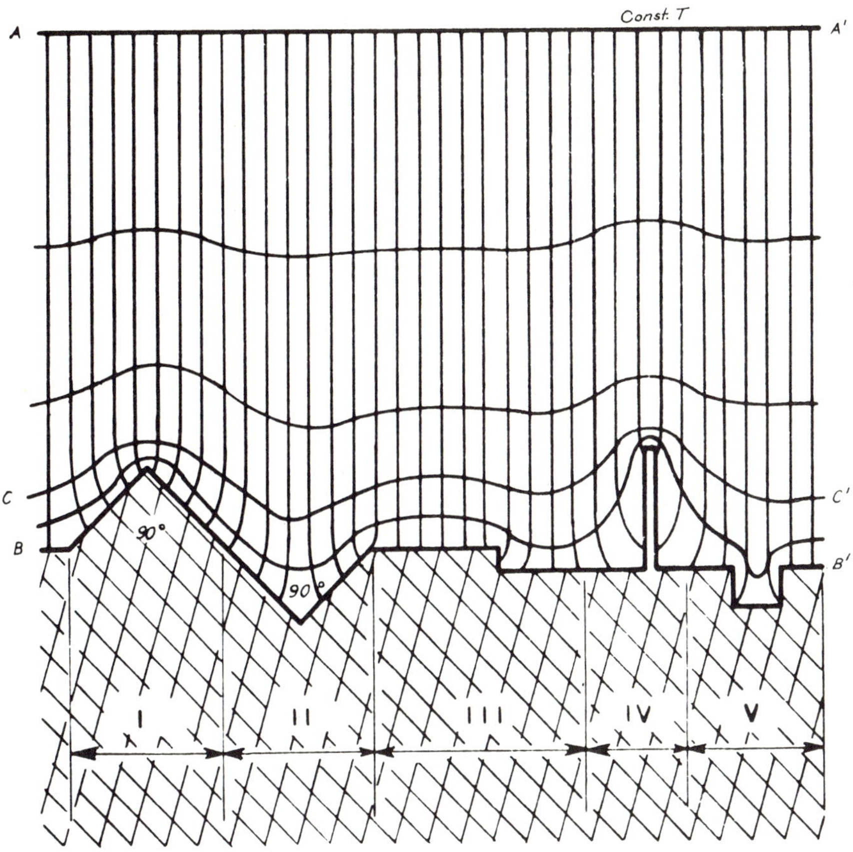

Figure 2

plate of constant voltage as used in an electroplating bath, a wall of constant temperature as in crystallizing by cooling, or a plane of constant concentration as in crystal growth by feeding from the outside. For convenience, the surface *BB'* is divided into sections which we will treat separately. They are:

I. A 90° protrusion.
II. A 90° trough.
III. A ledge.
IV. A rectangular groove.

The accurate solution of section I is shown in Figure 3. Notice that the streamlines shown as dotted lines are more concentrated at the corner than they are on the face. This is in agreement with Vogel's theory. Notice also that the solid lines representing the equipotentials are compacted near the corner and that the crystal would not have to grow as far to cross the first line at the corner as it would have to grow on the faces. In other words, this portion of the crystal, if being electroplated, will find that metallic ions are arriving at this corner faster than on the faces and that electrons are escaping fastest there, as shown by the high concentration of dotted lines at the corner. Or if heat is the controlling factor, more heat is escaping from the corner, or if concentration is the controlling factor, more molecules of solute can impinge upon the corner per square centimeter than can impinge upon the faces. Figure 4 shows the solution of the streamlines and equipotential lines about a square[7] having four corners, such as are illustrated in Figure 3 and section I of Figure 2.

We shall use section II of Figure 2 to illustrate the method of solution of these problems using conformal mapping. Figure 5 shows the construction of equipotential lines and streamlines for the simple case of two parallel plates. If AA' represents a plate at potential 5 volts and BB' a plate at zero voltage, the space between plates will have the equipotentials located on planes. Furthermore, the equal differences in voltage will be equally spaced if the space is a uniform resistor. These equipotentials are perpendicular to the flow lines or streamlines shown in dotted lines and the two sets of lines form a grid work, making the space appear as a coordinate system. Conformal mapping really amounts to a transformation of coordinates in a way that "bends" the coordinate system or distorts it.

Let the horizontal axis be the real, or x axis and the vertical axis of Figure 5 be the imaginary axis ty. Let the voltage be V and the streamlines be C, so that

$$C + tV = X + jy = Z \tag{1}$$

and call this space the Z space. Let us transform this to a new space, which we call the t space having coordinates ϕ and ψ. This transformation will be a solution of the Laplace equation in a complicated boundary val-

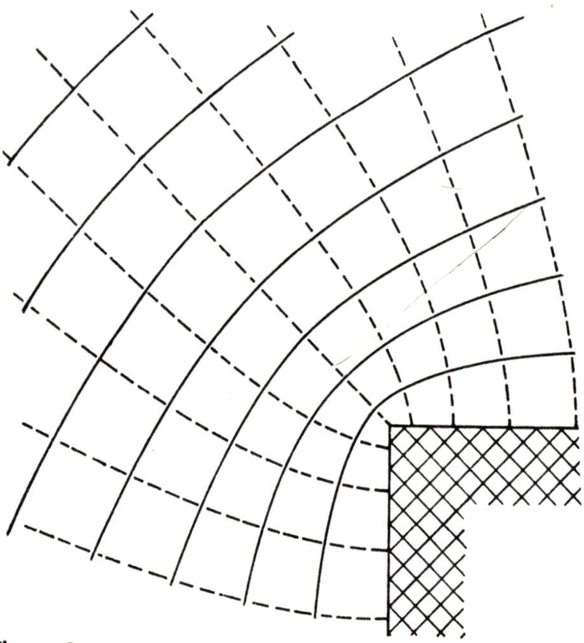

Figure 3

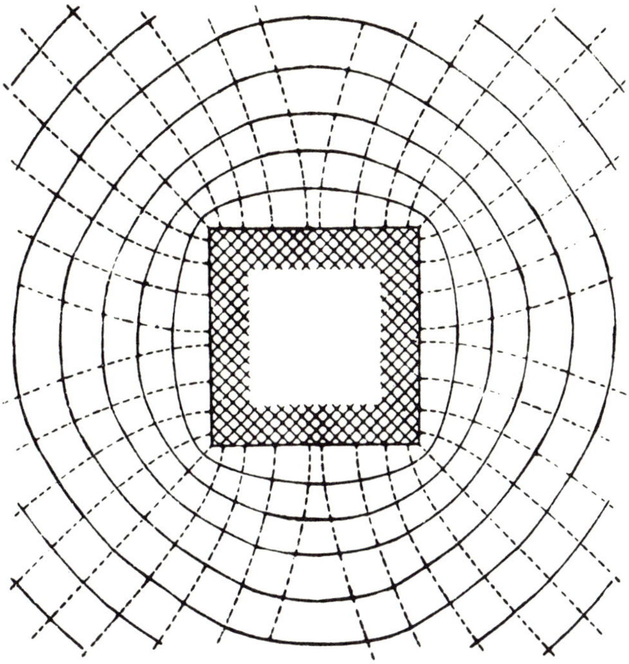

Figure 4

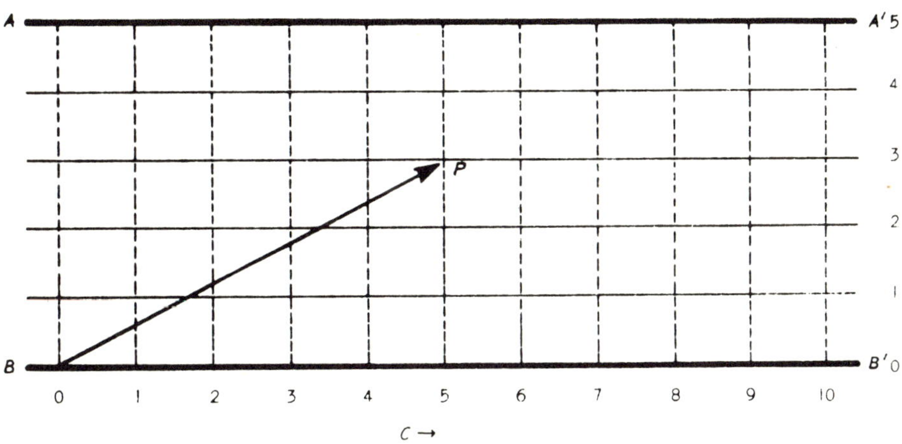

Figure 5

ue problem on the basis of our knowledge of the simple problem shown in Figure 5. For example, if $t = \phi + j\psi = Z^2$

$$t = x^2 = (x + jy)^2 = x^2 - y^2 - 2jxy$$

$$= \phi + j\psi \tag{2}$$

then we may equate the real parts to get

$$\phi = x^2 - y^2 \tag{3}$$

and the imaginary parts, getting

$$\psi = -2xy \tag{4}$$

Equation (3) may be plotted for constant values of ϕ. For example, $x^2 - y^2 = 1$ is a hyperbola asymptotic to the 45′ line $\phi = 0$ shown in Figure 6. The other lines at $\phi = 1, 2, 3, 4$, and 5 are shown and indicate how the streamlines of Figure 5 are distorted by the transformation. Equation (4) represents a set of hyperbolas asymptotic to the x, y axis, as shown in Figure 6. Figure 6 represents a depression as shown in section II of Figure 2, and shows that the streamlines are less congested at the bottom of the depression than at other parts of the crystal, indicating that this depression will become more pronounced as growth proceeds because of the faster growth of the surrounding areas of the crystal.

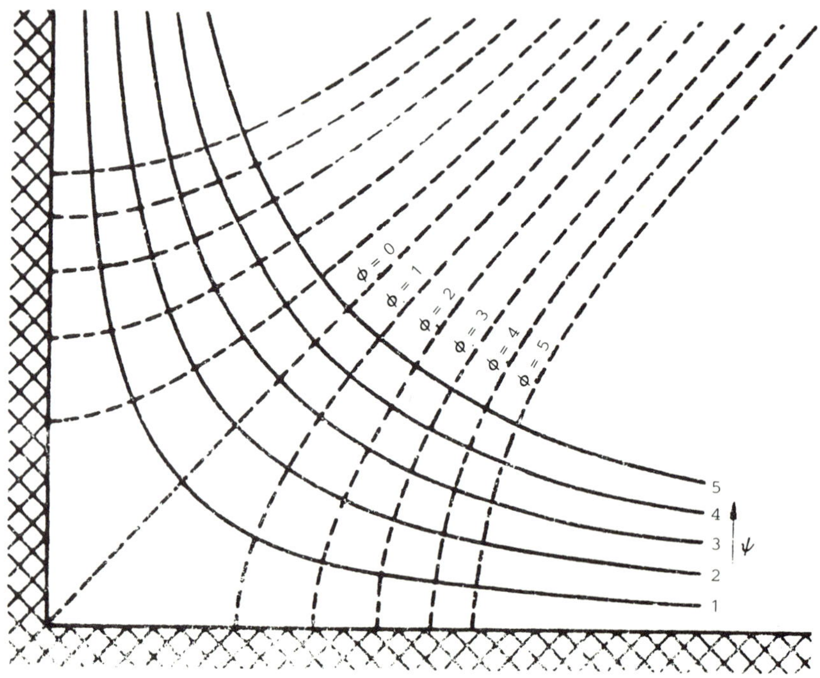

Figure 6

It may be noted that the last illustration was solved by choosing $t = z^2$ as a transformation equation, and it may be asked: How did we know that this was the proper transformation to choose for this particular problem? The answer is that this choice was based on experience. However, it is no longer necessary to rely upon experience since Schwartz and Christoffel[6] have put the transformation in a differential form:

$$\frac{dz}{dt} = K(t - a)^{[(\alpha/\pi)-1]}(t - b)^{[(\beta/\pi)-1]}(t - c)^{[(\gamma/\pi)-1]} \ldots \tag{5}$$

where α represents the angle (in radians) which the BB' axis or x axis is bent to produce another figure. For example, to get Figure 6, the zero-potential line is bent to form a right angle, i.e., $\alpha = \pi/2$, since there are no bends (or since all other angles are 180°), then $\beta = \gamma = \partial \ldots = \pi$. In our last example

$$\frac{dz}{dt} = K(t - a)^{[(\pi/2\pi)-1]} = K(t - a)^{-1/2} \tag{6}$$

The values $a, b, c, \ldots$ in equation (5) are the values along x at which the axes are bent. In our immediate illustration $a = 0$, so that

$$\frac{dz}{\partial t} = Kt^{-1/2} \tag{7}$$

Integrating equation (7)

$$Z = K\int \frac{dt}{t^{1/2}} = 2Kt^{1/2} \tag{8}$$

Since K is an arbitrary constant having magnifying properties only, it may be ½ if we choose. Then $z = t^{1/2}$ or $t = Z^{1/2}$, as we stated at the start.

Figure 7 shows a crystal surface having two bends and is the solution of section III of Figure 2. Note that corner S is in more rapid growth than corner L. The details of this solution are furnished by Walker.[6a]

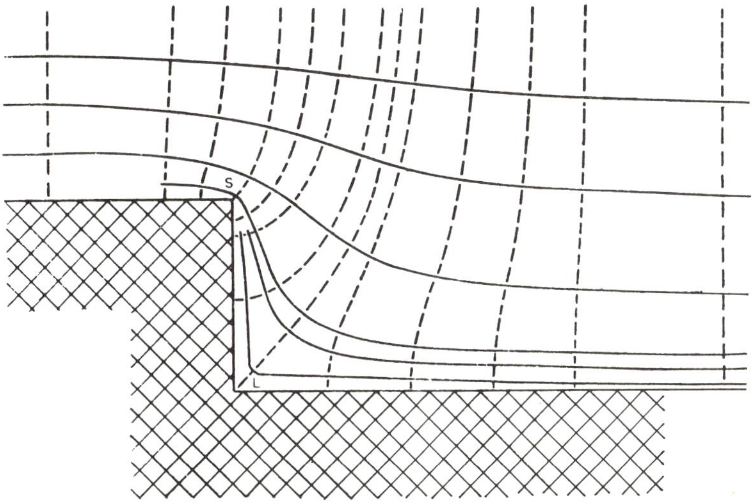

Figure 7

Figure 8 shows a solution of section IV of Figure 2 and clearly shows that a spine grows faster on its ends. This is the problem of the open-ended pipe.[6b]

Figure 9 shows the solution of section V of Figure 2, and suggests that if such a rectangular groove were to form on the face of a crystal it would very likely seal itself off, leaving a liquid cavity in the crystal.[6a]

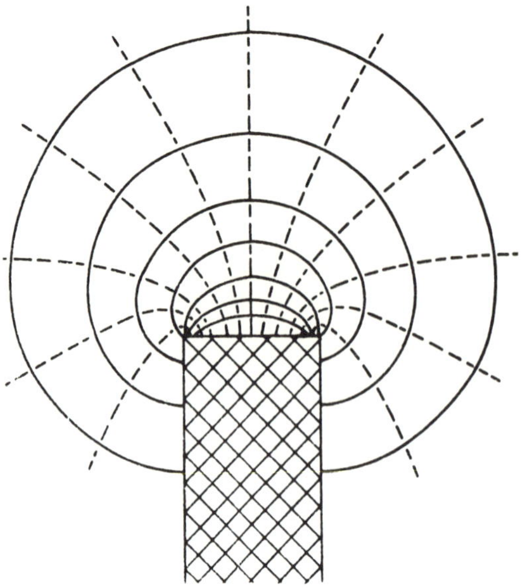

Figure 8

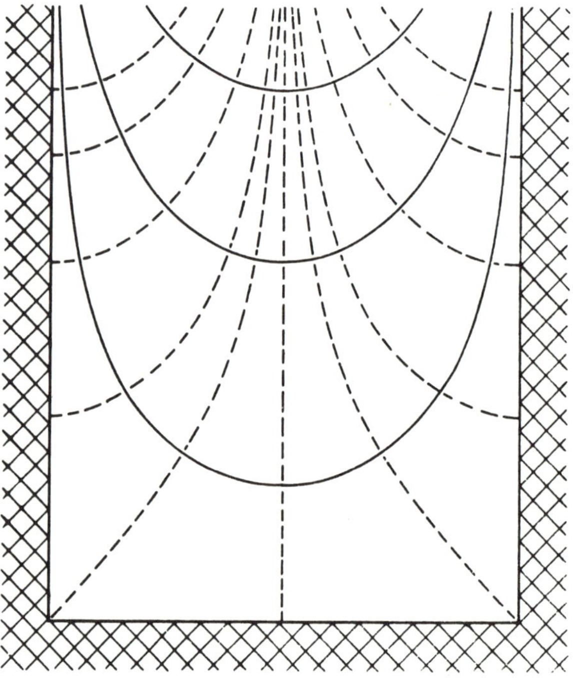

Figure 9

It should be pointed out that the complete problem of dendritic growth has not beeen considered in this section of the report. Dendrites are characterized by (A) long spines and fins which, when once initiated, continue to be enhanced when the controlling factors of growth are (a) the flow of heat, (b) the flow of electrons or ions, or (c) the diffusion of solute or solvent; and (B) by branching of the spines, which results in the characteristic tree-like formations. We feel that the pattern assumed by the branching is determined by other factors such as thermal and acoustical vibrations within the crystal. The latter problem is to be discussed in the section on snow crystals.

REFERENCES

1. H. E. Buckley, *Crystal Growth*, John Wiley & Sons, Inc., 1951, p. 505.
2. R. Vogel, *Z. Anorg. Chem.*, **116**, 21 (1921).
3. See reference (1), p. 488.
4. Kossel, *Nach. Ges. Wiss. Gutingen*, 135 (1927).
5. I. N. Stranski, *Discussions Faraday Soc.* **5** 13 (1949).
6. For discussions of conformal mapping the reader is referred to (a) Miles Walker, *Conjugate Functions for Engineers*, Oxford University Press, 1933; (b) I. S. Sokolnikoff, and E. S. Sokolnikoff, *Higher Mathematics for Engineers and Physicists*, McGraw-Hill Book Company, New York, 1934; (c) William R. Smythe, *Static and Dynamic Electricity*, McGraw-Hill Book Company, New York, 1939, p. 77.
7. Some very ingeneous experiments were set up by Bunn, and Bunn and Emmett to measure the concentration gradients in the neighborhood of growing crystals. For a description of the work see *Discussions Faraday Soc.* **5** 119–144 (1949). Their Figures 8 and 10, pp. 137 and 138, bear only a faint resemblance to our Figure 4. This disagreement is probably due to dislocation or strong points of nucelation (see Part II of this report).

Appendix

COMPUTER SIMULATION OF CRYSTAL GROWTH

STEREO PLATE: THE EQUILIBRIUM FORM OF BARITE

Cecil J. Schneer

This note was prepared expressly for this Benchmark volume

Friedel wrote, ". . . . give me the parameters of a crystal and I will tell you what are its forms and their relative development, I will draw the crystal for you . . . ."[1]

Since the habit of any real crystal is the resultant of not only its structure but the individual history of its growth, the statement would be wrong, Friedel wrote, and the Bravais rule expresses a statistical generality of mineralogical, rather than deterministic validity.

What are the consequences of a given set of assumptions of

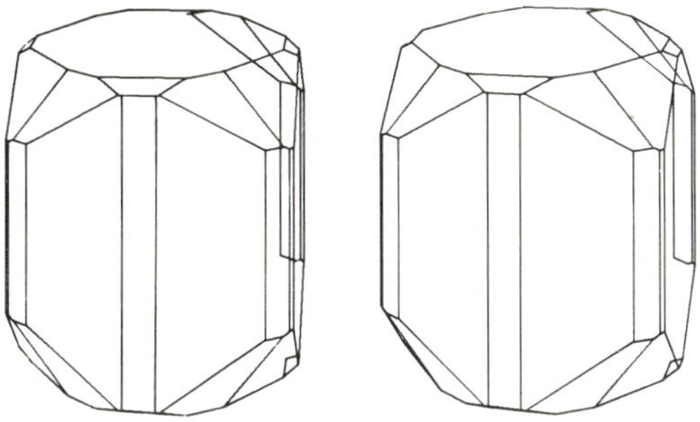

[1] This volume, p. 82.

the relationship between external morphology and internal structure? The computer offers a direct mathematical-mechanical transformation. The illustration here is of a stereo pair drawn in two-point perspective and representing, by the relative sizes of the faces, a statistical equilibrium barite crystal plotted at the Computation Center of the University of New Hampshire.

AUTHOR CITATION INDEX

SUBJECT INDEX

About the Editor

CECIL J. SCHNEER is Professor of Geology and the History of Science at the University of New Hampshire, where he teaches mineralogy and the history of science. He served as Chairman of the Faculty for 1975–1976. Dr. Schneer has been a mining geologist in Peru, Academic Guest at the Solid State Physics Institute (E.T.H.) of Zurich, and Visiting Professor at the Institute of Mineralogy, University of Milan.

Professor Schneer received his A.B. Magna cum Laude in 1943 at Harvard and his Ph.D. in 1954 at Cornell, where he studied mineralogy with W. T. Holser and history of science with Henry Guerlac. In the summer of 1967, he convened the New Hampshire Inter-Disciplinary Conference on the History of Geology and edited the *Proceedings* (1969). He directed the New Hampshire Bicentennial Conference on the History of Geology, 1976, and is editing the Proceedings. He is the author of *The Search for Order* (1960), and of *Mind and Matter* (1969), and serves as Associate Editor of the *Dictionary of Scientific Biography* and as Advisory Editor of *Isis*, the journal of the History of Science Society. He is President of the United States National Committee for the History of Geology and Vice President for America of the International Committee for the History of Geology of the International Union of Geological Sciences.